ADVANCES IN SEISMIC EVENT LOCATION

MODERN APPROACHES IN GEOPHYSICS

VOLUME 18

The titles published in this series are listed at the end of this volume.

ADVANCES IN SEISMIC EVENT LOCATION

edited by

CLIFFORD H. THURBER
University of Wisconsin-Madison,
Madison, WI, U.S.A.

and

NITZAN RABINOWITZ
Geophysical Institute of Israel,
Lod, Israel

SPRINGER-SCIENCE+BUSINESS MEDIA, B.V.

Library of Congress Cataloging-in-Publication Data

ISBN 978-90-481-5498-2 ISBN 978-94-015-9536-0 (eBook)
DOI 10.1007/978-94-015-9536-0

Printed on acid-free paper

Cover illustration: Convergence trajectory to the Tel-Aviv offshore event applying unconstrained Nelder and Mead algorithm using five P-readings and one S-reading from the five stations shown (filled circles). The S-reading is taken from station DLIA. Note the zig-zag pattern of the trajectory. Unlike many other linearized algorithms, this algorithm approaches the minimum by moving away from points with high RMS values (dashed contours).

Originally published by Kluwer Academic Publishers in 2000
Softcover reprint of the hardcover 1st edition 2000

Contents

List of Contributors

Richard Aster, Department of Earth and Environmental Science and Geophysical Research Center, New Mexico Institute of Mining and Technology, Socorro, New Mexico, USA

Catherine Berge-Thierry, Institut de Protection et de Sûreté Nucléaire, Fontenay-aux-Roses, Paris, France

E. Robert Engdahl, Department of Physics, University of Colorado-Boulder, USA

Petr Firbas, CTBTO, International Data Centre, Vienna Int'l Centre, P.O.Box 1250, A-1400 Vienna, Austria

Manfred Joswig, Tel Aviv University, Israel

Edi Kissling, Institute of Geophysics, Swiss Federal Institute of Technology (ETH) Zurich, Switzerland

Anthony Lomax, Géosciences-Azur, University of Nice - Sophia Antipolis, Valbonne, France

Jose Pujol, Center for Earthquake Research and Information, The University of Memphis, Memphis, TN 38152, USA

Nitzan Rabinowitz, Geophysical Institute of Israel, PO Box 182, Lod, Israel

Charlotte Rowe, Department of Earth and Environmental Science and Geophysical Research Center, New Mexico Institute of Mining and Technology, Socorro, New Mexico, USA

David M. Steinberg, Department of Statistics and Operations Research, Raymond and Beverly Sackler Faculty of Exact Sciences, Tel Aviv University, Tel Aviv 69978, Israel

Clifford H. Thurber, Department of Geology and Geophysics, University of Wisconsin-Madison, USA

Jean Virieux, Géosciences-Azur, University of Nice - Sophia Antipolis, Valbonne, France

Philippe Volant, Institut de Protection et de Sûreté Nucléaire, Fontenay-aux-Roses, Paris, France

Acknowledgements

We would like to acknowledge the Shalheveth Freier Center for Peace, Science, and Technology for their support and endorsement of the Dead Sea Workshop. We are grateful to Florian Haslinger and Renate Hartog for their careful reading of sections of the first draft of the book, and Megan Mandernach and Chad Trabant for helping to proofread the final version. We also thank Cathy Egan for her expert typing of numerous sections of the book.

Preface

This book is an outgrowth of the Shalheveth Freier First International Workshop on Advanced Methods in Seismic Analysis, held at the Dead Sea in January, 1998. The scientific steering committee consisted of Nitzan Rabinowitz, Clifford Thurber, Edi Kissling, and Wim Spakman. The workshop focused on the topics of seismic event location and seismic tomography. We would like to take this opportunity to thank all of the participants in that workshop.

This book presents papers from the Dead Sea workshop related to the topic of seismic event location. The first chapter provides an overview of global seismic event location, the development of one-dimensional travel time models, and efforts to incorporate lateral heterogeneity into global earthquake location. The second chapter presents a novel constrained non-linear earthquake location method. The third chapter examines the topic of optimal seismic network configuration. The next three chapters focus on earthquake location in three-dimensional (3-D) models, including methods for travel time determination, probabilistic location in 3-D models, and location calibration using 3-D models. The last three chapters focus on issues related to multiple-event location, including the joint hypocenter determination method, the use of pattern recognition and rule-based system techniques for event identification and location, and automated methods for arrival time refinement.

Chapter 1

ADVANCES IN GLOBAL SEISMIC EVENT LOCATION

Clifford H. Thurber
Department of Geology and Geophysics, University of Wisconsin-Madison, USA

E. Robert Engdahl
Department of Physics, University of Colorado-Boulder, USA

Key words: Earthquake location, velocity model, travel time, station correction, source correction, path correction, confidence region, depth phases, 3-D structure

Abstract: We review the fundamentals of earthquake location and document the evolution of global one-dimensional models for travel times and velocity structure. We discuss in detail the issues of location uncertainty, weighting, corrections (station, source, and path), and data quality. Nuclear monitoring concerns have brought about an increased need for improvements in global and regional seismic event location capability. Several recent studies highlight the value of utilizing information from multiple phases and global 3-D models to improve event locations. Further progress in the use of 3-D models for high-accuracy locations will require substantial efforts to improve the quality of the arrival time data used to determine the 3-D structure of the Earth.

1. INTRODUCTION

A recent surge in interest in improving global (teleseismic) event location can be attributed to two main factors: (1) the need for improved seismic monitoring due to the signing of the Comprehensive Test Ban Treaty (CTBT) (Husebye and Dainty, 1996), and (2) the need for improved source locations for use in seismic tomography (van der Hilst and Engdahl, 1992;

C.H. Thurber and N. Rabinowitz (eds.), Advances in Seismic Event Location, 3–22.

Engdahl et al., 1998). Potential avenues for progress include improved velocity models, improved data quality and data usage, better station and/or source-region travel-time corrections, and better location algorithms. Following a brief introduction to standard location techniques, we present some of the recent advances in teleseismic event location, focusing primarily on applications using one-dimensional (radially symmetric) velocity models, and highlighting areas where significant advancement remains to be made.

Standard location methods are based on Geiger's method (Geiger, 1912), which became practical with the advent of modern computers. The basic methodology is essentially unchanged from the classic papers of Bolt (1960), Flinn (1965), and Engdahl and Gunst (1966). Briefly, predicted arrival times for a trial hypocenter and origin time are calculated for the observing stations using the chosen velocity model. The arrival time residuals (observed minus calculated) are then related to hypocenter (latitude θ, longitude ϕ, depth z) and origin time (t_0) perturbations by a linearized equation of the form

$$r = -\sin\theta \sin\alpha \left(\frac{\partial T}{\partial \Delta}\right) \Delta\phi + \cos\alpha \left(\frac{\partial T}{\partial \Delta}\right) \Delta\theta + \left(\frac{\partial T}{\partial z}\right) \Delta z + \Delta t_0 \tag{1a}$$

where α is azimuth from the event to the station. This can be written in matrix form as

$$\mathbf{r} = \mathbf{A}\,\mathbf{x} \tag{1b}$$

where **r** is the vector of residuals, **A** is the matrix of derivatives, and **x** is the vector of origin time and hypocenter perturbations. Due to non-linearity, this system of equations is solved iteratively via matrix inversion until convergence is attained. Possible convergence criteria include reaching a specified level of either data misfit, data misfit change (variance reduction), perturbation step size, or reaching a set maximum number of iterations. Options for carrying out the matrix inversion include forming the normal equations

$$\mathbf{A}^T \mathbf{r} = \mathbf{A}^T \mathbf{A}\,\mathbf{x} \tag{2}$$

followed by the application of a standard solution algorithm for square symmetric matrices (such as Gaussian reduction (Bolt, 1960) or the Crout procedure (Flinn, 1960)), or using step-wise linear regression (Lee and Lahr, 1975), the QR algorithm (Buland, 1976), or singular value decomposition

(Klein, 1978). Alternative solution methods such as the grid search approach are discussed in detail in Chapters 2 and 4.

The matrix solution methods provide a means to estimate uncertainties in location and origin time. In the vicinity of the solution $\mathbf{x}^*$, the linear approximation of Equation 1a is generally relatively accurate, and the covariance matrix for the model parameters $\mathbf{C}_m$ can be calculated by

$$\mathbf{C}_m = \sigma_d^2 \, (\mathbf{A}^T \mathbf{A})^{-1} \, [(\mathbf{A}^T \mathbf{A})^{-1}]^T \qquad (3)$$

where σ_d^2 is an estimate of the data variance, usually based on the variance of the fit to the arrival times (the residuals). Individual parameter uncertainties are simply the square root of the corresponding diagonal elements of $\mathbf{C}_m$.

The covariance matrix can be used to estimate confidence regions (most generally a 4-dimensional ellipsoid) using

$$(\mathbf{x} - \mathbf{x}^*)^T \, \mathbf{C}_m^{-1} \, (\mathbf{x} - \mathbf{x}^*) = \kappa_p^2 \qquad (4)$$

where

$$\kappa_p^2 = M \, \sigma_d^2 \, [F_p(M, N\text{-}M)]/(N\text{-}M) \qquad (5)$$

and M is the total number of hypocentral parameters (nominally 4), N is the number of arrival-time data, 100p is the selected confidence level percentage, and F_p is the F-statistic at that confidence level. It is important to point out that this type of error estimate does not take into account location bias due to systematic model errors (that is, the deviation of the true earth from the model used to calculate travel times), so the actual error will generally exceed this estimate.

One of the key issues in stabilizing the solution is the use of weighting. Weighting is generally done based on reported arrival-time reading precision (Engdahl et al., 1998), reported arrival quality (Buland, 1976), phase variance as a function of distance (Buland and Engdahl, in preparation), residual size (Bolt, 1960), station distance (Klein, 1978) and/or azimuthal distribution of stations (Buland, 1976). Weighting is easily incorporated in the inversion by constructing a weight matrix W with diagonal elements equal to the square root of the weight value and modifying Equation 1b as follows:

$$\mathbf{W}\,\mathbf{r} = \mathbf{W}\,\mathbf{A}\,\mathbf{x} \qquad (6)$$

(e.g., Bolt (1960)) and then solving the weighted system as before. Uncertainty estimation and weighting are discussed further in a later section.

2. EVOLUTION OF ONE-DIMENSIONAL (1-D) MODELS

Prior to the 1980's, global seismic event locations were determined by major seismological agencies (e.g., NEIC, ISC) using the Jeffreys-Bullen (JB) Tables for P, S, and other later-arriving phases (Jeffreys and Bullen, 1940). We note that the JB Tables are still being used by NEIC and ISC in large part because of their complete representation of phases. An important alternative were the Herrin Tables for P waves (Herrin et al., 1968), which were developed based on data from nuclear explosions as well as earthquakes. These tables were constructed empirically using travel times from well-observed events. Both were known to have biases and limitations, but it was not until the 1980's and later that widely-accepted alternative models were developed.

The new generation of models were constructed in a totally different way; rather than simply establishing empirical tables for phase travel times, the new inverse modeling approach was to construct one-dimensional models for structure that fit the travel times (using data from the ISC Bulletin) and other parametric data. The Preliminary Reference Earth Model (PREM) of Dziewonski and Anderson (1981) was the most important member of this generation of new global 1-D models. However, PREM was constructed to fit both body wave travel time data and normal mode data (plus the Earth's mass and moment of inertia), so it was not generally considered to be a viable alternative for use in seismic event location. Soon afterwards, in fact, Dziewonski and Anderson (1983) published a separate analysis of just P waves in an effort to produce an improved travel time table. They included station corrections in their inversion following the "time term" approach of Cleary and Hales (1966), but incorporating azimuthally-dependent terms in their formulation (see the Station, Source, and Path Corrections section).

It was not until 1991 when the new body wave model iasp91 was published (Kennett and Engdahl, 1991) that the JB and Herrin Tables finally had a serious competitor for seismic event location. The iasp91 model was commissioned by the International Association of Seismology and the Physics of the Earth's Interior (IASPEI), after which the model was named. As in the case of PREM, iasp91 is a 1-D model of structure (P and S velocities versus radius) from which travel times can be calculated, using the highly efficient scheme of Buland and Chapman (1983).

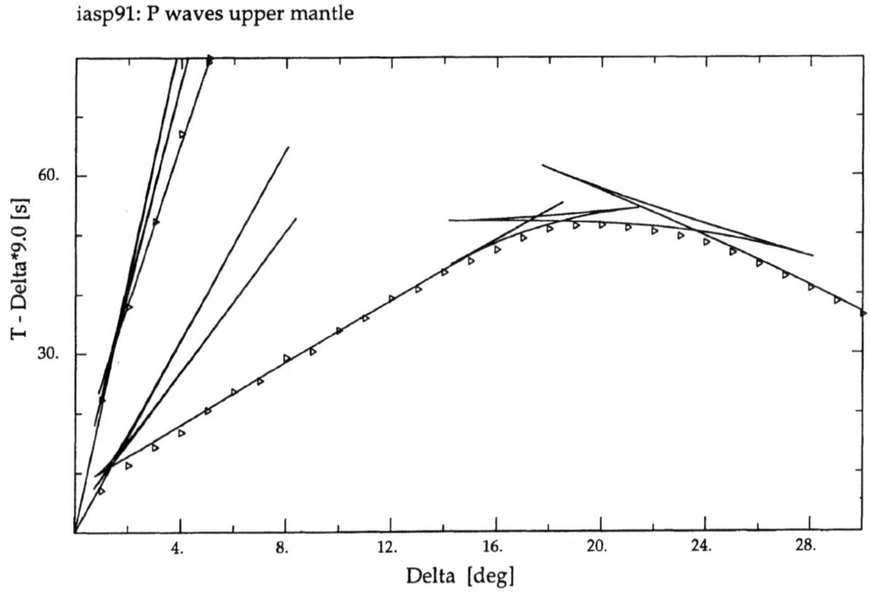

a)

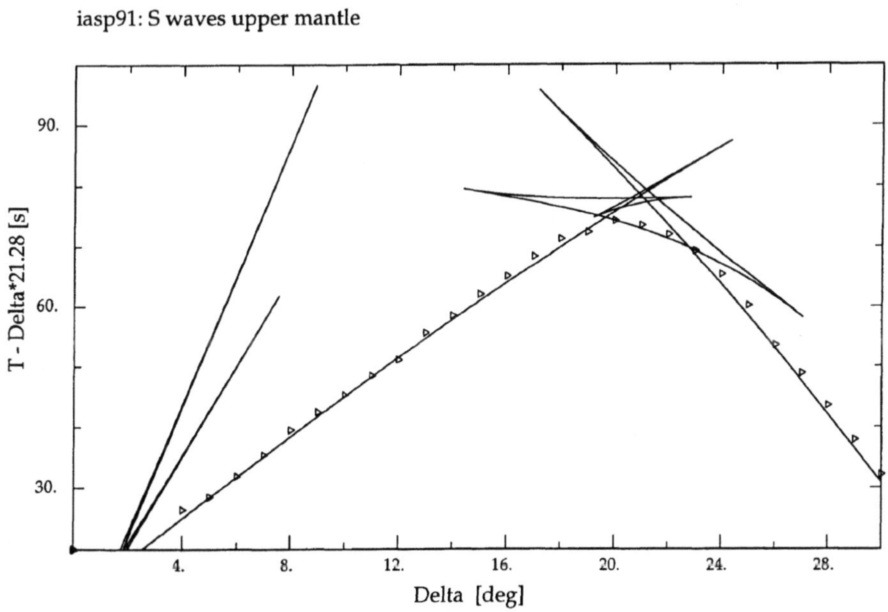

b)

Figure 1. Fit of model iasp91 as a function of epicentral distance for the upper mantle for a) P summary travel times and b) S summary travel times of Dziewonski and Anderson (1981, 1983) (from Kennett and Engdahl (1991)).

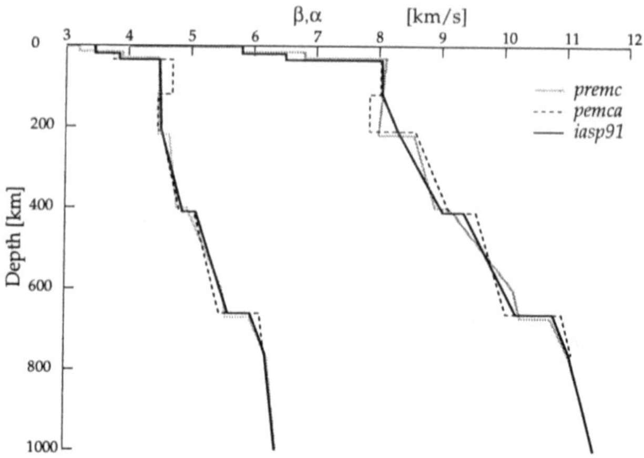

Figure 2. Comparison of the velocity models iasp91, pemca, and premc for the upper mantle (from Kennett and Engdahl (1991)).

The upper mantle part of the iasp91 model was fit to the "summary times" for P and S waves from Dziewonski and Anderson (1981, 1983) (Figure 1), with the latter adjusted by a 2 s baseline shift. As shown in Figure 2, the iasp91 model has substantial differences from PREM. There is no upper mantle low velocity zone for either P or S waves in iasp91, meaning that there were no shadow zones due to upper mantle structure. This is a practical advantage for locating events because travel times are not discontinuous (also because normally only first arriving times are reported so that it is not necessary to have discontinuities), but it ran counter to the prevailing ideas about upper mantle structure. The characteristics of the main upper mantle discontinuities are also different from previous models. The 210 km discontinuity is essentially absent in iasp91, and the 410 km and 660 km discontinuity velocity jumps are slightly greater in amplitude in iasp91 compared to PREM. Kennett and Engdahl (1991) emphasized that there is a geographic bias in the upper mantle structure due to the distribution of sources and stations, so that iasp91 should not be construed as representing a true average upper mantle model.

For the lower mantle, path coverage was generally more uniform, so the lower mantle parts of the iasp91 P and S models were considered to be more representative of the average Earth. Constraints on the P structure were reasonably good except near the core-mantle boundary, but for the S structure, the complication of interfering phases put a practical limit on the amount and quality of data (Kennett and Engdahl, 1991). On the whole, however, iasp91 does a good job of representing teleseismic travel times, as indicated by the analysis of "test event" data (Figure 3).

An alternative lower-mantle model was developed by Morelli and Dziewonski (1993) using the same model parameterization as Kennett and Engdahl (1991). Morelli and Dziewonski (1993) solved for multiple source-region station corrections averaged over 5° areas to account for lateral heterogeneity in an approximate manner. They then derived new sets of summary travel times for lower mantle and core P and S phases binned in 1° intervals of epicentral distance, and inverted those summary times for 1-D P and S velocity models. The resulting model, SP6, is generally comparable to iasp91 for lower mantle P and S. SP6 has slightly lower velocity gradients with depth and correspondingly higher velocity jumps at the 660 km discontinuity and the core mantle boundary, with the latter being enhanced by negative velocity gradients in SP6 for the D" region. The SP6 model fits the S data significantly better than iasp91, which appears to have a 2 s baseline error. The differences between iasp91 and SP6 are significant for the core, due to the addition of substantial core phase data for the derivation of SP6.

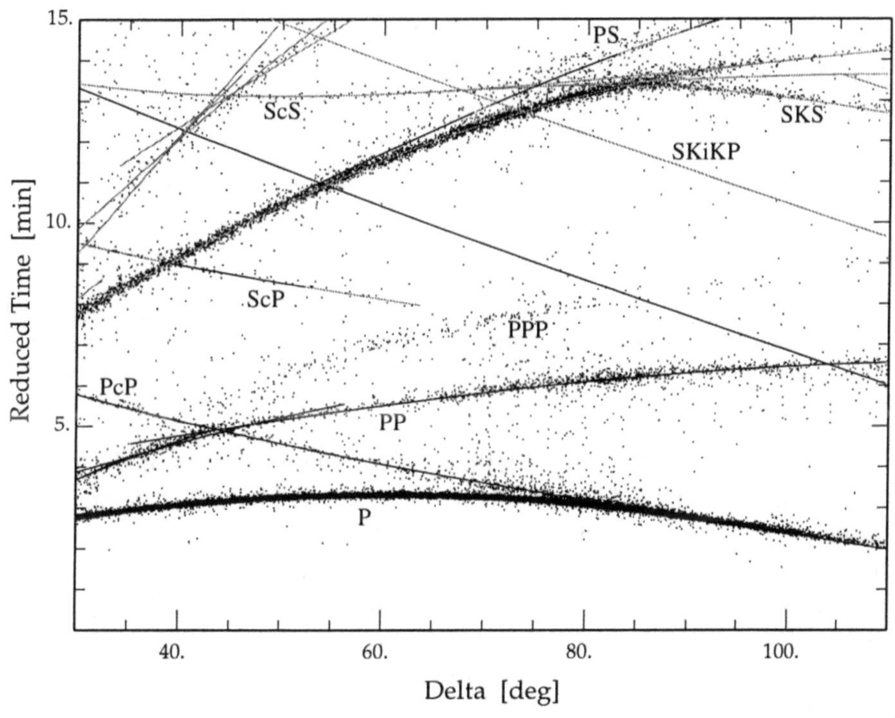

Figure 3. Comparison of iasp91 teleseismic travel times (reduced with slowness 6.84 s deg^{-1}) for teleseismic P, S, and associated phases for the test event data corrected to surface focus (from Kennett and Engdahl (1991)).

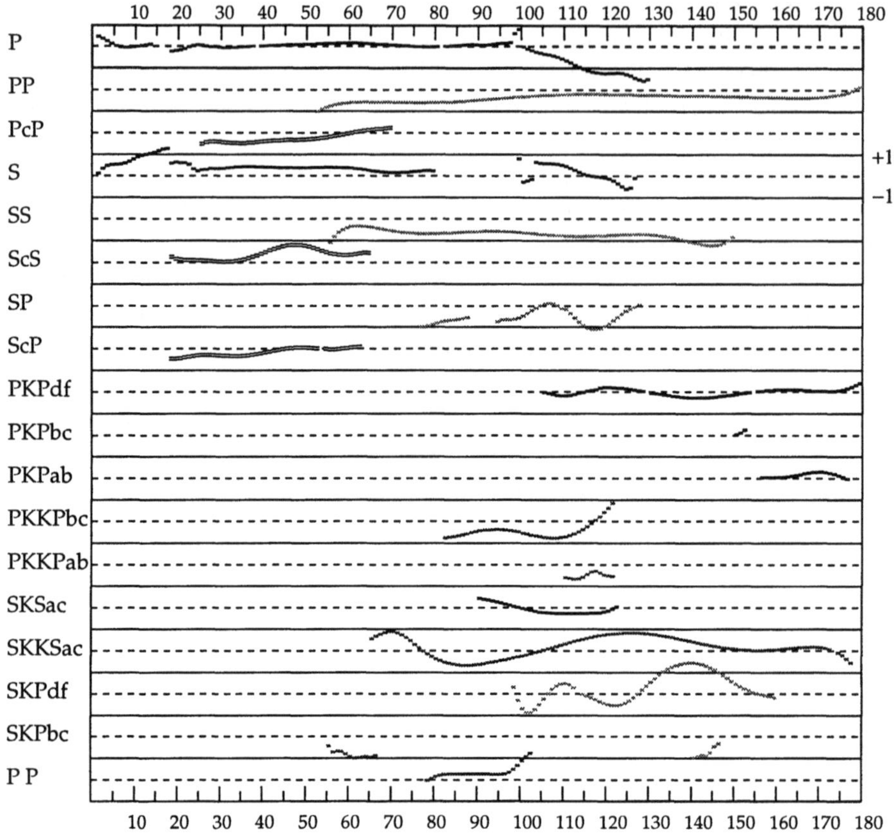

Figure 4. Display of the normalized residuals (normalized by the corresponding standard deviation estimates) between the ak135 travel times and the smoothed empirical travel times (from Kennett et al. (1995)). Note in particular the absence of any significant mantle S wave bias.

Kennett et al. (1995) started with the best characteristics of the iasp91 and SP6 models and added steps to improve data quality (see the later section on this topic) to derive model ak135. The main aspect of their strategy was to improve the locations of the seismic events used to create their set of travel time tables. They accomplished this by including depth phases and using iasp91 (and just P phases) in the relocation step. As shown in the composite residual plot in Figure 4, model ak135 provides a very good fit to a wide range of phases. The mantle S wave bias of iasp91 has been removed, and most core phase times are quite well matched. Thus, for global earthquake location there has been convergence on a global, radially symmetric, P- and S-velocity Earth model that provides a good average fit to reported phase arrival times.

3. UNCERTAINTY ESTIMATION AND WEIGHTING

Seismic event locations are most useful for subsequent analyses (e.g., discrimination, tomography, or tectonic interpretation) if they are of good quality. Therefore, the estimation of location uncertainty is of fundamental importance. The standard covariance matrix approach (Equation 3) is normally effective in evaluating relative location precision when events are well observed. This requires having observing stations that are well distributed in azimuth and (to a lesser degree) distance and having some control on source depth. Flinn (1965) presents an excellent discussion of confidence regions, comparing the linear estimate to grid-based and jackknife-type estimates. Some problems with the standard approach are that the estimate of data variance σ_d^2 becomes unreliable for relatively small numbers of observations (Evernden, 1969), that it neglects contributions to location error from inadequacies in the travel time model, and that the linear approximation is inadequate in some circumstances.

Buland (1976) noted that data variance estimated from either the residual variance or *a priori* experience will generally be too small due to the inaccuracy of the travel time model. He suggested that the σ_d^2 value should simply be increased to compensate for this problem. Jordan and Sverdrup (1981) developed an elegant statistical approach for making this compensation. Their philosophy is to replace the usual *a posteriori* estimate s_d^2 of the data variance based on the data misfit for an individual earthquake

$$s_d^2 = \| \mathbf{r} \|^2 / (N - M) \tag{7}$$

(where N is the number of data and M is the number of hypocenter parameters) by an estimate based on a combination of *a priori* and *a posteriori* information:

$$s_d^2 = \frac{K\, s_0^2 + \| \mathbf{r} \|^2}{(K + N \pm M)} \tag{8}$$

where the *a priori* variance estimate s_0 is given a relative weight of K. Thus the *a priori* information is treated as contributing K observations to the estimate of σ_d^2 and the data for the particular event contribute N observations. For self-consistency, given a large number of events, the mean of the estimated variances should approach s_0. Jordan and Sverdrup (1981) also show that the distribution of the estimated standard deviation needs to be consistent with the adopted value of K. In the limit of no prior information (K = 0) their estimate reverts to the standard formula (Flinn, 1965), whereas for infinite prior information ($K = \infty$) their estimate becomes equivalent to that of Evernden (1969).

One limitation of the Jordan and Sverdrup (1981) approach is that it assumes a Gaussian error distribution with a constant data variance for all observations. As shown in Equation 6, varying data uncertainty estimates can be incorporated simply by including a weighting matrix in the standard linearized analysis. However, the Gaussian data error assumption is inappropriate, as residual outliers are common in teleseismic arrival time data (Bolt, 1960; Buland, 1986).

The "uniform reduction" method (Jeffreys, 1939) is effective for treating the situation of normally-distributed errors plus a background of outliers. Bolt (1960) assumed a residual frequency distribution of the form

$$f(r) = a\,s + \frac{(1 \pm s)}{\sigma\sqrt{2\pi}} \exp\left(\pm \frac{(r \pm m)^2}{2\,\sigma^2}\right) \tag{9}$$

where a and s are constants and m and σ are the mean residual and its standard deviation. This corresponds to a residual weighting function of the form

$$w(r) = \left\{1 + \mu \exp\left(\pm \frac{(r \pm m)^2}{2\,\sigma^2}\right)\right\}^{\pm 1} \tag{10}$$

with $\mu = \sigma\,(2\pi)^{1/2}\,a\,s/(1 - s)$.

In one of the few recent studies of contributions to teleseismic location uncertainty, Billings et al. (1994) used a combination of Monte Carlo analysis and varying phase and station observations to assess hypocenter location errors. They investigated the role of picking errors by perturbing actual arrival time data multiple times using perturbation values drawn from a Gaussian distribution (assuming a mean of 0 s and a standard deviation of 0.25 s for P arrivals) and then relocating the event. The spatial distribution of the suite of Monte Carlo trial relocations provides an estimate of the influence of picking error alone on relative mislocation, as all other factors were held fixed. For the one event they analyzed in this way, they observed significant disagreements between scatter plots of location parameter values versus surface contours of data misfit, which essentially represents an *a posteriori* estimate of the probability density function for the location (with low misfit indicating high probability). However, much of the disagreement was apparently due to the problem of projecting a multi-dimensional distribution onto a 2-D plane. Correlation between parameters "tilt" the multi-dimensional distribution, which then biases the apparent distribution when it is projected to 2-D and compared to misfit contour/slices. For uncorrelated parameter projections, the comparison were quite favorable.

Billings et al. (1994) also attempted to examine the role of velocity model errors by comparing locations of an event derived using different combinations of phases and network geometries. The logic is that picking errors are random whereas model errors are systematic, so that locations including different phases or stations would only be affected by the systematic model errors. The flaw in this analysis, though, is the confounding effect of variable location constraint due to the use of different phases (e.g., pP) and distributions of stations (both in distance and azimuth). However, their conclusion that velocity-model errors will be mapped principally into depth errors for well-observed events is certainly valid. Such depth errors will generally be several times larger than picking-induced errors.

Finally, Billings et al. (1994), examined the combined effect of (large-scale) velocity heterogeneity and observation geometry on location. Previous studies of events in the Flores Sea indicated that paths from there to Australia stations were generally fast compared to iasp91 models. Thus for a small magnitude event observed by relatively few, nearby stations, the calculated location will be biased in a southerly direction towards Australia to compensate for the early arrivals there. For 2 larger events, observations at more distant stations, especially those with southerly azimuths, counteract the early Australia arrivals, and the larger events are located further to the north. In their example, the location shift was of the same order as the location shifts (for an event elsewhere) due to picking errors. Thus they conclude that magnitude-related bias can be a significant mislocation factor.

4. STATION, SOURCE, AND PATH CORRECTIONS

Station corrections, as well as source and path corrections, are a long-recognized mechanism for trying to compensate for velocity heterogeneity when 1-D velocity models are assumed. Station and source corrections are a logical extension of the "time term" method of refraction seismology; that is, to relate arrival time residuals δt_{ij} to a sum of correction terms

$$\delta t_{ij} = d_i + a_{ij}(\Delta) + b_j \qquad (11)$$

where d_i is a time correction for earthquake i, $a_{ij}(\Delta)$ is an average correction to the travel time table for distance Δ, and b_j is a time correction for station j. This approach was first used in a global study by Cleary and Hales (1966). Their analysis concentrated on travel times for 25 teleseismic events to stations in North America, but they also derived station corrections for worldwide stations assuming their adjusted travel time curve was globally applicable. They found significant systematic average travel time corrections relative to the J-B tables in the 33° to 99° range, with a mean difference of about -1.5 seconds. Later studies confirmed this bias.

The station correction values of Cleary and Hales (1966) showed systematic patterns both within North America and worldwide. In the central U.S. and southern Canadian shield, corrections are on the order of -1.0 s, indicating early arrivals, while for the tectonically active Basin and Range area, average corrections are approximately +0.5 s, indicating late arrivals. Remarkably, this general trend of negative corrections in shield areas and positive residuals in areas of recent tectonic uplift extended to many areas worldwide. Also, Pacific margin regions with subduction zones showed large negative corrections. These station corrections trends were significant, as standard errors for the corrections averaged about 0.25 s worldwide. Cleary and Hales (1966) concluded by pointing out the need for higher-precision data to improve studies of this kind, calling for the reading of P times to 0.1 s, and by indicating the potential for further studies of regional crust and upper mantle structure based on the station correction values.

As part of their development of a new travel time table for P (Herrin et al., 1968), Herrin and Taggart (1968) produced a set of station corrections for 321 worldwide stations and over 400 events that included an azimuthally-varying term, using the form

$$C_{ij} = A_i + B_i \sin(\alpha_{ij} + E_i) \qquad (12)$$

where i is the station and α_{ij} is the azimuth to event j. The station correction parameters were calculated by least squares for the final residuals from the Herrin et al. (1968) analysis. Estimated standard errors averaged about 1 s, much higher than the Cleary and Hales (1966) study, although this is not surprising given the more than order of magnitude increase in the number of data used. Their mean station corrections (A_i terms) correlated moderately well with those of Cleary and Hales (1966), but the inclusion of the azimuthally varying term (often with an amplitude exceeding the mean correction) resulted in many substantial differences between the two studies. However, the general trend of negative corrections in shield and island arc areas and positive residuals in other tectonically active areas remained. The azimuthal variations were not interpreted.

The next quantum leap in the amount of data used for a global station correction study was that of Dziewonski and Anderson (1983), who used over 3000 earthquakes and nearly 1000 seismic stations in their analysis. As in the study of Herrin and Taggart (1968), Dziewonski and Anderson (1983) included azimuthally varying terms (up to 2, depending on the station's and distribution of observations), and like Cleary and Hales (1966), they adjusted the teleseismic travel time curve during the inversion process. Despite the huge increase in the number of data, the data fit was reasonably good, with misfit averaging less than 1 s for stations without azimuthal terms and less than 0.5 s for stations with azimuthal terms.

The spatial patterns in mean station corrections were slightly different in the Dziewonski and Anderson (1983) study. For North America, coastal regions (extending several hundred kilometers inland) showed systematically large positive corrections (late arrivals, often exceeding 1 s) while the continental interior showed primarily small negative anomalies. In Eurasia, the Alpine-Himalaya belt was dominated by large positive corrections, again with generally small negative corrections in stable shield areas (Baltics, Siberia). The Australian shield stands out as having the largest systematic negative corrections.

The azimuthally varying station corrections, where determined, had terms in α and 2α. The α terms tended to show an intriguing pattern of having the slow direction point towards the ocean for many stations near a coast, perhaps due to thickening crust. The 2α term showed stronger regional trends. In many areas, the slow direction is oriented roughly parallel to the greatest principal stress direction of Zoback and Zoback (1980), perhaps indicating stress-or flow-induced alignment of crystals.

Recent station correction studies have evolved in opposite directions. Engdahl et al. (1998) adopted a simpler 'patch correction' approach, determining from teleseismic residuals a single median correction for all stations within 5 x 5 degree regions. Patch medians derived separately from P and PKP data agreed well with each other, but Engdahl et al. (1998) did not provide standard error or misfit values. In contrast, Morelli and Dziewonski (1993) derived source-region-specific corrections in 5°-sized regions for each individual station, meaning that each station could have hundreds of correction values for the various source regions worldwide. Not surprisingly, the use of such a complex station correction strategy leads to a substantial reduction in residual variance.

A potential problem with azimuthally-varying station corrections stems from the fact that the corrections are derived from paths that do not uniformly sample the 3-D source region, and as a result are biased. Tests carried out using 3-D ray tracing in the Bijwaard and Spakman (1999) model produced by non-linear global P-wave tomography confirm that for many stations, source-region upper mantle and lower mantle path structure maps into the azimuthal terms. This may result in bias in the location of events in the same direction but at different epicentral distances.

5. IMPROVING QUALITY AND USAGE OF DATA

One direct method to improve seismic event locations is by improving the quality and utilization of the data. Hypocenter determination is significantly improved by using, in addition to direct P and S phases, the arrival times of PKiKP, PKPdf, and the teleseismic depth phases (pP, pwP and sP) in the relocation procedure. However, using later phases from catalog arrival times requires a method for improving the identification of the phases (Engdahl et al., 1998). Alternatively, if digital data are available, cross-correlation techniques (VanDecar and Crosson, 1990) can be used to determine relative arrival times with much better precision.

Early teleseismic location work relied essentially exclusively on the use of first arriving P phases (Bolt, 1960; Flinn, 1965). Standard teleseismic catalogs (ISC, NEIC) still rely almost entirely on first arriving P phases for locating events (Engdahl et al., 1998). Many studies have shown that the inclusion of later arriving phases can provide greater constraints on hypocenter parameters, especially depth (for example, Buland (1976), Engdahl and Billington (1986), Gomberg et al. (1990), Billings et al. (1994)). Epicenter constraints are improved by the inclusion of S wave and P core phases because their travel-time derivatives (Equation 1a) differ significantly in magnitude from direct P, while depth-origin trade-off is ameliorated by the inclusion of depth phases (pP, pwP, sP) because their travel time derivatives are opposite in sign to direct P (Engdahl et al., 1998).

A "chicken and egg" problem with the use of later phases is that their correct identification often requires knowledge of the event depth and distance. Recognizing this problem, Kennett et al. (1995) and Engdahl et al. (1998) included a phase re-association step in their location procedure. In their approach, phase arrivals were re-identified after each iteration using a statistically-based association algorithm. An example of how this approach can be implemented is illustrated in Figure 5, from Engdahl et al. (1998). Probability density functions (PDFs) for relevant phases, centered on their theoretical relative travel times for a given hypocenter, can be compared to the observed phase arrivals. When PDFs overlap for a particular phase, the idea is to assign a phase identification in a probabilistic manner based on the relevant PDF values, making sure not to assign the same phase to two different arrivals. Engdahl et al. (1998) applied this procedure to depth phases in their study.

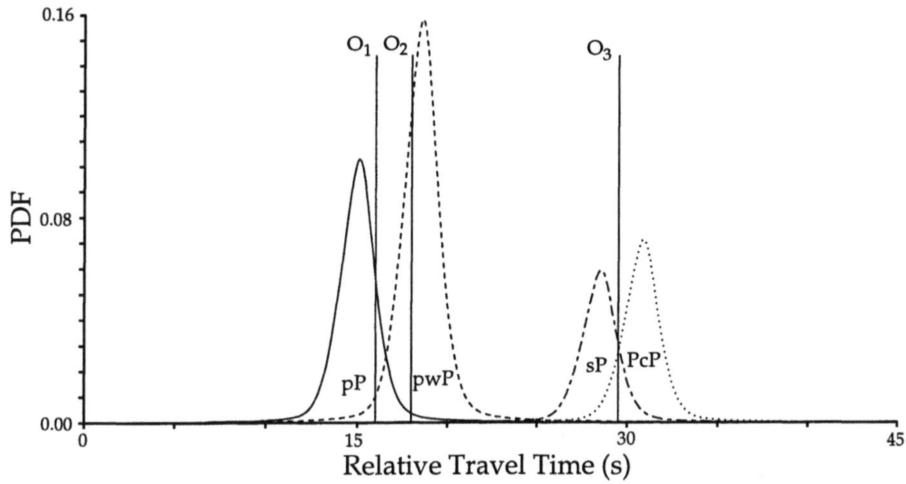

Figure 5. Probabiliy density functions (PDF's) for four phases (pP, pwP, sP, and PcP) centered at their theoretical relative travel times from a hypothetical deep event and hypothetical observed phases (O1, O2, and O3) with unknown identifications (from Engdahl et al. (1998)).

6. EFFECTS OF ASPHERICAL EARTH STRUCTURE

The travel times predicted by recently developed, radially symmetric, Earth models (such as ak135) are extremely valuable for earthquake location and phase identification. Nevertheless, most earthquakes occur in or near subducted lithosphere where aspherical variations in upper mantle seismic wave velocities are large (i.e., on the order of 5 - 10%; Bijwaard et al., 1998). Such lateral variations in seismic velocity, the uneven spatial distribution of seismological stations, and the specific choice of seismic data used to determine the earthquake hypocenter can still easily combine to produce bias in earthquake locations of several tens of kilometers (van der Hilst and Engdahl, 1992) . Kennett and Engdahl (1991) have shown that a set of reference events (events for which we have well-constrained hypocenters, such as nuclear explosions or earthquakes located within a local network) are on average mislocated by about 14 km using standard procedures. Tests of global location bias using a new archive of reference event information (Engdahl, 1998) and the Engdahl et al. (1998) location algorithm show that most explosions and earthquakes are mislocated by less than 20 km if the azimuthal gap to observing stations at all distances is less than 120° (Figure 6). Thus, at least in the case of events well constrained

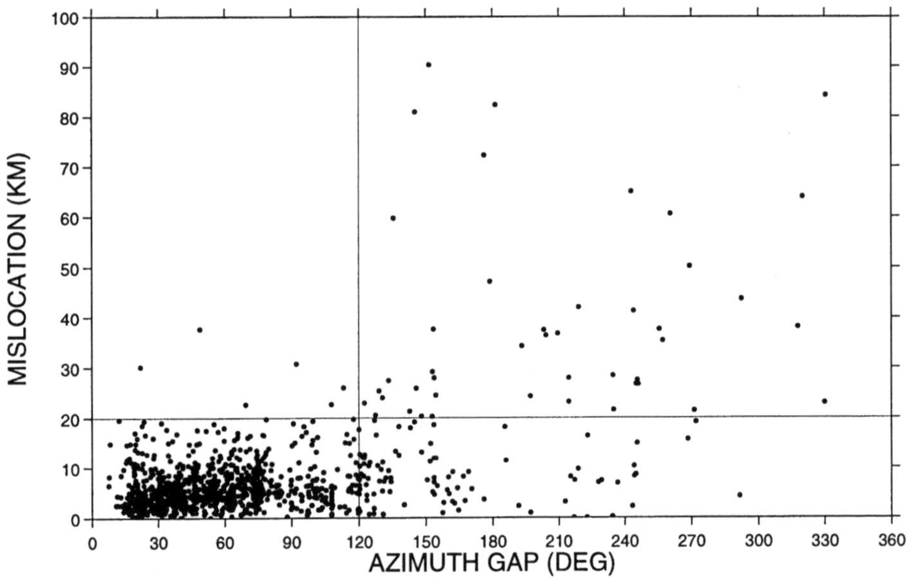

Figure 6. Mislocation of events (in kilometers) as a function of azimuthal gap (in degrees). Note that for azimuth gaps below 120°, almost all test events are mislocated by less than 20 km.

azimuthally by reporting stations, mislocation errors introduced by lateral heterogeneity can be minimized. For smaller and/or poorly recorded events, however, there is not much hope of significantly reducing the resulting mislocation error until we can somehow account for aspherical Earth structure in routine earthquake location procedures.

7. SEISMIC MONITORING UNDER THE CTBT

Operational monitoring and international verification of the CTBT will be the responsibility of an International Data Center (IDC) in Vienna, Austria. However, an enhanced seismic monitoring capability of the International Monitoring System (IMS) is inherently limited by the systematic error of event locations. The CTBT will limit On-Site Inspections to a 1000 km^2 region. This is interpreted as a minimum performance requirement for the IMS/IDC to locate seismic events on land throughout the world. As defined by a 90% coverage ellipse, this would apply to all seismic events above the IDC event definition threshold (currently $m_b \sim 3.75$). A powerful and direct method to address the problem of errors in event location is the development of high-quality reference event databases (including but not limited to origin times and hypocenters for seismic sources) with validated travel-time information for a wide range of seismic phases over regional and teleseismic paths. These databases can be used to support the calculation of regional travel-time curves and source-specific station corrections, referenced to the IMS seismic network. They can also be used to estimate the probable magnitude of mislocation in different regions of the world and to identify specific regions where calibration experiments are necessary to correct these problems. Such information is needed to increase the capability to locate small seismic events throughout the world.

8. FUTURE DIRECTIONS

The current "state-of-the-art" in teleseismic event location is well represented by three papers, the work of Engdahl et al. (1998) using 1-D velocity structure and those of Smith and Ekstrom (1996) and Antolik et al. (2000) using 3-D velocity structure. Engdahl et al. (1998) combined the use of regional and teleseismic P and S phases, selected core phases, and teleseismic depth phases in their location algorithm. As discussed above, a statistically-based association algorithm was used to automatically identify depth phases, resulting in significant improvement in depth estimation. Use of the Engdahl et al. (1998) algorithm to locate events with reported arrival times from seismic stations well distributed in azimuth and distance teleseismically does appear to enhance the non-random structural component of travel time residuals. In principle, this means that higher resolution imaging of 3-D global P- and S-wave structure by tomographic inversions of hypocenters and phase data (van der Hilst et al., 1997; Bijwaard et al., 1998) can be improved by providing improved hypocenters, especially depths.

Smith and Ekstrom (1996) compared Kennett and Engdahl (1991) reference event locations to locations derived from standard 1-D Earth models (JB, PREM, iasp91) and a recent Earth model (S&P12/WM13) which incorporates large scale heterogeneity of P and S wave velocities in the mantle. The 3-D model S&P12/WM13 improved event locations over all three 1-D models, with or without station corrections. For explosion reference events the average mislocation is reduced by 40%; for earthquake reference events the improvements are smaller. However, it is clear that where slabs with strong small-scale heterogeneities occur (e.g., the Aleutians), modeling only large-scale heterogeneity cannot significantly improve teleseismic event location. Furthermore, the location problem is exacerbated when an event is poorly recorded as well, suggesting that for events recorded primarily at regional distances (where outlier observations are commonly reported) some kind of grid search method may be more appropriate for their location (Boeve et al., 1999).

Antolik et al. (2000) extended the study of Smith and Ekstrom (1996), using ISC data for the same data set augmented with events from Lop Nor, by testing the location capability of four 3-D mantle P velocity models (two spherical harmonic, two block models). They found no significant differences in location accuracy between the low-resolution spherical harmonic models and the (presumably) higher resolution block models. To mimic a CTBT scenario, they also assessed location accuracy using just 30 P arrivals per event. They found that 66% of the time, locations fell within the 1000 km^2 target region, and when calibration information was available, this improved to 75%. The authors suggested that increased use of Pn (including Pn anisotropy) and azimuth data might provide further improvements. However, they also noted that poor phase picks in the ISC data might be a limiting factor. This possibility is strongly supported by the recent studies of Thurber et al. (1999) and Röhm et al. (1999). Thurber et al. (1999) showed that re-picked arrival-time data using a waveform cross-correlation technique yielded significantly improved relative event locations compared to ISC data, even with more than an order of magnitude fewer arrival times. Röhm et al. (1999) demonstrated the existence of systematic phase timing variations at many stations in data reported to the ISC. Their tests showed that most of these temporal variations must be due to timing errors or picking bias. The authors noted that these problems may lead to bias in tomography studies or studies involving temporal changes (e.g., inner core rotation).

In conclusion, we envision continued and substantial progress in the future on global and regional seismic event location, driven in part by the CTBT. It also seems clear that improvements in global and regional 3-D models hinge on improved source locations. The ISC database remains the primary data source for many of these studies, but it appears that limitations of the ISC database may limit potential improvements to both locations and structure. Thus it is important to improve the reporting of data to the ISC, such as in the ISOP program (e.g., Engdahl and Bergman, 1992), but it will also be necessary to utilize global waveform data to make a quantum improvement in the quality of available data.

ACKNOWLEDGEMENTS

The first author (CHT) acknowledges the Defense Threat Reduction Agency, U.S. Department of Defense, and the U.S. National Science Foundation for support of research related to this chapter. We thank Brian Kennett for assistance with the figures.

REFERENCES

Antolik, M., G. Ekström, and A. Dziewonski (2000) Global event location with full and sparse datasets using three-dimensional models of mantle P wave velocity, *J. Geophys. Res.*, submitted.

Bijwaard, H. and W. Spakman (1999) Fast kinematic ray tracing of first- and later-arriving global seismic phases, *Geophys. J. Int.* **139**, 359-369.

Bijwaard, H., W. Spakman, and E.R. Engdahl (1998) Closing the gap between regional and global travel time tomography, *J. Geophys. Res.* **103**, 30055-30078.

Billings, S.D., M.S. Sambridge, and B.L.N. Kennett (1994) Errors in hypocenter location: picking, model, and magnitude dependence, *Bull. Seismol. Soc. Am.* **84**, 1978-1990.

Boeve, R., W. Spakman, H. Bijwaard, and E.R. Engdahl (1999) 3-D earthquake location with grid search methods: Application to nuclear explosions (abstract), *International Workshop on Tomographic Imaging of 3-D Velocity Structures and Accurate Earthquake Location (PAPHOS99), Platres, Cyprus, 5-9 July 1999.*

Bolt, B.A. (1960) The revision of earthquake epicentres, focal depths and origin times using a high-speed computer, *Geophys. J. Roy. Astr. Soc.* **3**, 433-440.

Buland, R. (1976) The mechanics of locating earthquakes, *Bull. Seism. Soc. Am.* **66**, 173-187.

Buland, R. (1986) Uniform reduction error analysis, *Bull. Seism. Soc. Am.* **76**, 217-230.

Buland, R., and C.H. Chapman (1983) The computation of seismic travel times, *Bull. Seism. Soc. Am.* **73**, 1271-1302.

Cleary, J., and A.L. Hales (1966) An analysis of the travel times of P waves to North American stations, in the distance range 32° to 100°, *Bull. Seism. Soc. Am.* **56**, 467-489.

Dziewonski, A.M. and D.L. Anderson (1981) Preliminary reference Earth model, *Phys. Earth Planet. Inter.* **25**, 297-356.

Dziewonski, A.M. and D.L. Anderson (1983) Travel times and station corrections for P waves at teleseismic distances, *J. Geophys. Res.* **88**, 3295-3314.

Engdahl, E.R. (1998) Development of an archive of seismic ground truth events globally in support of monitoring under the CTBT, *Proc. 20th Annual Seis. Res. Symp. on Monitoring a Comprehensive Test Ban Treaty (CTBT), Santa Fe, NM, Sept. 1998*, 11-18.

Engdahl, E. R., and R.H. Gunst (1966) Use of a high speed computer for the preliminary determination of earthquake hypocenters, *Bull. Seism. Soc. Am.* **56**, 325-336.

Engdahl, E. R., and S. Billington (1986) Focal depth determination of central Aleutian earthquakes, *Bull. Seism. Soc. Am.* **76**, 77-93.

Engdahl, E. R., and E.A. Bergman (1992) The International Seismological Observing Period in Africa: *Tectonophys.* **209**, 1-16.

Engdahl, E.R., R. van der Hilst, and R. Buland (1998) Global teleseismic earthquake relocation with improved travel times and procedures for depth determination, *Bull. Seism. Soc. Am.* **88**, 722-743.

Evernden, J.F. (1969) Precision of Epicenters obtained by small numbers of world-wide stations, *Bull. Seism. Soc. Am.* **59**, 1365-1398.

Flinn, E.A. (1960) Local earthquake location with an electronic computer, *Bull. Seism. Soc. Am.* **50**, 467-470.

Flinn, E.A. (1965) Confidence regions and error determinations for seismic event location, *Rev. Geophys.* **3**, 157-185.

Geiger, L. (1912) Probability method for the determination of earthquake epicenters from the arrival time only, *Bull. St. Louis Univ.* **8**, 60-71.

Gomberg, J.S., K.M. Shedlock, and S.W. Roecker (1990) The effect of S-wave arrival times on the accuracy of hypocenter estimation, *Bull. Seism. Soc. Am.* **80**, 1605-1628.

Herrin, E. and J. Taggart (1968) Regional variations in P travel times, *Bull. Seism. Soc. Am.* **58**, 1325-1337.

Herrin, E., W. Tucker, J. Taggart, D.W. Gordon, and J.L. Lobdell (1968) Estimation of surface focus P travel times, *Bull. Seism. Soc. Am.* **58**, 1273-1291.

Husebye, E.S. and A.M. Dainty (1996) *Monitoring a Comprehensive Test Ban Treaty*, Kluwer Academic Publishers, Dordrecht, 864 pp.

Jeffreys, H. (1939) *Theory of Probability*, Oxford University Press, London.

Jeffreys, H., and K.E. Bullen (1940) *Seismological Tables*, British Association for the Advancement of Science, London.

Jordan, T.H. and K.A. Sverdrup (1981) Teleseismic location techniques and their application to earthquake clusters in the south-central pacific, *Bull. Seism. Soc. Am.* **71**, 1105-1130.

Kennett, B.L.N. and E.R. Engdahl (1991) Traveltimes for global earthquake location and phase identification, *Geophys. J. Int.* **105** 429-465.

Kennett, B.L.N., E.R. Engdahl, and R. Buland (1995) Constraints on seismic velocities in the Earth from traveltimes, *Geophys. J. Int.* **122**, 108-124.

Klein, F.W. (1978) Hypocenter Location Program HYPOINVERSE, *U.S. Geological Survey Open-File Report 78-694*, 113 pp.

Lee, W.H.K., and J.C. Lahr (1975) HYPO71 (revised): A computer program for determining hypocenter, magnitude, and first motion pattern of local earthquakes, *U.S. Geological Survey Open-File Report 75-311*, 114 pp.

Morelli, A. and A.M. Dziewonski (1993) Body wave traveltimes and a spherically symmetric P- and S-wave velocity model, *Geophys. J. Int.* **112**, 178-194.

Röhm, A.H.E., J. Trampert, H. Paulssen, and R.K. Snieder (1999) Bias in reported seismic arrival times deduced from the ISC Bulletin, *Geophys. J. Int.* **137**, 163-174.

Smith, G.P. and G. Ekstrom (1996) Improving teleseismic earthquake locations using a three-dimensional Earth model, *Bull. Seism. Soc. Am.* **86**, 123-132.

Thurber, C., F. Haslinger, and C. Trabant (1999) Testing event location capability with ground truth events in Kazakstan, *Proc. 21st Seismic Research Symposium*, 283-293.

VanDecar, J.C., and R.S. Crosson (1990) Determination of teleseismic relative phase arrival times using multi-channel cross-correlation and least squares, *Bull. Seism. Soc. Am.* **80**, 150-159.

van der Hilst, R.D., and E.R. Engdahl (1992), Step-wise relocation of ISC earthquake hypocenters for linearized tomographic imaging of slab structure, *Phys. Earth Planet. Inter.* **75**, 39-54.

van der Hilst, R.D., S. Widiyantoro, and E.R. Engdahl (1997). Evidence for deep mantle circulation from global tomography, *Nature* **386**, 578-584.

Zoback, M.L. and M.D. Zoback (1980) State of stress in the conterminous United States, *J. Geophys. Res.* **85**, 6113-6156.

Chapter 2

HYPOCENTER LOCATION USING A CONSTRAINED NONLINEAR SIMPLEX MINIMIZATION METHOD

Nitzan Rabinowitz
Geophysical Institute of Israel, P.O. Box 182, Industrial Zone North, Lod 71100, Israel

Key words: constrained hypocenter location, Simplex optimization, Nelder and Mead minimization, Flexible Tolerance Method

Abstract: We present a powerful new hypocenter location method based on an extension of the unconstrained Nelder and Mead Simplex method. This method incorporates linear and nonlinear constraints into the location scheme in an efficient manner. We demonstrate the merit of the method by locating events incorporating constraints given by P arrival order, S-P times, and back azimuths.

1. INTRODUCTION

In this study, we present a specific constrained search algorithm as an effective hypocenter location procedure. The motivation to develop constrained search location algorithms arises in the context of automated location, where only a small number of detecting stations may be used to estimate a preliminary location (for example, Alpha arrays from the International Monitoring System (IMS) network within the CTBT framework). In cases where only P-waves are detected, there is a strong risk that the location algorithm will improve the misfit function at the expense of moving further and further away from the feasible region where the true hypocenter occurred. There are several natural constraints that can provide effective bounds to the epicenter position, namely the azimuthal estimates of the detecting stations (three component stations and arrays are capable of

C.H. Thurber and N. Rabinowitz (eds.), Advances in Seismic Event Location, 23–49.

providing back azimuthal estimations with varying degrees of uncertainty) and, depending on the network geometry with respect to the epicenter, the P-arrival orders. Thus, a location strategy is required that will insure that:

1) the initial hypocenter will be placed close enough to the true hypocenter;
2) improvements to the values of the misfit function will be carried out while insuring that the amount of "constraints violation" is gradually diminished.

The proposed method is a constrained simplex approach that falls into the class of grid search methods. In a scenario where only a few P-waves are detected, this method is much more efficient than the standard derivative-based, Gauss-Newton location algorithms. It is commonly found that in cases where the arrival data poorly constrain the parameters, a derivative-based algorithm may encounter severe numerical instability, caused by the fact that a similar dependence of derivatives at different stations on the hypocenter parameters may cause the partial derivative matrix to become rank deficient (see Chapter 1).

Apart from being a fully nonlinear hypocenter location algorithm thus avoiding any matrix inversions, the proposed algorithm has another important advantage: the ability to incorporate into its optimization scheme any type of constraint, both in linear and nonlinear forms. Such an optimization procedure falls into the class of Nonlinear Programming Methods (NLP). The goal of NLP is to find an extremum of an objective function subject to equality and/or inequality constraints. No general method exists to solve nonlinear problems in the sense of linear programming theory, for which the existence of a "general all purpose algorithm" is guaranteed, provided that the objective function and the constraints are expressed in linear terms. A major difficulty in solving an NLP problem in which the constraints cannot be easily eliminated is the need to balance the aims of reducing the objective function and staying inside, or close to, the feasible region. Therefore, in many NLP problems, a considerable part of the computation time is spent on satisfying rather rigorous feasibility requirements. Thus, an effective NLP strategy must enable the objective function to be minimized while controlling the constraint violations.

Our elected NLP algorithm for earthquake location tasks is referred to as the Flexible Tolerance Method (FTM) and is described in detail by Himmelblau (1972, Chapter 8). The FTM is essentially based on the Nelder and Mead simplex optimization technique (Nelder and Mead, 1965). It attempts to reduce simultaneously both the objective function and the constraint violations using an "improvement-correction" scheme. The objective function is improved at the first step and, in the second step, the resulting improved value is corrected in the sense that it is transferred to a point "close enough" to the feasible region in the event that the improved value fell "reasonably far" from the feasible region.

2. THE NELDER AND MEAD SIMPLEX OPTIMIZATION

A fundamental key for understanding the concepts underlying the FTM is the Nelder and Mead Optimization method. We describe briefly how it works in a non-formal way (see Appendix A for a formal description): recalling that the definition of a simplex is a set of $n+1$ points in n-dimensional space E^n, then, according to the Nelder and Mead method, the minimization of the objective function is carried out by forming successive series of simplexes in E^n. The procedure starts with an initial simplex having the initial guess as one of its vertices or, alternatively, as its center of gravity. Usually, a regular polyhedron is constructed (i.e., for two independent variables a regular simplex is an equilateral triangle, for three variables the regular simplex is a tetrahedron and so forth). In Appendix A, we show a method for obtaining the coordinates of the vertices of a regular simplex.

Once the simplex is formed, the objective function is evaluated at each of the vertices of the simplex and a projection is made from the point yielding the highest value of the objective function (termed "worst vertex") through the center of gravity of the simplex (termed the "centroid"). Such an operation is referred to as a **reflection**.

If the reflection yields a point that is better than all the vertices of the simplex, an **expansion** operation is undertaken in an attempt to continue the search along the same direction. The reflected point is placed further away from the simplex centroid on the projection line connecting the worst vertex and a new simplex is formed. If the expanded point is better than some but not all of the simplex vertices, the reflected point replaces the worst vertex and a new simplex is formed.

If the reflected point is found to be better ONLY than the worst vertex, a **contraction** operation is performed in an attempt to alter the search direction. The reflected point is moved slightly back along the projection line and placed closer to the simplex centroid. The contracted point replaces the worst vertex and a new simplex is formed. If the reflected point is even worse (i.e., higher objective function value) than the worst vertex, a reduction operation takes place, indicating that the current simplex is in close vicinity to the global optimum. The simplex edges are reduced by half. Termination takes place when the simplex radius defined as the average distance of the simplex vertices from its centroid, decreases beyond a prescribed tolerance level.

Figure 1 illustrates how the algorithm works. It is applied to the hypocenter location of an ML=3.9 offshore event that occurred some 40 km west of Tel-Aviv in the Mediterranean Sea on September 13, 1990 (coordinates x=60, y=160). Starting from coordinate x=100, y=0, this figure traces the series of simplexes as constructed by the Nelder and Mead algorithm. The search starts with a series of reflections and expansions until a contraction takes place around the coordinates x=-50 y=140. At this point the simplex changes its direction eastward and continues with a series of reflections and expansions. In the vicinity of the epicenter, a series of reductions takes place and the simplex collapses on the epicenter. In this scheme, we used only five P-arrivals at Israel Seismic Network (ISN) stations (filled circles), and allowed only two independent variables, x and y (the travel time RMS misfit contour map is superimposed).

As a purely unconstrained minimization algorithm, the simplex algorithm has been implemented in several hypocenter location studies. Rabinowitz and Kulhanek (1988) studied its effectiveness in locating teleseismic events using a sparse network (the Swedish Seismograph network). The problem of locating teleseismic events involved fitting of tabular data. In the context of searching for an efficient algorithm for fitting tabular data, Olsson and Nelson (1975) compared the Nelder and Mead algorithm against several nonlinear optimization methods. They found that the Nelder and Mead algorithm shows a convincing capability to handle a wide variety of optimization problems without requiring any modification due to special features of the problem studies.

Rabinowitz (1988) and Prugger and Gendzwill (1988) studied its properties in the context of locating local events. Prugger and Gendzwill (1988) compared the performance of the Nelder-Mead simplex algorithm using an L1 norm misfit function against a standard Gauss-Newton location procedure in the context of local events and found that the simplex algorithm out-performs the other. The L1 norm misfit function generally has been found to be more robust (i.e., less sensitive) to the presence of arrival time outlier values (Anderson, 1982; Claerbout and Muir, 1973; Shearer, 1997).

3. THE FLEXIBLE TOLERANCE METHOD

Formally, the FTM can be described as follows:

Problem I:
Minimize: $f(\mathbf{x}),\ \mathbf{x} \in E^n$

Subject to: $h_i(\mathbf{x})=0, i=1,\ldots,m$
$g_i(\mathbf{x}) \geq 0, i=m+1,\ldots,p$

where n is the number of independent variables, m and p-m are the numbers of equality and inequality constraints, respectively; and $f(\mathbf{x})$, $h_i(\mathbf{x})$, and $g_i(\mathbf{x})$ may be linear or nonlinear functions.

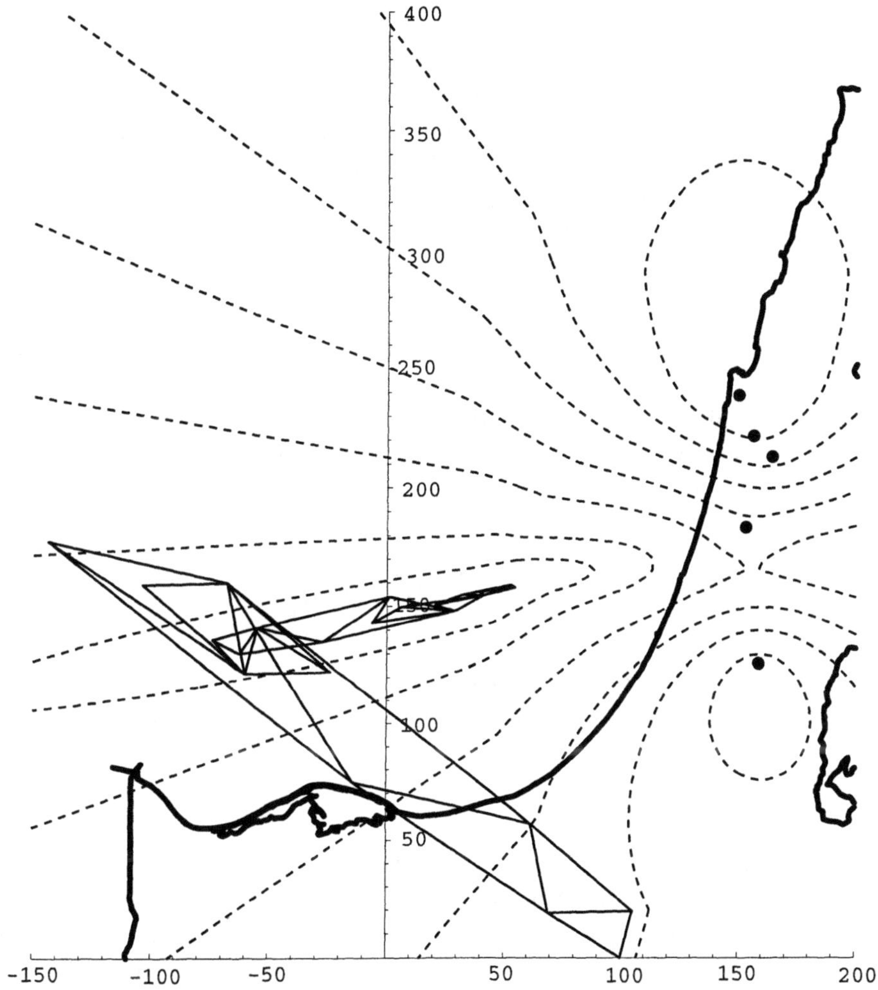

Figure 1. Convergence scheme of the Nelder and Mead Simplex algorithm applied to a Tel-Aviv offshore earthquake of September 1990. The search progresses by forming a sequence of simplexes starting near x=100, y=0. Note the change in search direction around point x=-50, y=140, owing to a contraction operation. RMS contour map (dashed lines) associated with the five P-readings of the stations shown in this figure (filled circles) is superimposed. Bold lines are coastlines.

We define two functionals $T(\mathbf{x})$ and ϕ. The first functional measures the extent of constraints violation. The algorithm will attempt to minimize this functional while simultaneously trying to reduce the objective function. $T(\mathbf{x})$ is defined as

$$T(\mathbf{x})=\left\{\sum_{i=1}^{m} h_i^2(\mathbf{x})+\sum_{i=m+1}^{p} U_i g_i^2(\mathbf{x})\right\}^{1/2} \tag{1}$$

where U_i is the Heaviside operator such that U_i=0 for $g_i \geq 0$ and U_i=1 for $g_i(\mathbf{x})<0$. Thus, $T(\mathbf{x})$ is the sum of the positive squared values of all the violated constraints. Note that a solution vector $\mathbf{x}$ is defined as feasible if $T(\mathbf{x})=0$ and non-feasible if $T(\mathbf{x})>0$.

The second functional, ϕ, controls the extent of constraints violation and also serves as a termination criterion. ϕ is defined as

$$\phi^{(k)} = \min\ \{\phi^{(k-1)}, \mathit{SimplexRadius}^{(k)}\} \tag{2}$$

where the simplex radius of the k-th stage is defined as the average distance of each simplex vertex $\mathbf{x}_i$ from the simplex centroid (usually defined as $\mathbf{x}_{(n+2)}$, see Appendix A):

$$\mathit{SimplexRadius}^{(k)} = \frac{n-m+1}{r+1}\sum_{i=1}^{n+1}\left\|\mathbf{x}_i^{(k)} - \mathbf{x}_{n+2}^{(k)}\right\| \tag{3}$$

where r is the number of simplex vertices.

The initial value $\phi^{(0)}$ is determined using the empirical formula:

$$\phi^{(0)} = 2(m+1)t \tag{4}$$

where t is the initial size of the simplex. To make it easier for the algorithm to avoid converging to local minima, it is desirable to select the value of t such that the initial simplex will spread widely out over the topology of the objective function. The value of t should be selected as a function of the expected range of variation of the independent variables. One practical suggestion (see Himmelblau p. 357) is the following:

$$t = \min\left\{[\frac{0.2}{n}\sum_{i=1}^{n}(U_i - L_i)]\ (U_1 - L_1),\ldots,(U_n - L_n)\right\} \tag{5}$$

where $(U_i - L_i)$ is the difference between the upper and lower bounds of the independent variables.

Constraints are incorporated into the minimization scheme by introducing the concept of "intermediate region" or a "near-feasible region" within which minimization of the objective function is allowed even though the accepted solution (i.e., points which yield lower value of the objective function) "slightly" violates the constraints. Choosing the near-feasible region definition to be a function of a simplex radius, expressed recursively as the function $\phi^{(k)}$, ensures that as the search proceeds to the solution of Problem I, the near-feasible region "shrinks" (in the Nelder and Mead sense, i.e., the simplex radius tends to zero) towards the feasible region. Then the resulting final solution that minimizes the objective function also satisfies the constraints. As a result of this strategy, Problem I can be replaced by a simpler problem having the same solution:

Problem II:
Minimize: $f(\mathbf{x}),\ x \in E$
Subject to: $\phi^{(k)} - T(\mathbf{x}) \geq 0$

Using the above definitions of $\phi^{(0)}$ and $T(\mathbf{x})$, the sharp definition of a point as belonging to either a feasible or a non-feasible region is now relaxed and another possibility, that of "near-feasibility" is introduced. The advantage of introducing the near-feasible concept lies in the fact that minimization of the objective function is allowed even at non-feasible points (actually at near-feasible points) while ensuring that the final solution will be a feasible one. The definition of feasible, near-feasible and non-feasible are expressed by the following conditions:

A vector $\mathbf{x}^{(k)}$ in E^n is said to be:
1) Feasible, if $T(\mathbf{x}^{(k)}) = 0$
2) Near-feasible, if $0 \leq T(\mathbf{x}^{(k)}) \leq \phi^{(k)}$
3) Non-feasible if $T(\mathbf{x}^{(k)}) > \phi^{(k)}$

The basic structure underlying the FTM algorithm can now be outlined as follows:

1) Set up the initial tolerance criterion ϕ^0. Note that it decreases during the progression of the search.
2) Moving from stage (k) to stage (k+1):
   I. Improve the value of $f(\mathbf{x})$ using the Nelder and Mead method to obtain $\mathbf{x}^{(k+1)}$ from $\mathbf{x}^{(k)}$.

II. If the resulting $\mathbf{x}^{(k+1)}$ is feasible or near-feasible ($0 \leq T(\mathbf{x}^{(k+1)}) \leq \phi^{(k)}$), then the move from $\mathbf{x}^{(k)}$ to $\mathbf{x}^{(k+1)}$ is permitted, return to Step I.
III. If the resulting $\mathbf{x}^{(k+1)}$ is non-feasible, then minimize $T(\mathbf{x})$ using the Nelder and Mead algorithm to obtain a new vector in the feasible region, $0 \leq T(\mathbf{x}^{(k+1)}) \leq \phi^{(k)}$ and return to Step I.

Note that the tolerance criterion $\phi^{(k)}$ defined in Equation 2 is a positive non-increasing function of the sequence of points $\mathbf{x}^{(0)},\mathbf{x}^{(1)},\ldots,\mathbf{x}^*$, generated during the progression of the search. The value of $\phi^{(k)}$ defined as a "simplex radius" *does not depend on the value of the objective function or on the constraint values and decreases only when a* **contraction** *operation is encountered.* This, in turn, guarantees that the extent of constraints violation in Problem II is progressively decreased as the search moves toward the solution of Problem I. Because the constraints are loosely satisfied in the early stages of the search and more tightly satisfied only as the search approaches the solution of Problem II, the overall computation effort required in the optimization is considerably reduced (Himmelblau, Chapter 8).

4. APPLICATION OF P-ARRIVAL ORDER CONSTRAINTS TO EARTHQUAKE LOCATION

The specific handling of the constraints as implemented by the FTM has important implications for the problem of automated seismic events location and the problem of location using spatially sparse networks. As all location algorithms require initial guessing of the hypocentral parameters, it is vitally important that the search for the global solution start from a region in the parameter space that is close enough to the true hypocenter to allow convergence. The P-wave arrival order at the different network stations constitutes a natural bound on the epicentral position (Anderson, 1981; Anderson, 1982; Sambridge and Kennett, 1986). Each pair of stations provides a geometrical constraint on the position of the epicenter. If we take the two earliest stations, say Stations 1 and 2, then the epicenter must lie closer to 1 than 2 and must lie below the perpendicular bisector of the line joining 1 and 2 closer to Station 1. All the other possible combinations of pairs of stations constitute n(n-1)/2 geometrical constraints on the epicenter location. If we run the FTM using these constraints and start the search from a non-feasible point, the FTM will not start reducing the misfit travel-time function, but rather will search for another starting point close enough to the feasible region which is either a feasible or near-feasible point. Such a new starting point will be obtained by minimizing the functional $T(\mathbf{x}^{(k)})$ until the condition $\phi^{(k)} - T(\mathbf{x}) \geq 0$ is satisfied.

We illustrate these ideas by analyzing a small earthquake (M_L=2) that occurred in the northern part of the Jordan rift on 24 January 1999. For our analysis, we used the five closest stations from the ISN. Figure 2(a-b) shows the set up of this event. The depth of this event is well constrained; using only stations within 100 km of the epicenter, it was estimated at 17.6 km.

Figure 2a shows the five stations (gray circles), the epicenter (dark circle) and the associated set of perpendicular bisector lines induced by the arrival times order. The stations are numbered in accordance with the order of their P-arrivals. There are 5(5-1)/2=10 bisector lines shown in the figure. Superimposed on this figure is the RMS contour map derived from the five P-readings. The RMS is defined as:

$$RMS = \left(\frac{1}{n} \sum_{i=1}^{n} (t_i^{obs} - t_i^{cal})^2 \right)^{1/2} \tag{6}$$

where t_i^{obs} and t_i^{cal} are the observed and calculated travel times to station i respectively, and n is the number of observations. Figure 2b depicts the feasible region (filled polygon) formed by the intersection of the 10 perpendicular bisector lines.

We first examine the sensitivity of the unconstrained version of the FTM (i.e., the Nelder and Mead algorithm) to the initial hypocenter in the context of the four dimensional problem. Figure 3 shows a set of 1000 random initial epicenters sampled inside a 200 km long square centered at the epicenter. Using a four parameter space (origin time and hypocenter), setting the initial simplex size at 10 km starting from an initial focal depth of 10 km and using the P-arrivals for the five selected stations, we ran the Nelder Mead algorithm for each initial epicenter in Figure 3. The overall majority of the runs converged to the "true" estimated hypocenter. Figures 4 and 5 depict the set of initial epicenters for which the Nelder Mead algorithm converged to incorrect solutions. Figure 4 shows a set of 36 initial epicenters for which the converged epicentral solutions lie at distances greater than 10 km from the "true" estimated epicenter. Figure 5 is even more striking. It shows a list of 13 (only!) initial epicenters for which the focal depth differences (with respect to the "true" estimated focal depth) are greater than 5 km. Most of the Gauss-Newton location algorithms are much more sensitive to the position of the initial guess. A common technique employed by derivative-based location programs such as HYPOINVERSE (Klein, 1978) to attempt avoiding divergence due to a poor starting point, is to begin the search with a linearized system with two degrees of freedom only (x and y variables) and increase it to four degrees of freedom once convergence to the global

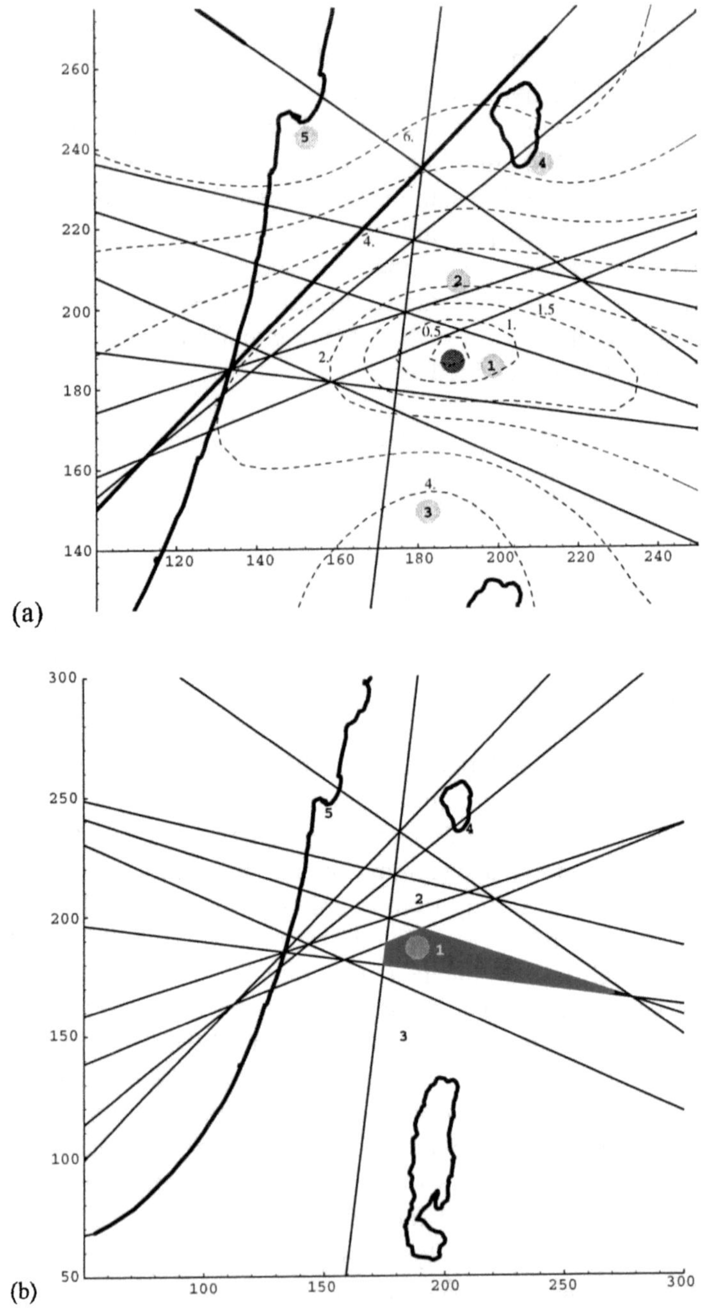

Figure 2. (a) Stations (gray circles) epicenter (black circle) and 10 perpendicular bisector lines associated with the 24 January 1999 event. Stations are marked in accordance with their distances from the epicenter. RMS contour map associated with the P-readings to these five stations is superimposed. (b) The feasible region (dark polygon) formed by intersections of the 10 perpendicular bisector lines obtained from the arrival times

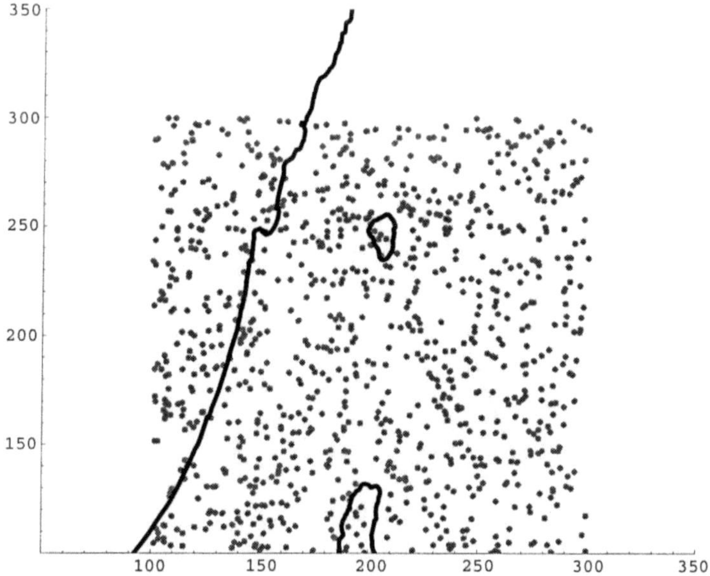

Figure 3. Set of 1000 random initial epicenters associated with the event shown in Figure 2.

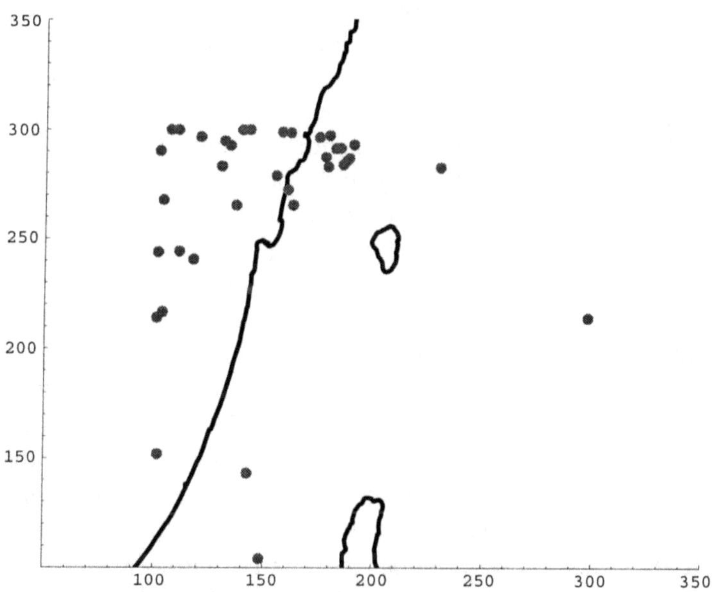

Figure 4. Set of 36 random initial epicenters associated with the event in Figure 2 for which the Nelder and Mead algorithm with 4 degrees of freedom converged to an incorrect epicenter (epicentral shift greater than 10 km).

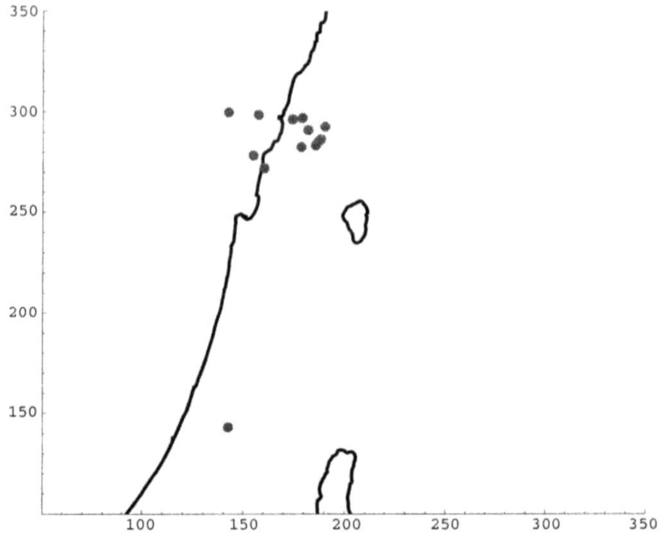

Figure 5. Set of 13 random initial epicenters associated with the event in Figure 2 for which the Nelder and Mead algorithm with 4 degrees of freedom converged to an incorrect focal depth (estimated depth shift greater than 5 km).

optimum is approached. This strategy is, of course, recommended for all location algorithms and was adopted for the FTM. It is worth noting that once the number of degrees of freedom is set to four (i.e., at the vicinity of the solution), the Nelder Mead algorithm performs substantially better in estimating the origin time and focal depth parameters, as it is designed to adapt itself effectively to the topology of the objective function.

The problematic nature of origin time estimation is connected with the existence of a strong correlation between origin time and focal depth. While the epicentral parameters are largely constrained by the station geometry, the focal depth can be traded off almost freely with the origin time with little effect on the misfit function. To overcome this problem, several authors suggested the use of a spatial-temporal separation in hypocenter location. Sambridge and Kennett (1986) and Kennett (1992) used golden section search (Whittle, 1971) in the origin time while minimizing the spatial location on a series of grids. Billings (1994) used a similar approach in a search algorithm that is based on the simulated annealing technique, and Billings et al. (1994) incorporated the spatial-temporal separation in a location procedure based on a genetic algorithm. We did not implement this idea into the FTM.

How does the FTM work subject to the constraints that are induced by the order of P-arrival times? First we select the value of t (initial simplex size) as 10 km, which implies $\phi^{(0)} = 2*10=20$ km (Equation 4).

Figure 6 is a contour map of the values of the functional $T(\mathbf{x})$, which measures the amount of constraints violation. The filled polygon indicates the feasible region (the intersections of the 10 bisector lines induced by the P-arrival order to the 5 stations). The contour labelled 20 delineates the initial near-feasible region for our problem. As the search progresses, this region shrinks towards the feasible region while ensuring that only points inside the sequence of decreasing near-feasible regions (as well as feasible regions) are accepted as reducing the value of the travel time misfit function.

Given any initial hypocenter, the first task of the algorithm is to check whether or not this initial point is qualified to start the search. It qualifies only if it lies inside the near-feasible region. Points that are located in the non-feasible region are transferred to the near-feasible region using the Nelder and Mead algorithm. Figure 7 shows the projection of the 1000 randomly picked initial epicenters onto the initial near-feasible region. The task of determining an initial starting point that lies inside the feasible region can also be carried out using standard linear programming techniques (see Fletcher, 1980, Vol. 2, Sections 8.2 and 8.4; Anderson, 1981).

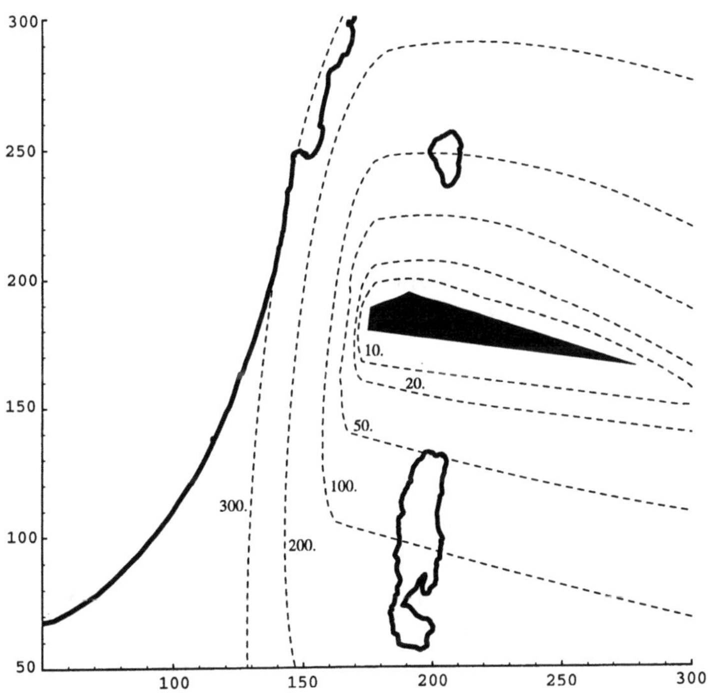

Figure 6. Feasible region (dark polygon) and various contours (dashed lines) indicating near-feasible regions induced by P-arrival orders associated with the event in Figure 2. The contour map of the near-feasible regions is calculated from the positive squared values of all the violated constraints.

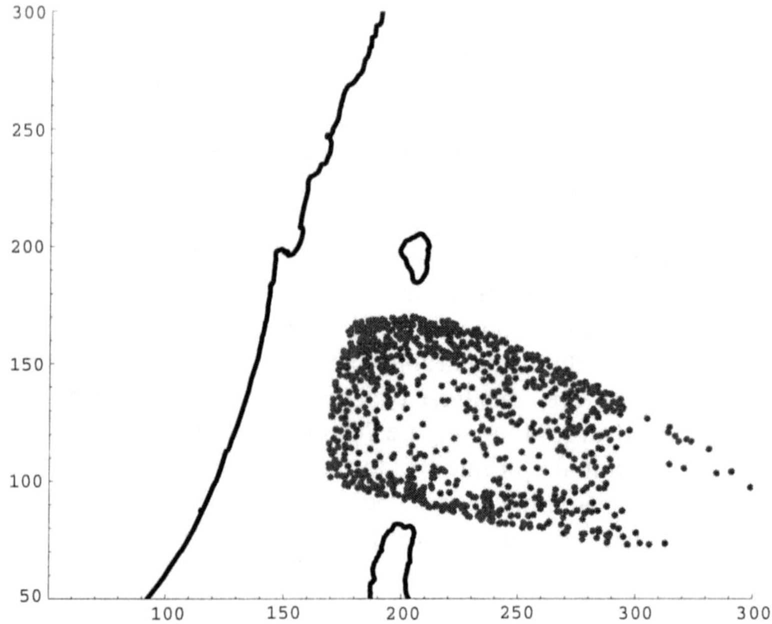

Figure 7. Transformation of the set of 1000 initial epicenters shown in Figure 3 onto the near-feasible region using tolerance level 20. In the constrained location scheme, this transformation ensures that every initial guess starts from a feasible point.

Once an initial epicenter is found inside the near-feasible or the feasible region, the search continues as follows:

1) Form an initial simplex where one of the simplex vertices (or, alternatively, its centroid) is the initial point found using the above procedure.
2) Reflect the worst vertex of the simplex through its centroid.
3) If the reflected vertex falls in the non-feasible region, transfer this point to a point in the near-feasible region by minimizing the functional $T(\mathbf{x})$ using the Nelder and Mead algorithm until $\phi^{(k)} - T(\mathbf{x}) \geq 0$ is satisfied, where the reflected point serves as the starting point.
4) If the above reflected/transferred point yields a point that is better than the best vertex of the simplex, expand this point (to keep the momentum of the search along the same direction) and check whether or not this expanded point lies in the near-feasible region. If it lies in the non-feasible region, transfer it to a point in the near-feasible region using the procedure described in Step 3 above. If the expanded/transferred point is better than the best vertex in the simplex, replace the worst vertex of the simplex by this (expanded) new point, declare a new simplex and go to Step 2. Otherwise,

replace the worst vertex of the simplex by the reflected point (obtained in Step 3) and go to Step 2.

5) If the above reflected/transferred point yields a point that is worse than all the other vertices in the simplex, perform a contraction operation (in an attempt to change the search direction) and transfer to the near-feasible region if necessary (using the procedure mentioned in Step 3). If the resulting contracted/transferred point is better than the worst vertex in the simplex, replace the worst vertex of the simplex by this new point, declare a new simplex and go to Step 2. Otherwise, perform a reduction operation (reduce the simplex size by half) which means that at this point the simplex starts collapsing toward the global minimum and go to Step 2.

6) Termination: the procedure terminates when $\phi^{(k)} \leq \varepsilon$ for a specified ε.

Figure 8 (a, b) shows two convergence trajectories to this epicenter. The one on the top resulted from the unconstrained Nelder and Mead algorithm (8a) and that on the bottom (8b) was obtained by FTM. Both trajectories start in the lower left-hand corner (coordinates x=0, y=0). As may be observed from these figures, the FTM makes a "large detour" in the course of its progression, caused by starting from a near-feasible point, whereas the Nelder and Mead algorithm approaches the epicenter more or less along a straight line. However, as shown in these Figures, the Nelder and Mead trajectory, being an unconstrained algorithm, oscillates back and forth between feasible and non-feasible regions, while the FTM trajectory oscillates mainly inside the region confined to the near-feasible region.

An interesting question that arises in the context of using the FTM for earthquake location subject to constraints induced by P-arrival order, is how sensitive the arrival order zone is to errors in picking the arrival times. Sambridge and Kennett (1986) addressed this question by suggesting a scheme for sampling grid points on either side of the boundary of the P-arrival zone and finding all those that satisfy a perturbed constraint. This constraint is generated by looking at a set of points along the boundary, calculating the P-wave travel times from the point to both stations and then selecting those epicentral points whose travel times would lie in a specified range about the travel times for the boundary point.

The FTM offers a different approach for handling the problem of a "perturbed constraint." For example, we can interpret the near-feasible region associated with tolerance level 10 (Figure 6) as a P-arrival zone associated with picking errors of about 2 seconds. In assessing the impact that the arrival time picking errors have on the P-arrival zone, it is important to check whether a given level of error causes a change in the order of P-arrival. If the P-arrival order associated with a given error is maintained, then the relevant near-feasible region can represent a "perturbed constraint."

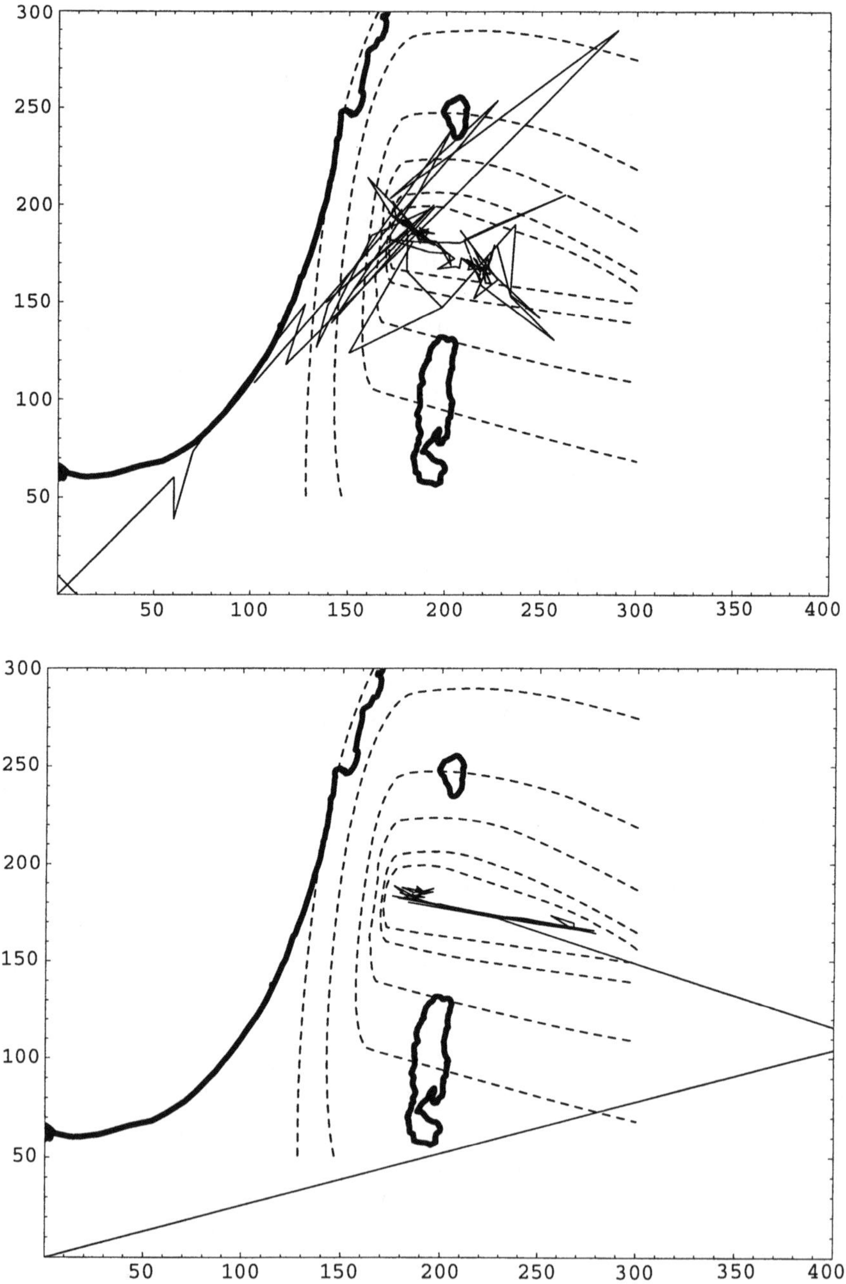

Figure 8. Convergence trajectories associated with the event in (2): (a) top - employing an unconstrained Nelder and Mead algorithm; (b) bottom - employing FTM algorithm subject to the inequality constraints induced by P-arrival orders. Contours of inequality constraints are superimposed on both figures. Note how the path oscillations associated with the FTM are confined inside the smallest near-feasible region.

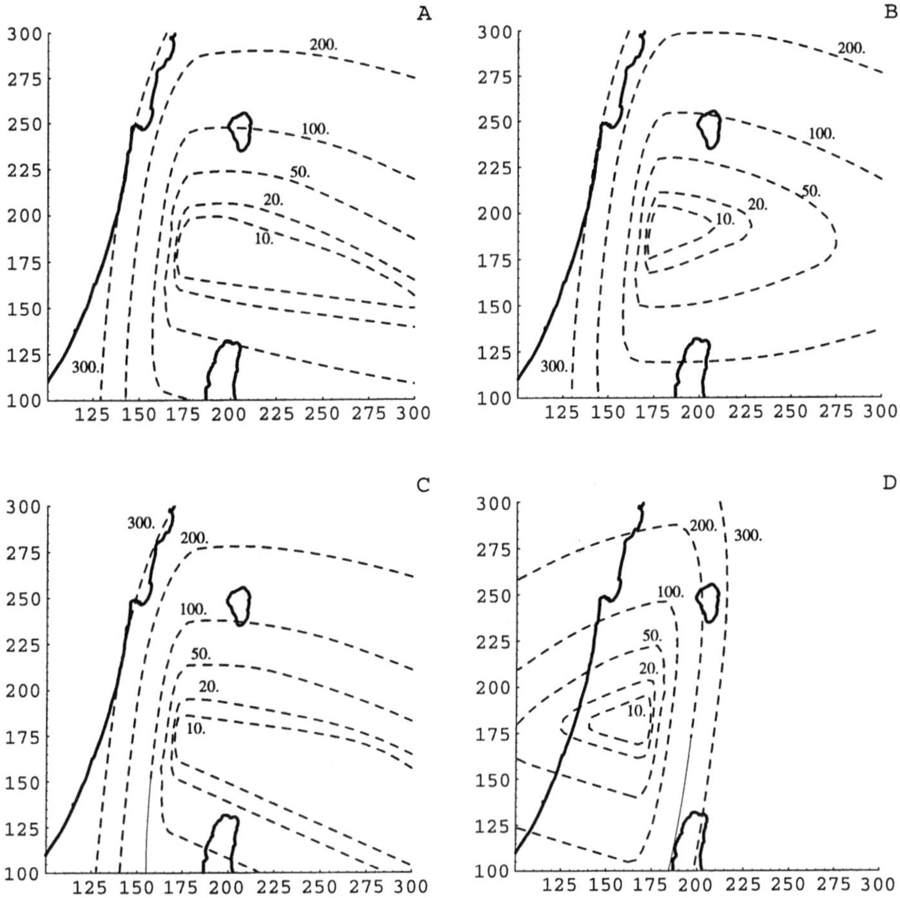

Figure 9. Changes in the feasible regions caused by changing one pair of arrival times order associated with the event in Figure 2. (a) The original contour map of inequality constraints, (b-d) three different near-feasible regions generated by altering one pair of arrival times order each time. Note that substantial changes in the original near-feasible zone (9a) may be due to picking errors on the order of two seconds (see text).

It is instructive to see what happens if the P-arrival order is altered. Figure 9 shows four different near-feasible regions associated with the P-arrival order of events shown in Figure 2. Compared to the original order (Figure 9a), Figures 9b to 9d show substantial changes in the P-arrival zones (and the associated near-feasible regions) introduced by an error level of about 2 seconds. Each figure represents a change in only one pair of arrivals.

These examples can guide us in estimating a perturbed-constraints P-arrival zone. From the complete set of P-arrivals associated with a given event, we can select a sequence of P-arrivals (set S) whose arrival order can be established with a high degree of confidence (in a noisy data set, finding the time shift using a cross correlation can do the trick). This sequence of arrivals $\{t_i\}$, $i \in S$, constitutes a set of inequality constraints that defines a P-arrival zone that is associated with time picking errors in the order of the minimum time difference of the set of successive pairs from this sequence:

$$\Delta t = \min_{i \in S} |t_{i+1} - t_i| \qquad (7)$$

In other words, since perturbations in the arrival time picking that are bounded by $|\Delta t|$ do not change the order of the P-arrival of set S, the above P-arrival zone is insensitive to a travel time picking error in the order of $|\Delta t|$.

5. APPLICATION TO AN OFFSHORE EVENT LOCATION

The task of locating offshore events using continental stations poses severe difficulties for any location algorithm. In this section, we show how the FTM can be effectively used to locate offshore events when "natural" simple constraints are incorporated into the location scheme. In order to illustrate our idea we refer again to the Tel-Aviv, M_L=3.9, offshore event of 13 September 1990, mentioned in Section 3. In our analysis we used five P-wave and one S-wave (at station DLIA) readings of ISN stations shown in Figure 10. For location purposes, we applied the standard travel time routines for a 1-D layered model (Lee and Stewart, 1981) and used a three-layered velocity model employed by the ISN.

Figure 10 depicts the convergence trajectory of the unconstrained location method (i.e., Nelder and Mead) to this event in the x-y plane (RMS contour map superimposed). Note the decaying zig-zag pattern of the trajectory. Unlike many optimization procedures, this algorithm approaches the minimum by moving away form high RMS values rather than by trying to move along a line (Gauss-Newton) toward the minimum.

At the next stage, we incorporate constraints into this location scheme. Figure 11 depicts the epicentral convergence trajectory obtained by the FTM, subject to the constraints that force the epicenter to lie inside the ring shown in the figure. The ring is formed out of two epicentral distance estimates from station DLIA (based on S-P times). Starting from a non-feasible initial epicenter far to the southwest, the FTM first maps the initial epicenter to a point inside the feasible region (around point x=130, y=125). The search then progresses westward toward the epicenter. Observe that this trajectory crosses the feasible region several times. The amount of constraints violation is gradually diminished until convergence is achieved.

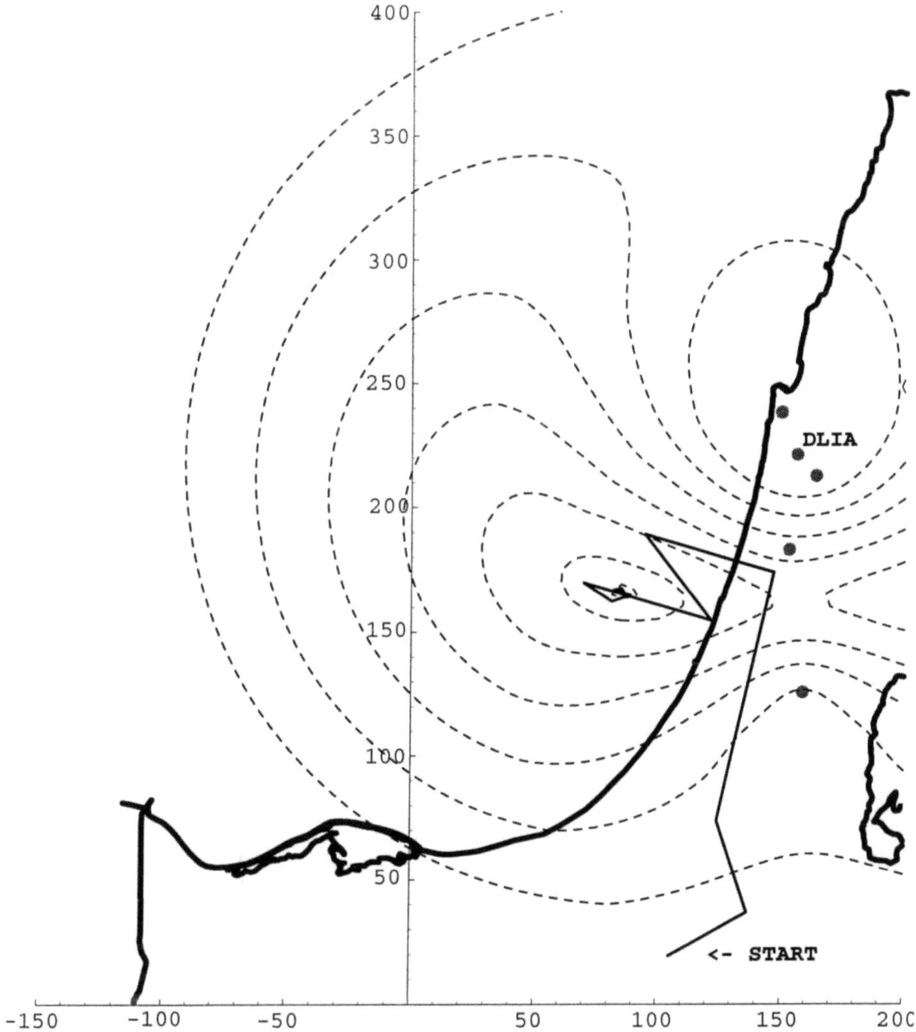

Figure 10. Convergence trajectory to the Tel-Aviv offshore event applying unconstrained Nelder and Mead algorithm using five P-readings and one S-reading from the five stations shown (filled circles). The S-reading is taken from station DLIA. Note the zig-zag pattern of the trajectory. Unlike many other linearized algorithms, this algorithm approaches the minimum by moving away from points with high RMS values (dashed contours).

Such a location scheme may look particularly attractive if we are specifically interested in robust location. We may, for example, suspect that our set of arrivals contains one or more outlying arrival times, but our knowledge of the velocity structure is insufficient to detect these outliers using standard statistical techniques (see Cook and Weisberg, 1982; Rabinowitz, 1986; Rabinowitz and Steinberg, 1988). Assuming that we have

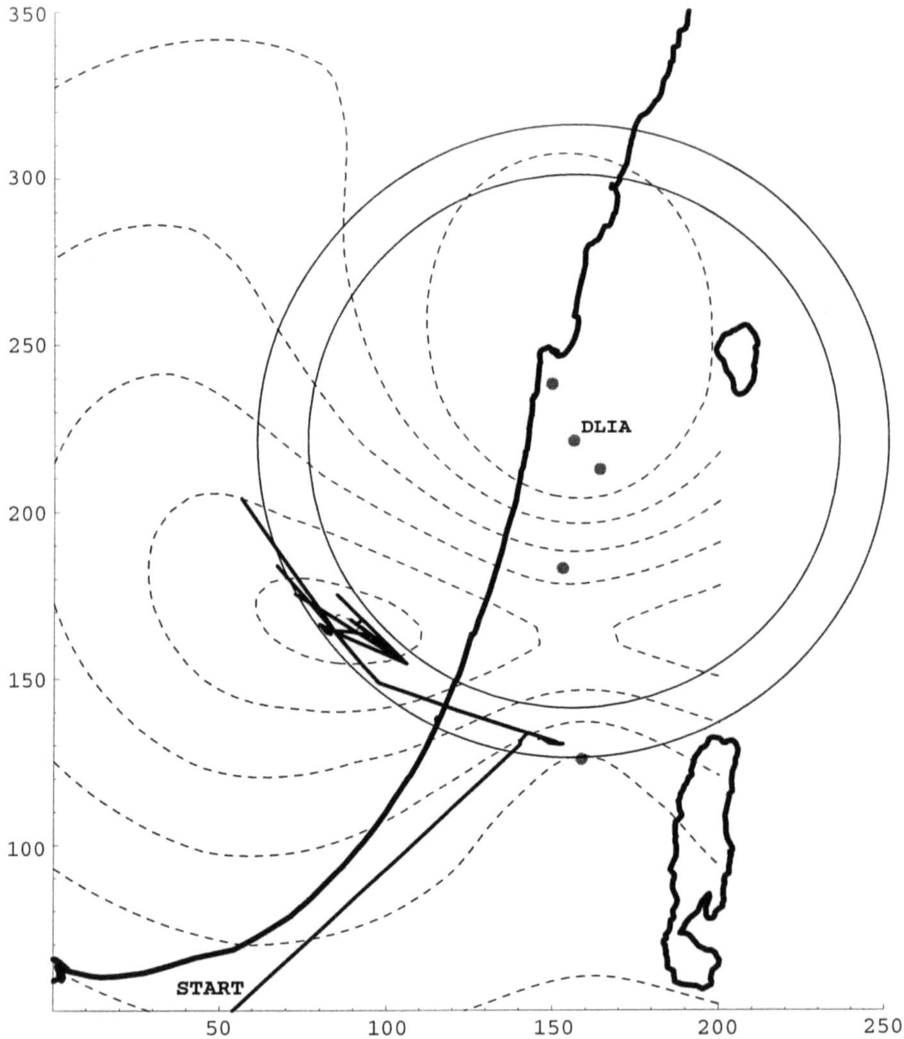

Figure 11. Convergence trajectory of the offshore event in Figure 10 when applying the FTM subject to two inequality constraints formed by two epicentral distances from station DLIA.

some a-priori knowledge of S-P travel time with respect to a given station, then constraining the solution inside a desirable ring estimated from a given S-P value may yield a robust estimate that substantially down-weights the outlying arrivals. The following is a simple illustration of this idea.

We have successively introduced outlying values of 2 seconds to each of the P-readings of the above event and then, for each outlier, we have located the event using first the unconstrained option and then the constrained option (i.e., the FTM), where we have constrained the solution to lie inside a ring formed out of the distances of 90-100 km from the epicenter to station DLIA. We have obtained the following results: for two stations the outlier caused a shift in the epicenter of 40 and 30 km respectively (relative to the assumed "true" epicenter) for the unconstrained option, while the constrained FTM version, as expected, fixed the estimated solutions within distances of 5 and 10 km from the "true" epicenter. The epicentral location was relatively insensitive to the presence of 2-second outlier values at the rest of the stations.

Azimuthal constraints are also used by many current location programs (Leinert et al., 1986; Roberts et al., 1989; Ruud et al., 1988). The incorporation of azimuthal constraints in the FTM seems to work relatively well. Figures 12a and 12b show an implementation of inequality azimuthal constraints to the FTM algorithm for the above offshore event of September 13 1990. The four rays drawn from the two marked stations (filled circles) form a feasible region (gray polygon). Each back-azimuth is associated with an error of ±5 degrees with respect to the "true epicenter". In order to illustrate this idea, the two marked stations estimated the epicenter station back-azimuth under the assumption that the location of the "true epicenter" coincides with the calculated epicenter using the unconstrained simplex algorithm. Normally, azimuth calculations will be evaluated independently from three-component single stations using a polarization technique (Ruud et al., 1988) or from an array using standard beam forming methods. Thus, we have four inequality constraints, two for each station (forming an angle of 10 degrees).

Figure 12a depicts the feasible region (gray polygon) and several near-feasible regions induced by these constraints. Figure 12b shows the RMS contour map associated with five P-readings of this event, superimposed on the feasible regions. When we run the unconstrained algorithm using the five P-readings and one S-reading (from station DLIA) the estimated epicenter is x=81.4, y=162.4, with an RMS value of 0.25. Omitting the S-reading, the unconstrained algorithm shifts the epicenter some 30 km to the west (x=52.4, y=156.6, RMS=0.16). Constraining these data using the above four inequality azimuthal constraints (by means of the FTM) locates the event at the left-hand edge of the feasible region (coordinates x=58, y=156, RMS=0.8). Of course, the quality of the resulting constrained solution may depend heavily on the reliability of the azimuth estimates. However, there are indications that azimuthal errors estimated by arrays or even small arrays (such as tripartite arrays), may be realistically restricted to about ±5 degrees. If we have good reason to believe that the azimuth estimate is sufficiently accurate, then we can use equality constraints.

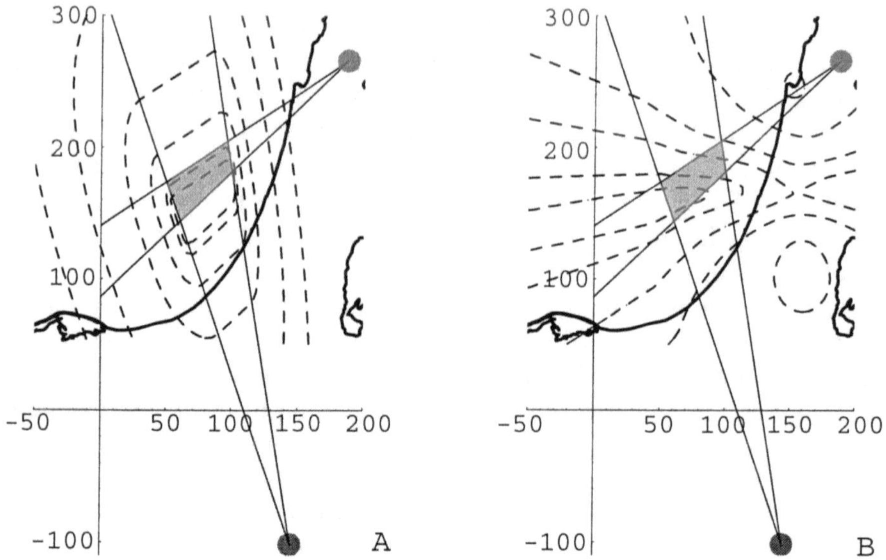

Figure 12. Constraining the offshore event in Figure 10 using four inequality constraints formed by back azimuths from two stations (gray circles): (a) feasible region (gray polygon) and several near-feasible regions induced by these constraints; (b) RMS contour map associated with five P-readings of this event, superimposed on the feasible region.

6. STATISTICAL UNCERTAINTY

A major drawback of the FTM is that it does not provide information on errors associated with statistical estimates. In their original paper, Nelder and Mead (1965) pointed out that such errors could be evaluated by adding a few selected points and generating the Hessian matrix. The error estimation associated with the FTM is performed using the standard approach, which entails calculations of the travel time derivatives with respect to the hypocentral parameters. Confidence ellipses are computed via:

$$(\mathbf{h}-\overline{\mathbf{h}})^T\,\mathbf{C}^{-1}(\mathbf{h}-\overline{\mathbf{h}}) \le k_\alpha^2 \tag{8}$$

where $\overline{\mathbf{h}}$ is the estimated hypocenter, $\mathbf{C}$ is the covariance matrix associated with the matrix of the partial derivatives evaluated at the estimated hypocenter. The value of k_α depends upon the distribution of the residuals, usually assumed to be normally distributed. Under the assumption that picking and velocity model errors are normally distributed, Flinn (1965) proposed the following formula for k_α:

$$k_{\alpha}^{2} = 4\, s^{2} F_{\alpha}(4, n-4) \tag{9}$$

where $F_{\alpha}(4, n\text{-}4)$ is an F statistic with 4 and n-4 degrees of freedom for the critical level α and

$$s^{2} = \frac{1}{(n-4)}\left(\frac{1}{n}\sum_{i=1}^{n} r_i^{2}\right)^{1/2} \tag{10}$$

where n is the number of observations and $r_i = t_i^{obs} - t_i^{cal}$.

The confidence ellipsoids that are calculated using this relation may be unrealistically large in case of a small number of observations (Evernden, 1969). Jordan and Sverdrup (1981) suggested a modification in this relation by taking account of a priori information on the residual distribution:

$$k_{\alpha}^{2} = \frac{4K + 4s^{2}(n-4)}{K+(n-4)} F_{\alpha}(4, K+n-4) \tag{11}$$

where K is the parameter that controls the amount of a priori information. For $K = 0$ Flinn's relation is retained, while for $K = \infty$, it is assumed that there is exact prior information. Jordan and Sverdrup (1981) recommended a value of $K = 8$. At this writing, this value is also adopted by the IMS.

7. SUMMARY AND CONCLUSIONS

In this study we have presented the constrained optimization method referred to as the Flexible Tolerance Method, as an efficient location algorithm. The algorithm may be considered a natural extension of the Nelder and Mead algorithm; it is a derivative-free algorithm and, as such, function evaluation is the only type of repetitive information used. The concept of a near-feasible region introduced in the algorithm enables the search process to adapt itself to the constraints before they are "severely" enforced as the search approaches the global minimum.

For the location problem, the order of P-arrival times constitutes an important set of inequality constraints on the epicentral position. These constraints are easily incorporated into the FTM. They may play an important role in an automatic location scheme, since the FTM accepts only initial epicenters that fall inside a near-feasible region (induced by the order of P-arrival times), and if a trial epicenter falls in the non-feasible region, a built-in procedure transfers this point to a near-feasible region.

We have presented three cases commonly encountered in practical earthquake location in which there is a real need to incorporate nonlinear constraints into the location scheme. These constraints are the time order of P-arrivals, back azimuths to some stations, and source-to-station distances derived from S-P information. Such constraints can be easily and effectively incorporated into the FTM algorithm and are likely to improve its convergence properties by controlling the convergence domain and down-weighting possible outlying observations. The use of nonlinear constraints is especially valuable under severe unfavorable conditions, such as the Tel-Aviv offshore event discussed in Section 5.

Most of the standard linearized location algorithms currently in use in many seismological centers do not use nonlinear constraints in routine analyses. Under unfavorable conditions, these algorithms usually truncate small eigenvalues and many iterations update the adjustment solution vector in less than the full parameter space dimension, which may very often lead to an incorrect solution. The proposed FTM location algorithm may complement these algorithms in analyzing poorly constrained events.

ACKNOWLEDGEMENTS

The author thanks Anthony Lomax and Clifford Thurber for helpful comments.

REFERENCES

Anderson, K.R. (1981) Epicentral location using arrival time order, *Bull. Seism. Soc. Am.* **71**, 541-546.

Anderson, K.R. (1982) Robust earthquake location using M-estimates, *Phys. Earth Plan. Int.* **30**, 119-130.

Billings, S.D. (1994) Simulated annealing for earthquake location, *Geophys. J. Int.* **118**, 680-692.

Billings, S.D., B.L.N. Kennett, and M.S. Sambridge (1994) Hypocenter location: genetic algorithms incorporating problem-specific information, *Geophys. J. Int.* **118**, 693-706.

Claerbout, F. and F. Muir (1973) Robust modeling with erratic data, *Geophysics*, **38**, 826-844.

Cook, R.D. and S. Weisberg (1982) *Residuals and influence in regression*, Chapman and Hall, New York, 230 pp.

Evernden, J.F. (1969) Precision of epicenters obtained by a small number of world wide station, *Bull. Seism. Soc. Am.* **59**, 1365-1398.

Fletcher, CR. (1981) *Practical methods of optimization*, Vol. 2, John Wiley, New York, 224 pp.

Flinn, E.A. (1965) Confidence region and error determinations for seismic event location, *Rev. Geophys.* **3**, 157-185.

Himmelblau, D.M. (1972) *Applied nonlinear programming*, McGraw-Hill Book Company, 498 pp.

Jordan, T.H. and K.A. Sverdrup (1981) Teleseismic location techniques and their applications to earthquake clusters in the south-central Pacific, *Bull. Seism. Soc. Am.* **71**, 1105-1130.

Kennett, B.L.N. (1992) Locating oceanic earthquakes – the influence of regional models and location criteria, *Geophys. J. Int.* **108**, 848-854.

Klein, R.W. (1978) Hypocenter location program HYPOINVERSE, U.S.G.S. Open File Report, 78-694.

Lee, W.H.K. and S.W. Stewart (1981) *Principles and application of microearthquake networks*, Academic Press, New York, 293 pp.

Leinert, B.R., E. Berg, and L.N. Frazer (1986) HYPOCENTER: an earthquake location method using centered, scaled and adaptively damped least squares, *Bull. Seism. Soc. Am.* **76**, 771-783.

Nelder, J.A. and R. Mead (1965) A simplex method for function minimization, *Computer J.* **7**, 308-313.

Prugger, A.F. and D.J. Gendzwill (1988) Microearthquake location: a nonlinear approach that makes use of simplex stepping procedures, *Bull. Seism. Soc. Am.* **78**, 799-815.

Olsson, D.M. and L.S. Nelson (1975) The Nelder-Mead Simplex procedure for function minimization, *Technometrics* **17**, 45-51.

Rabinowitz, N. (1988) Microearthquake location by means of nonlinear simplex procedure, *Bull. Seism. Soc. Am.* **78**, 380-384.

Rabinowitz, N. (1986) On identifying influential arrivals by means of density information matrix, *Bull. Seism. Soc. Am.* **76**, 1817-1821.

Rabinowitz, N. and O. Kulhanek (1988) Application of a nonlinear algorithm to teleseismic locations using P-wave readings from the Swedish seismographic network, *Phys. Earth Plan. Int.* **50**, 111-115.

Rabinowitz, N. and D.M. Steinberg (1988) Effect of single arrival times on the hypocentral focal depth determination, *Geophys. J.* **94**, 567-570.

Roberts, R.G., A. Christofferson, and F. Cassidi (1989) Real time events detection, phase identification and source location estimation using single station component seismic data and a small PC, *Geophys. J, Int.* **97**, 471-480.

Ruud, B.O., E.S. Huesebye, S.F. Ingate, and A. Christofferson (1988) Event location at any distance using seismic data from a single, three-component station, *Bull. Seism. Soc. Am.* **78**, 308-325.

Sambridge, M.S. and B.L.N. Kennett (1986) A novel method of hypocenter location, *Geophys. J. Roy. Astr. Soc.*. **87**, 679-697.

Shearer, P.M. (1997) Improving local earthquake location using the L1 norm and waveform cross correlation: application to Whittier Narrows, California, aftershock sequence, *J. Geophys. Res.* **102**, 8269-8283.

Wittle, P. (1971) *Optimization under Constraints*, Wiley, London.

APPENDIX A – NELDER AND MEAD SIMPLEX ALGORITHM

This algorithm minimizes a function of n independent variables using $(n+1)$ vertices of a polyhedron in E^n. Each vertex can be defined by a vector $\mathbf{x}$. The vertex in E^n that yields the highest values of $f(\mathbf{x})$ is projected through the center of gravity (centroid) of the remaining vertices. Improved values of the objective function are found by successively replacing the point with the highest value of $f(\mathbf{x})$ by better points until the minimum of $f(\mathbf{x})$ is found.

The procedure starts by constructing an initial regular simplex. The coordinates of the vertices of the initial regular simplex can be extracted from the following $n(n+1)$ matrix $\mathbf{D}$, in which the columns represent the components of the vertices and the rows represent the coordinates, i=1 to n:

$$\mathbf{D} = \begin{bmatrix} 0 & d_1\, d_2 \ldots d_2 \\ 0 & d_2\, d_1 \ldots d_2 \\ 0 & d_2\, d_2 \ldots d_2 \\ \ldots & \ldots\ldots \ldots \ldots \\ 0 & d_2\, d_2 \ldots d_1 \end{bmatrix} \tag{A1}$$

where

$$d_1 = \frac{t}{n\sqrt{2}}\left(\sqrt{n+1}+n-1\right) \tag{A2}$$

and

$$d_2 = \frac{t}{n\sqrt{2}}\left(\sqrt{n+1}-1\right) \tag{A3}$$

with t = the distance between two vertices. Define:

$$f(\mathbf{x}_h) = \max\ \{f(\mathbf{x}_1),\ldots,f(\mathbf{x}_{n+1})\} \tag{A4}$$

and

$$f(\mathbf{x}_l) = \min\ \{f(\mathbf{x}_1),\ldots,f(\mathbf{x}_{n+1})\} \tag{A5}$$

where $\mathbf{x}_1,\ldots,\mathbf{x}_{n+1}$ are the $(n+1)$ vertices of the polyhedron in E^n. Define $\mathbf{x}_{n+2}$ as the gravity center (centroid) of these vertices. The procedure for finding the vertex in E^n at which $f(\mathbf{x})$ has a better value involves four operations:

1. Reflection

Reflect $\mathbf{x}_h$ through the centroid by computing

$$\mathbf{x}_{n+3} = \mathbf{x}_{n+2} + \alpha(\mathbf{x}_{n+2} - \mathbf{x}_h) \tag{A6}$$

where $\alpha>0$ is the reflection coefficient, $\mathbf{x}_{n+2}$ is the centroid, $\mathbf{x}_h$ is the vertex at which $f(\mathbf{x})$ is the largest of $(n+1)$ values of $f(\mathbf{x})$.

2. Expansion

If $f(\mathbf{x}_{n+3}) \leq f(\mathbf{x}_l)$, expand the vector $(\mathbf{x}_{n+3} - \mathbf{x}_{n+2})$by computing

$$\mathbf{x}_{n+4} = \mathbf{x}_{n+2} + \gamma(\mathbf{x}_{n+3} - \mathbf{x}_{n+2}) \tag{A7}$$

where $\gamma > 1$ is the expansion coefficient used to elongate the search vector if the reflection has produced a vertex with a value of $f(\mathbf{x})$ smaller than the smallest $f(\mathbf{x})$ obtained prior to the reflection.

If $f(\mathbf{x}_{n+4}) < f(\mathbf{x}_l)$, replace $\mathbf{x}_h$ by $\mathbf{x}_{n+4}$ and continue from Step 1. Otherwise, replace $\mathbf{x}_h$ by $\mathbf{x}_{n+3}$ and continue from Step 1.

3. Contraction

If $f(\mathbf{x}_{n+3}) > f(\mathbf{x}_i)$for all $i \neq h$, contract the vector $(\mathbf{x}_h - \mathbf{x}_{n+2})$ by computing

$$\mathbf{x}_{n+5} = \mathbf{x}_{n+2} + \beta(\mathbf{x}_h - \mathbf{x}_{n+2}) \tag{A8}$$

where $0 < \beta < 1$ is the contraction coefficient used to reduce the search vector if the reflection has not produced a vertex with a value of $f(\mathbf{x})$ smaller than the second largest value of $f(\mathbf{x})$ obtained prior to the reflection. Replace $\mathbf{x}_h$ by $\mathbf{x}_{n+5}$ and return to Step 1.

4. Reduction

If $f(\mathbf{x}_{n+3}) > f(\mathbf{x}_h)$reduce all the vectors $(\mathbf{x}_i - \mathbf{x}_l)$ $i=1,\ldots n+1$ by one half from $\mathbf{x}_l$ by computing

$$\mathbf{x}_i = \mathbf{x}_l + 0.5(\mathbf{x}_i - \mathbf{x}_l) \quad i = 1,\ldots,n+1 \tag{A9}$$

and return to Step 1.

Termination Criterion:

The search is terminated when:

$$\left(\frac{1}{n+1} \sum_{i=1}^{n+1} \left[f(x_i) - f(x_{n+1}) \right]^2 \right)^{1/2} \leq \varepsilon \tag{A10}$$

where ε is an arbitrarily small number. The values of α, β and γ are chosen experimentally; recommended values are: $\alpha=1$, $\beta=0.5$ and $\gamma=2$.

Chapter 3

A STATISTICAL OUTLOOK ON THE PROBLEM OF SEISMIC NETWORK CONFIGURATION

Nitzan Rabinowitz
Geophysical Institute of Israel, PO Box 182, Lod, Israel

David M. Steinberg
Department of Statistics and Operations Research, Raymond and Beverly Sackler Faculty of Exact Sciences, Tel Aviv University, Tel Aviv 69978, Israel

Key words: Comprehensive Nuclear Test Ban Treaty (CTBT), seismic network, *D*-optimal design, hypocenter location, outlier detection, robust configuration, station importance.

Abstract: We examine the problem of optimal seismic network configuration. We apply ideas from the statistical literature on optimal experimental design and influential observations. We review several different criteria that have appeared in the literature and propose a new criterion based on equalizing arrival time importance. All the configuration criteria achieve networks that improve the resolution for determining hypocenters. Our new criterion enhances the robustness of the network to outlying observations at the expense of a slight reduction in resolution. The network configuration problem has taken on special importance in the context of monitoring the Comprehensive Nuclear Test Ban Treaty (CTBT) and we discuss the application of the configuration criteria to CTBT monitoring and to on site inspection search areas.

C.H. Thurber and N. Rabinowitz (eds.), Advances in Seismic Event Location, 51–69.

1. INTRODUCTION

Seismic networks are used to monitor seismicity, locate seismic events, and to determine their magnitudes and source parameters. Technological progress in instrumentation and communications has substantially increased the use of seismic networks, in particular small, mobile networks that can be rapidly deployed in a specific area. The configuration of the network has important implications for the quality of the information that will be obtained; therefore, careful study of the configuration is essential in order to maximize the information-to-cost ratio. In this article, we will review ideas and methods for optimizing network configuration decisions.

The question of optimal network configuration has important implications in the context of the Comprehensive Nuclear Test Ban Treaty (CTBT), an international agreement banning nuclear tests signed in 1996. Seismic monitoring is the most important technology that will be used to verify compliance with the treaty. The CTBT Organization (CTBTO) is in the process of deploying a global seismic network (the International Monitoring System - IMS) that should be able to detect waves generated by nuclear tests. Seismic data from the IMS is recorded on-line by the International Data Center (IDC) in Vienna, which processes and analyzes the data and prepares several bulletins of seismic events.

One of the important elements in the verification regime of the CTBT is the concept of On Site Inspection (OSI). If an event is suspected of being a nuclear test in violation of the treaty, the CTBTO is authorized to send a team of experts to the suspect region to conduct a detailed scientific investigation within days of the event. Data from the global seismic network can be used to make an initial estimate of the epicenter of the event, including a 90% confidence ellipsoid. However, due to the sparse nature of the IMS network, it is difficult to obtain a precise location estimate. Therefore, one of the first tasks of the OSI team is to deploy a small mobile seismic network that will provide much more accurate estimates of epicenters as well as event depths. Rapid deployment of such a network is essential as aftershock sequences from an earthquake or an explosion may decay in strength soon after the main event. Optimal configuration of the network is critical to achieve high resolution power for hypocenter locations.

Network configuration must also be considered in deploying permanent networks for seismic monitoring, for example a national or regional seismic network. Typically, such a network must monitor a rather wide area subject to limitations such as budget, geography, access, etc. Rational allocation of resources requires the use of formal criteria such as those that we will describe here to quantify the quality of a proposed seismic network.

Informed decisions can then be made as to the final size of the network and the most advantageous locations for the stations.

The main focus of this chapter will be to present, analyze and discuss criteria related to optimal network design for event location. Many of the ideas will be drawn from the statistical theory of optimal experimental design. We will also explore the concept of site influence on the location and show how it is related to the design criteria. We will consider an additional criterion based directly on the station influence.

2. CRITERIA FOR OPTIMAL CONFIGURATION

Criteria for optimal network configuration are based primarily on the confidence ellipsoid of the location estimate. The confidence ellipsoid has the form

$$(\theta - \hat{\theta})^T M(\theta - \hat{\theta}) \leq Const \tag{1}$$

where $\theta = (X_0, Y_0, Z_0, t_0)$ contains the hypocenter and origin time, $\hat{\theta}$ is the least squares estimate of θ, $M = A^{\mathrm{T}}A$ is called the information matrix in the statistical literature and A is the n x 4 matrix of partial derivatives of the n arrival times with respect to θ evaluated at $\hat{\theta}$. The value of the constant depends upon the distribution of the residuals, usually assumed to be Gaussian, and on the method used to estimate their variance. Under the assumption that picking and velocity-model errors are normally distributed and that the root mean square error of the residuals s^2 is used to estimate the variance, the constant will be $4s^2F_{\alpha,4,n\text{-}4}$ where n is the number of phase arrivals and $F_{\alpha,4,n\text{-}4}$ is an F statistic with 4 and n-4 degrees of freedom for the critical level α (Flinn, 1965). The above results can be easily extended to the case of weighted least squares estimation, which will be appropriate when the arrival time errors are assumed to have unequal variances (Jordan and Sverdrup, 1981, Bratt and Bache, 1988).

With a small number of observations, the quantile from the F distribution becomes large and the confidence ellipsoid calculated using the above relation might be unrealistically large (Evernden, 1969). Jordan and Sverdrup (1981) suggested a method to solve this problem by taking account of *a priori* information on the residual variance. They proposed a modified estimator of the residual variance

$$s_{JS}^{\;2} = \frac{K s_0^{\;2} + (n-4)s^2}{K + (n-4)} \tag{2}$$

where s_0^2 is a prior estimate of the error variance perhaps based on past events and K controls the amount of *a priori* information. For $K = 0$, Flinn's relation is retained, while setting $K = \infty$ corresponds to the assumption that the error variance is known to be exactly s_0^2. Jordan and Sverdrup recommended a value of $K = 8$. The constant is modified accordingly to $4\, s_{JS}^2\, F_{\alpha,4,n\text{-}4}$. At this writing the IDC follows the Jordan-Sverdrup method with $K = 8$.

2.1 The D-criterion for a Single Source

The volume of the confidence ellipsoid is proportional to $\det(M)^{-1/2}$. An obvious optimality criterion is therefore to make the ellipsoid as small as possible by configuring the network to maximize $\det(M)$. A configuration that maximizes $\det(M)$ is called *D*-optimal. It is common to use $(\det(M))^{1/4}$ as the *D*-criterion rather than the determinant itself. The exponent corresponds to the fact that there are four parameters in the location problem and permits a rough interpretation of the criterion values in terms of sample sizes (see Steinberg et al., 1995 on this issue). The *D*-criterion has been studied at length in the statistical literature (Box and Lucas, 1959, Atkinson and Donev, 1992, Pukelsheim, 1993, Atkinson, 1996).

The information matrix M depends, of course, on both the network configuration and the location of the hypocenter. Thus, the *D*-criterion above is in fact formulated only for a single source location. It can be fruitfully applied in settings where it is desired to monitor a specific location that is suspected as a likely source or, perhaps, is the planned site of a critical facility. The former context is highly relevant to OSI monitoring where the initial analysis of the IMS data will provide a tentative event location.

D-optimal seismic networks can be characterized theoretically using a fundamental result known as the equivalence theorem of Kiefer and Wolfowitz (1961). In the context of seismic network configuration, this result shows that the Fréchet derivative of $\det(M)$ with respect to the information from an additional station is proportional to $v = f^{T} M^{-1} f$, where f is the vector of partial derivatives of the arrival times to the new station with respect to θ. With this key result, they were able to show that a configuration is *D*-optimal if and only if $v \leq p$ for all possible sites, where p is the number of parameters in θ (4 for the standard location case). Further, they showed that any *D*-optimal network will have $v = p$ at every station in the network.

Steinberg and Rabinowitz (1999) used the equivalence theorem to prove the following properties of *D*-optimal networks for a single source in a layered velocity model:

1. A *D*-optimal network configuration must place all stations on concentric circles about the epicenter, with the stations on each circle equidistant from one another. Thus, the *D*-optimality criterion agrees with seismological intuition that a good configuration must "surround" the epicenter.

2. A triangular quadripartite network, which places one station directly above the hypocenter and three stations at equidistant points on a circle about the epicenter, is *D*-optimal for a hypocenter in the half-space. The maximal value of the *D*-criterion is obtained as the radius of the circle tends to infinity. For typical crustal models, a moderate radius will yield a value near the maximum. This type of network was recommended by Lilwall and Francis (1978) and Uhrhammer (1980).

3. For a hypocenter in a layer above the half-space, a *D*-optimal configuration includes a station above the hypocenter and stations on two concentric rings around the epicenter. Each ring includes at least three stations at equidistant locations. The radii of the two rings must be chosen so that waves arrive at stations on the near circle via direct ray paths and reach stations on the far circle via refraction paths. The triangular quadripartite network, which uses only one ring, can be quite inefficient for a hypocenter above the half-space. The superior resolution power of a network with mixed arrivals, in particular for depth resolution, was noted by Peters and Crosson (1972).

4. Network configurations that are *D*-optimal for P-wave arrivals only will also be *D*-optimal when both P- and S-wave arrivals are available. The *D*-criterion improves markedly from adding the S arrivals, which corroborates the observations by Buland (1976) and Gomberg et al. (1990) that adding even a single S arrival can substantially improve location estimates.

Mitchell (1974) used the ideas in the equivalence theorem to develop an effective algorithm, known as DETMAX, for constructing *D*-optimal experimental designs. Rabinowitz and Steinberg (1990) applied DETMAX in the network configuration context. The algorithm is based on a simple rule that guides the choice of which station to add to, or remove from, a given network. The algorithm adds stations from a finite list of possible candidate sites. Adding the *i*-th candidate to a network with information matrix M increases the determinant to

$$det(M)\,(1+v_i) \tag{3}$$

where $v_i = f_i^T M^{-1} f_i$. Thus the greatest improvement in the *D*-criterion is achieved by adding a station at the site that maximizes the quantity v_i. The best station to remove from an augmented network is the site at which v_i (with the current information matrix *M*) is minimal. The DETMAX algorithm works by applying this result to iteratively add and subtract stations from a network, progressively improving the criterion value. The optimization problem typically has many local maxima, so a number of random starting designs are used to increase the probability of finding the global maximum.

A practical problem with the *D*-criterion for a single source is that it tends to locate stations at the same site, which does not make sense for practical seismology. Collocated stations will provide identical signals and redundant information. The *D*-criterion, as presented here, is based on the assumption that each station provides statistically independent arrival time data and therefore does not reflect the redundancy. Several solutions to this problem have been considered. Rabinowitz and Steinberg (1990) proposed a correlated error model for the arrival times in which arrivals at collocated stations are perfectly correlated rather than independent. The *D*-criterion then automatically excludes collocated stations because they do not improve the criterion value. Steinberg et al. (1995) extended the *D*-criterion to multiple sources (see section 2.3). The need to simultaneously monitor many sources tends to spread out the stations and thus prevents collocation unless a very large network is considered. Shimshoni et al. (1992) developed the OPTINET program for *D*-optimal configurations and included a restriction that stations not be located within 5 km of each other.

2.2 Configuration for a Subset of the Parameters

Interest often focuses on a subset of the parameters rather than on the full vector θ. Focal depth might be of prime interest. For example, in the context of the CTBT, an accurate focal depth can screen out a deep event as clearly being a natural event. Similarly, interest might focus on the epicenter, for example to limit the initial search area for an OSI team.

The *D*-criterion can be simply adapted to parameter subsets by using the reduced information matrix for the relevant parameters, which is defined as follows. Let $\theta = (\theta_1, \theta_2)$ where the relevant parameters are those in θ_1. Partition the information matrix accordingly as

$$M = \begin{pmatrix} M_{1,1} & M_{1,2} \\ M_{2,1} & M_{2,2} \end{pmatrix} \qquad (4)$$

where $M_{1,1}$ is the block corresponding to θ_1. If $M_{2,2}$ is non-singular, the reduced information matrix for θ_1 is

$$M(\theta_1) = M_{1,1} - M_{1,2}M_{2,2}^{-1}M_{2,1} \tag{5}$$

The definition of the reduced information matrix can also be extended to cover cases in which $M_{2,2}$ is singular. The *D*-criterion can then be applied to $M(\theta_1)$. This criterion is known in the statistical literature as the D_S (for subset) criterion.

Steinberg and Rabinowitz (1999) derived some results on D_S-optimal networks, which we summarize here.

1. The determinant of the reduced information matrix for the hypocenter parameters (ignoring origin time) is proportional to the determinant of the full information matrix. Therefore, any *D*-optimal network is also D_S-optimal for the hypocenter.

2. A D_S-optimal network for the epicenter must place all stations symmetrically around the epicenter but will have no stations at the epicenter. As a result, this network will have no ability to determine focal depth and requires the extended definition of the reduced information matrix.

2.3 Criteria for Multiple Sources

Seismic sources are typically distributed along faults and most networks must monitor several, often geographically diverse, systems of faults. Thus, it is important to extend the single source criteria discussed in the last two sections to the multiple case setting. We review here several extensions that have been proposed.

All the extensions represent the multiple sources by a discrete grid of *k* isolated point sources. Steinberg et al. (1995) showed that even a rather coarse grid is sufficient to give an adequate representation. Denote by A_i the matrix of partial derivatives of the arrival times with respect to source *j* and by $M_i = A_i^T A_i$ the corresponding information matrix. The extensions are all based on different ideas for assessing the collection of *k* information matrices.

2.3.1 The DMS Criterion

Steinberg et al. (1995) proposed maximization of the *DMS*-criterion (D-Multiple Source)

$$DMS = \sum_{1}^{k} \alpha_j \ln[\det(M_j)] \tag{6}$$

where α_j reflects the relative importance attached to source *j* by the designer. They recommended that large weights be assigned to those sources where effective monitoring is critical and to sources that have demonstrated a high level of seismicity. They studied several examples and found that the *DMS*-criterion gave efficient networks.

2.3.2 Kijko's Criterion

The paper by Kijko (1977) was the pioneering work to apply statistical design criteria to the problem of seismic network configuration. He considered multiple sources and proposed maximization of

$$\sum_{1}^{k} \alpha_j[\det(M_j)] \tag{7}$$

Kijko's criterion treats the determinants as additive, in contrast to the *DMS*-criterion, which treats the log determinants as additive. Steinberg et al. (1995) showed that Kijko's criterion is likely to generate networks that do not effectively distribute information across the potential sources. Instead, the stations will provide excellent coverage for the most resolvable source and poor coverage for other sources. The *DMS*-criterion generates networks with much more balanced coverage of the sources.

2.3.3 Average Information Criterion

An alternative to the *DMS*-criterion and to Kijko's criterion is to average the information matrices prior to computing the determinant. The resulting design functional is

$$D_{AI} = \det\left(\sum_{1}^{k} \alpha_j M_j\right) \tag{8}$$

This criterion would be useful for problems in which singular information matrices might arise, say, due to incomplete data. The *DMS*-criterion cannot be applied in that case.

2.3.4 The Minimax Approach of Bartal et al.

Bartal et al. (2000) considered the problem of multi-source network configuration for adding a small number of national stations (known as Cooperating National Facilities, or CNF) in the context of the CTBT. The CNF stations would then be available to augment IMS data for events in or near that nation. Bartal et al. argued that the main consideration for CNF stations should be to guarantee that no IDC epicentral error ellipse should exceed a specified threshold. Thus, they were led to propose a criterion of the form

$$\max_j \det(M_j(\theta_1)) \tag{9}$$

where θ_1 is the epicenter. The optimal network would then be one that minimizes this criterion. The minimax approach does best for the most difficult to cover epicenter whereas the previous approaches all find networks that perform well in an average sense.

3. INFLUENTIAL OBSERVATIONS AND NETWORK CONFIGURATION

An important issue in hypocenter location is to identify phase arrivals that may have unusually strong effects on the determined hypocenter. This problem is especially important in small seismic networks, where experience has shown that hypocenter location can be very sensitive to dropping or adding individual stations. In practice, seismic analysts often drop phase arrivals with large residuals that would otherwise shift the hypocenter location. We think that this practice is unwise and leads to excessively optimistic fit statistics in location.

In this section we describe and explain the use of statistical criteria that have been developed to identify influential observations. We also discuss the relation of these influence measures to network configuration and suggest an alternative configuration criterion based on them.

3.1 Influence, Importance, and Outliers

Cook (1977) studied the problem of identifying which observations in a linear statistical model have the greatest effect on the estimated parameter vector. Cook's approach can also be applied in nonlinear models, such as hypocenter location, by using the linear approximation to the model at the point of convergence to the solution. The basic notion underlying Cook's approach is to determine how the estimate changes when an observation (or a set of observations) is omitted. The same ideas can be applied to assess the influence of phase arrivals that might be added to a given location problem.

The idea of station influence could be effectively applied by the IDC. The IMS seismic stations are divided into two categories, 50 primary (alpha) stations and 120 auxiliary (beta) stations. Initial event locations are made automatically by the IDC using data from the primary stations only. Additional information is then automatically retrieved from those auxiliary stations which, based on the initial location, appear most likely to improve the location accuracy. Use of station influence should help guide the decision as to which auxiliary stations should be added.

Cook's statistic is most often applied to measuring the influence of a single observation and we consider that case in the context of hypocenter location, as first developed by Rabinowitz (1986). As before, let θ denote the unknown hypocenter and origin time and let $\hat{\theta}$ denote the estimated parameters using all phase arrivals. Now, suppose the i-th phase arrival is deleted and the resulting solution is denoted by $\hat{\theta}_{(i)}$. The Cook statistic then computes the distance between $\hat{\theta}_{(i)}$ and $\hat{\theta}$ in the metric that corresponds to the covariance matrix of the complete solution,

$$D_i = (\hat{\theta} - \hat{\theta}_{(i)})^T (A^T A)(\hat{\theta} - \hat{\theta}_{(i)})/(4s^2), \tag{10}$$

where, as before, A is the matrix of partial derivatives evaluated at $\hat{\theta}$. The rationale for this choice of metric, rather than, say, Euclidean distance, is that the information matrix $M=A^TA$ measures changes in the solution relative to the resolution provided by the network. For example, if the network has little ability to determine focal depth but provides good precision for the epicenter, changes in the former will contribute much less to the influence than changes in the latter.

The measure D_i can also be converted to a standardized probability scale, as suggested by Cook (1977). His idea was to relate D_i to the contours of the confidence ellipsoid for the solution vector $\hat{\theta}$. As we noted earlier, the contour of the 100(1-α)% confidence ellipsoid will be the solution of the equation

$$(\theta-\hat{\theta})^T M(\theta-\hat{\theta})/(4s^2)=F_{1-\alpha,4,K+n-4} \tag{11}$$

where K controls the trade-off in estimating σ^2 between the root mean square error of the event and prior information. The value of α determines the size of the ellipsoid and so we can think of generating a family of expanding ellipsoids indexed by decreasing values of α. We can then measure the effect of dropping a phase arrival by computing the value of α that corresponds to the confidence ellipsoid for $\hat{\theta}_{(i)}$. The computation amounts to substituting $\hat{\theta}_{(i)}$ for θ in the last equation, which gives D_i on the left-hand side, and computing the corresponding value of α for the F distribution on the right-hand side. Values of α close to 0 imply that dropping the i-th phase arrival has moved the solution to the edge of a confidence ellipsoid that is quite far removed from the full data solution.

In the context of CTBT monitoring, the influence measure has immediate practical importance. The confidence ellipsoid for the solution may be used to screen out events as either too deep or too far off-shore to be an explosion. More important, in the case of an accusation and subsequent OSI, the confidence ellipsoid is used to limit the search area for the OSI team. Thus, it is crucial to know if single phase arrivals have a strong effect on the solution.

Cook's influence measure for the i-th observation can be decomposed into a product of two terms, one of which depends on the so-called studentized residual r_i and one of which depends only on the matrix of partial derivatives. The studentized residual is defined by $r_i = e_i /[s(1\text{-}h_{ii})^{1/2}]$ where $e_i = t_i - t_i$ is the difference between observed and fitted arrival time and $h_{ii} = f_i^T M^{-1} f_i$ is the i-th diagonal entry of the "hat matrix," which is defined by $H = A(A^TA)^{-1}A^T$. The simple residual e_i is divided by $s(1\text{-}h_{ii})^{1/2}$ to reflect the fact that the standard deviation of e_i is $\sigma(1\text{-}h_{ii})^{1/2}$. The singular value decomposition (SVD) of A is typically used to compute H. The SVD is given by $A=USV^T$, where U and V^T are orthonormal matrices and S is a diagonal matrix of the singular values of A. Then $H=VV^T$. We note that the presence of a column of 1's in A guarantees that $1/n \le h_{ii} \le 1$.

The decomposition formula derived by Cook (1977), applied to hypocenter location, gives

$$D_i = \frac{r_i^2}{4}\frac{h_{ii}}{1-h_{ii}}. \tag{12}$$

The second term depends on the geometry of the network and on the particular types of phases that were recorded. It thus measures the inherent strength of the associated phase arrival to dictate the solution. Cook called this term the potential of the observation and it is obviously a monotone increasing function of h_{ii}. Many standard location programs include h_{ii} as part of their output to help analysts identify high leverage arrivals. This measure is referred to as "importance" in Hypoinverse (Klein, 1978) and we will adopt this terminology hereafter.

Applying Cook's (1977) analysis to hypocenter location shows that the phase arrivals with the most dominance in determining the hypocenter will be those with large importance and large residuals. The above equation shows that the overall station influence is a product of the stochastic error (as reflected in the residual) and the network geometry (as reflected by the station importance). Analysts would be wise to examine importances and Cook's statistics along with residuals in deciding whether a phase arrival should be dropped or added.

3.2 Influence and *D*-optimality

The phase arrival importance is also closely related to the volume of the error ellipsoid and thus to the *D*-optimality criterion. For the full set of phase arrivals, the ellipsoid volume is proportional to $(\det(M))^{-1/2}$. If the *i*-th phase arrival is omitted, the volume of the error ellipsoid will be proportional to $(\det(M_{(i)}))^{-1/2}$, where $M_{(i)}$ is the reduced information matrix. The matrix identity (Cook and Weisberg, 1982, p. 210)

$$\det(M) = \det(M_{(i)})/(1 - h_{ii}) \tag{13}$$

shows that the effect of a single phase arrival on the error ellipsoid is a direct function of its importance. Thus arrivals with high importance not only have the greatest potential effect on the solution, they also have the greatest effect on its assessed precision.

We pointed out in Section 2 that the DETMAX algorithm generates *D*-optimal networks from an initial network by successively adding or subtracting stations. We see here that the contribution of a station to the *D*-criterion is an increasing function of its importance. Thus the best station to subtract from a given network is the station with the smallest importance. Now consider what happens when we augment a network with a single station. For a given network and a list of candidate stations, the DETMAX algorithm adds the station that maximizes $v_i = f_i^T M^{-1} f_i$. Note that this expression has the same form as the importances h_{ii} for the stations that are already in the network. The difference here is that we apply the formula also to candidate sites not currently in the network. We can use additional matrix identities to link the value v_i at the design stage to the importance h_{ii} that the added station will have in the augmented network, $v_i = h_{ii}/(1-h_{ii})$. Since the added station maximizes v_i, it must have the largest importance in the augmented network.

3.3 Importance-Based Configuration Criteria

Alternative network configuration criteria can be based directly on the station importances. We propose here such a criterion and show that it has valuable robustness properties. By robustness we mean here limited sensitivity to the presence of unusually large phase picking errors (outliers). We also discuss the relation of the criterion to *D*-optimality.

For any network configuration, the sum of the importances must equal 4. We noted above that the ability of a phase arrival to influence the estimated hypocenter is an increasing function of its importance. Thus, a reasonable goal is to ensure that no station has excessive influence. Since the sum of the importances is constant, a good network should minimize the maximal station importance. Thus, we propose finding a network by minimizing the robust importance (*RImp*) configuration criterion

$$RImp = \max h_{ii} \tag{14}$$

A robust importance optimal network will have importances that are nearly equal, which was suggested by Box and Draper (1975) as a means to limit sensitivity to outliers. Further, as noted by Huber (1975) and Davies and Hutton (1975), if a station importance is close to 1, it is very likely that the station will have a small residual even if it is subject to a large picking error and the error will thus go undetected. Limiting the maximal importance in the network will thus also improve the ability to detect outliers if they are present.

We have already noted the close relationship between station importances and the *D*-optimality criterion, as seen in the equivalence theorem of Kiefer and Wolfowitz (1980) and the DETMAX algorithm of Mitchell (1974). It might appear from those results that the *RImp*-optimality criterion is thus equivalent to the *D*-optimality criterion. We stress here that the criteria are not equivalent and can lead to quite different configurations. The difference, although subtle, is important. The Kiefer-Wolfowitz theorem is phrased in terms of sites, not stations, and allows multiple stations at each site. Each site is weighted by the number of stations placed there. The term v that we used in describing the theorem for a particular site thus equals the sum of the h_{ii} for all the stations located at the same site. Furthermore, all the h_{ii} at the same site will be equal to one another. Thus the *D*-optimality criterion will generate designs in which these site-wise sums are (nearly) equal whereas the *RImp*-optimality criterion will generate designs in which the individual station importances are (nearly) equal. The *RImp* criterion automatically filters out collocated stations.

We illustrate the difference between the *RImp*-optimality criterion and the *D*-optimality criterion with a simple example. Figure 1 shows *RImp*-optimal (panel *a*) and *D*-optimal (panel *b*) networks with 6 stations for locating a source just northeast of a 100 km square region and 10 km deep. Stations could be located anywhere inside the region and were determined for both criteria using a genetic algorithm (see below). The *RImp*-optimal network uses 6 distinct sites and has maximal importance of 0.679, slightly higher than the theoretical optimum of 0.667. The *D*-optimal network has two pairs of stations that are essentially collocated. The sum of the importances at each distinct site is 0.999 so the network achieves the site-wise equality of the Kiefer-Wolfowitz theorem, but has individual station importances that differ greatly from one another. Thus, it is not a desirable configuration by the *RImp*-criterion. In terms of the *D*-criterion, though, the latter network achieves a criterion value 36% higher than does the former network.

We show a second example in Figure 2, in which a network of 30 stations is configured to precisely locate an event with epicenter at the center of a 300-km square grid and at depth 10 km. The *RImp*-optimal design found by a genetic algorithm places 15 stations on a ring about 25 km from the source. All of these stations receive *P*-wave arrivals by direct paths. The 15 remaining stations are spread out quite symmetrically about the source at much larger distances and receive *P*-wave arrivals by refracted paths.

Our experience in working with the *RImp*-criterion is that convergence to a solution occurs much more rapidly than with the *D*-criterion. Thus, we conjecture that the objective function for the former criterion is more peaked and has fewer local optima than the latter function.

We add a few words on the optimization algorithm used here. Genetic Algorithms (GAs) are global optimization and search procedures that have advantages in solving problems with many local extrema (Goldberg, 1989). These algorithms generate new solutions (offspring) from previous solutions (parents) in a way that mimics the biological process of evolution. The process proceeds via three basic operations, reproduction, crossover and mutation. See Goldberg (1989) for a thorough description of GAs. We have used a versatile, modern FORTRAN implementation of GA (Caroll, 1997). The FORTRAN code and further details are available from Caroll's web site, http://www.staff.uiuc.edu/~carroll/ga.html.

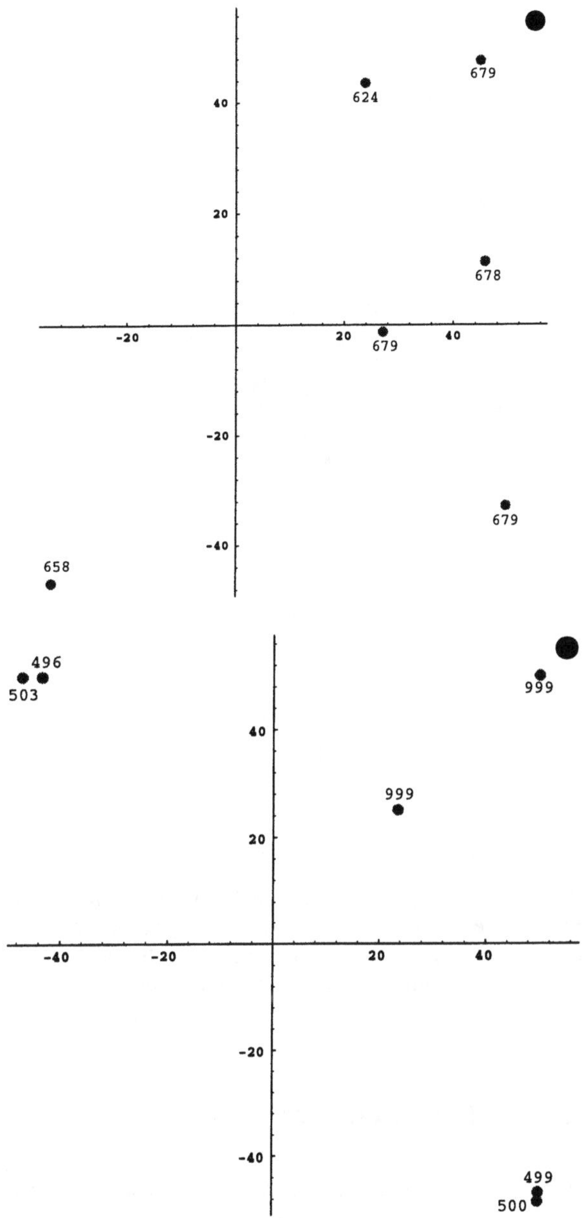

a)

b)

Figure 1. Network configurations with 6 stations for locating a source just northeast of a 100-km square region and 10 km deep. The optimal network using the *RImp*-criterion is shown in (a) and the *D*-optimal network is shown in (b). The *D*-optimal network achieves a smaller error ellipsoid but suffers from collocation of stations. The numbers indicate the station importances times 1000.

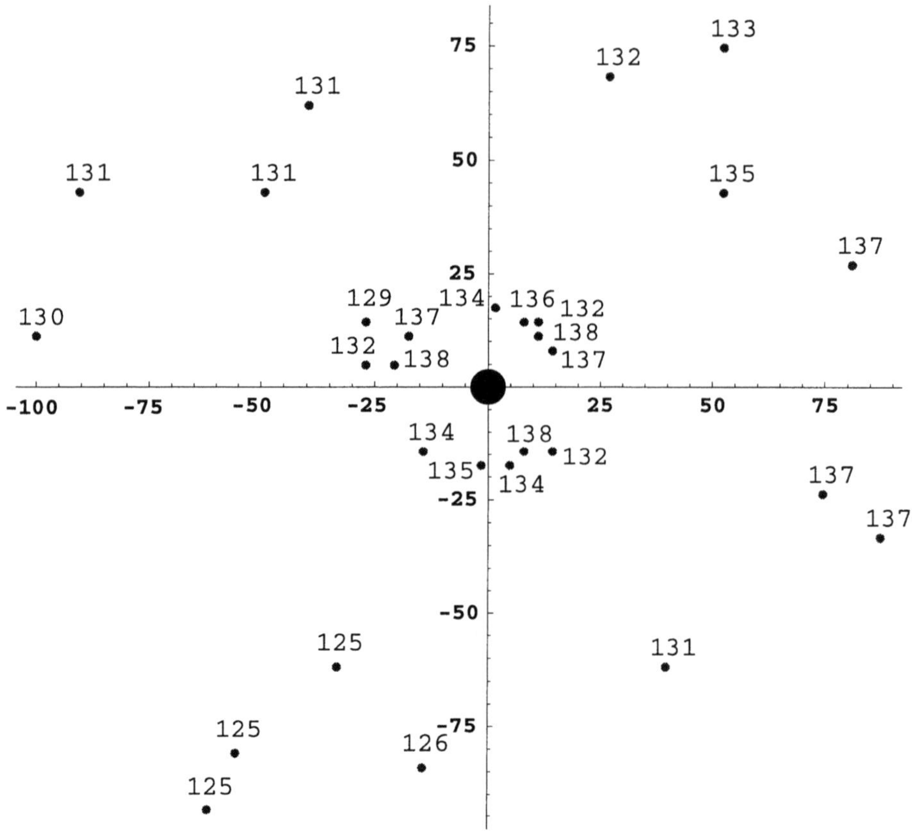

Figure 2. Optimal network configuration derived from the *RImp*-criterion for locating an event with epicenter at the center of a 300 km square grid and at depth 10 km using a 30 station network. The network has 15 stations almost evenly spaced on a ring about 25 km from the source and the remaining 15 stations spread out around the source near the boundaries of the grid. The numbers indicate the station importances times 1000 and range from 125 to 138.

3.4 Specialized Influence Measures

As we noted in the last section, interest often focuses on a subset of the parameters rather than on the full vector θ. Influence measures and the associated configuration criterion can also be adapted to parameter subsets.

Rabinowitz and Steinberg (1988) showed how to compute station importances specifically for focal depth, applying the general theoretical ideas of Cook (1977). Similar ideas can be used to derive importances for the epicenter. The *RImp*-optimality criterion can then be applied to the subset specific importances rather than to the importances for the full parameter vector.

4. SUMMARY AND CONCLUSIONS

In this chapter, we have studied the seismological problem of optimal network configuration using ideas from the statistical theory of optimal design and influential observations. The issue of network configuration has assumed far greater importance in recent years because of its role in monitoring the Comprehensive Test Ban Treaty. The CTBT has led to a fundamental shift in the significance attached to location errors and their assessment in the form of an error ellipsoid. Traditionally attention has focused on the estimated hypocenter itself and the error ellipsoid has taken a secondary role. For monitoring the CTBT, the error ellipsoid has become the primary concern, because of its role in screening out natural events, and even more so because of the direct use of the epicentral error region in defining the search area of an OSI team. Questions regarding the size of error ellipsoids are of great relevance. Thus, for example, determining the effect of adding or removing a station on the size of the ellipsoid is not just of theoretical interest, but has immediate practical, and even political, implications. Similarly, the goal of the CTBT to achieve search areas as small as possible (at most 1000 square km) has posed seismologists with the problem of how to configure a network that will meet this goal. The global network of primary stations will not always be sufficient and it is then crucial to add data from the most useful auxiliary stations and, if available, CNF stations. The ideas that we have discussed in this chapter are immediately applicable to this problem.

The configuration criteria that we have presented and discussed are all related to the basic connection between the size of the error ellipsoid, the determinant of the information matrix, and their dependence on the geometry of the network. The D-optimality criterion directly applies this connection and finds networks that minimize the size of the 4-dimensional ellipsoid. Related criteria can be used to generate networks that are optimal for subsets of the parameters, for example the epicenter or the focal depth. We have also shown how the determinant of the information matrix is related to arrival time importances and have proposed here a new configuration criterion that seeks networks with equal importances. This criterion prevents collocation of stations and improves robustness to outliers.

Implementation of any of the criteria that we have discussed requires the use of a numerical optimization algorithm. The criteria typically have many local optima and so present difficult optimization problems. However, advances in recent years in computing capability and the development of efficient global optimization algorithms such as genetic algorithms and simulated annealing have largely eliminated difficulties in the practical implementation of the criteria. Many of these algorithms are easily and freely available via the Internet.

We have limited our presentation to the optimal configuration of a seismic network for locating events with a known travel-time model. The same basic ideas can be applied to more general geophysical inversion problems. For example, one could optimize a network for determining the moment tensor and source function of an earthquake. One could also derive an optimal network for joint hypocenter determination in which the arrival times from a suite of events are used to simultaneously determine a layered velocity model and to locate the hypocenters, or in full 3-D simultaneous inversion. Curtis (1999a, 1999b) has done initial work on configuration for inversion problems in the context of borehole experiments. This is a promising direction for future research.

REFERENCES

Atkinson, A.C. (1996) The usefulness of optimum experimental designs, *J. Roy. Stat. Soc.*, Series B **58**, 59-76.

Atkinson, A.C. and A.N. Donev (1992) *Optimum Experimental Designs*, Oxford: Oxford University Press.

Bartal, Y., Z. Somer, G. Leonard, D.M. Steinberg, and Y Ben Horin (2000) Optimal seismic networks in Israel in the context of the Comprehensive Test Ban Treaty, *Bull. Seism. Soc. Am.* **90**, in press.

Box, G.E.P. and N.R. Draper (1975) Robust designs, *Biometrika* **62**, 347-352.

Box, G.E.P. and H.L. Lucas (1959) Design of experiments in nonlinear situations, *Biometrika* **46**, 77-90.

Bratt, S. and T.C. Bache (1988) Locating events with a sparse network of regional arrays, *Bull. Seism. Soc. Am.* **78**, 780-798.

Buland, R. (1976) The mechanics of locating earthquakes, *Bull. Seism. Soc. Am.* **66**, 173-187.

Caroll, L.D. (1997) FORTRAN Genetic Algorithm (GA) Driver, Department of Aeronautical and Astronautical Engineering, University of Illinois at Urbana-Champaign, Urbana, Illinois, WWW page http://www.staff.uiuc.edu/~carroll/ga.html.

Cook, R.D. (1977) Detection of influential observations in linear regression, *Technometrics* **19**, 15-18.

Cook, R.D. and S. Weisberg (1982) *Residuals and Influence in Regression*, New York: Chapman and Hall.

Curtis, A. (1999a) Optimal experiment design: Cross-borehole tomographic examples, *Geophys. J. Int.* **136**, 637-650.

Curtis, A. (1999b) Optimal design of focussed experiments and surveys, *Geophys. J. Int.* **139**, 205-215.

Davies, R.B. and B. Hutton (1975) The effects of errors in the independent variables in linear regression, *Biometrika* **62**, 383-391.

Evernden, J.F. (1969) Precision of epicenters obtained by small numbers of world-wide stations, *Bull. Seism. Soc. Am.* **59**, 1365-1398.

Flinn, E.A. (1965) Confidence region and error determinations for seimic event location, *Reviews of Geophysics* **3**, 157-185.

Goldberg, D.E. (1989) *Genetic Algorithms in Search, Optimization, and Machine Learning*, Reading, Massachusetts: Addison-Wesley.

Gomberg, J., K. Shedlock, and S. Roecker (1990) The effect of *S*-wave arrival times on the accuracy of hypocenter estimation, *Bull. Seism. Soc. Am.* **80**, 1605-1628.

Huber, P. (1975) Robusteness and designs, in Srivastava, J.N. (ed), *A Survey of Statistical Design and Linear Models*, Amsterdam: North-Holland.

Jordan, T.H. and K.A. Sverdrup (1981) Teleseismic location techniques and their applications to earthquake clusters in the south-central Pacific, *Bull. Seism. Soc. Am.* **71**, 1105-1130.

Kiefer, J. and J. Wolfowitz (1960) The equivalence of two extremum problems, *Can. J. Math.* **12**, 363-366.

Kijko, A. (1977) An algorithm for optimum distribution of a regional seismic network-I, *Pageoph.* **115**, 999-1009.

Klein, R.W. (1978) Hypocenter location program hypoinverse, *U.S. Geological Survey Open File Rep.* **78-694**.

Lilwall, R.C. and T.J.G. Francis (1978) Hypocentral resolution of small ocean bottom seismic networks, *Geophys. J. Roy. Astron. Soc.* **54**, 721-728.

Mitchell, T.J. (1974) An algorithm for construction of *D*-optimal experimental design, *Technometrics* **16**, 203-210.

Peters, D.C. and R.S. Crosson (1972) Application of prediction analysis to hypocenter determination using local array, *Bull. Seism. Soc. Am.* **62**, 775-788.

Pukelsheim, F. (1993) *Optimal Design of Experiments*, New York: John Wiley.

Rabinowitz, N. (1986) On identifying influential arrivals by means of density information matrix, *Bull. Seism. Soc. Am.* **76**, 1817-1821.

Rabinowitz, N. and D.M. Steinberg (1988) Effect of single arrival time on the hypocentral focal depth determination, *Geophys. J.* **94**, 567-570.

Rabinowitz, N. and D.M. Steinberg (1990) Optimal configuration of a seismographic network: a statistical approach, *Bull. Seism. Soc. Am.* **80**, 187-196.

Shimshoni, Y., D. Mizrachi, D.M. Steinberg, and N. Rabinowitz (1992) OPTINET: A computer program for designing the optimal configuration of a seismographic network, *Inst. Petrol. Res. Geophys. Report* **321/47/91(4)**.

Steinberg, D.M and N. Rabinowitz. (1999) Optimal seismic monitoring for event location with application to on site monitoring of the Comprehensive Test Ban Treaty, submitted for publication.

Steinberg, D.M, N. Rabinowitz, Y. Shimshoni, and D. Mizrachi (1995) Configuring a seismographic network for optimal monitoring of fault lines and multiple sources, *Bull. Seism. Soc. Am.* **85**, 1847-1857.

Uhrhammer, R.A. (1980) Analysis of small seismographic station networks, *Bull. Seism. Soc. Am.* **70**, 1369-1379.

Chapter 4

ADVANCES IN TRAVEL-TIME CALCULATIONS FOR THREE-DIMENSIONAL STRUCTURES

Clifford H. Thurber
Department of Geology and Geophysics, University of Wisconsin-Madison, USA

Edi Kissling
Institute of Geophysics, Swiss Federal Institute of Technology (ETH) Zurich, Switzerland

Key words: Ray tracing, travel time, shooting, bending, finite-difference, three-dimensional structure, tomography, accuracy.

Abstract: An accurate and efficient technique for determining travel times and ray paths in a heterogeneous medium is an essential part of methods for seismic event location and seismic tomography using three dimensional velocity models. A wide variety of methods has been developed over the last few decades. We present a review of the major classes of methods for travel time and ray path calculation. They can be categorized as either exact or approximate, and the computational strategy involved usually can be classified as shooting, bending, perturbation, or grid-based. Many methods determine solutions using the ray equations, but some methods rely on Fermat's principle, Huygen's principle, or the eikonal equation. Relative advantages and disadvantages of the various techniques are discussed. We conclude by highlighting some recent comparative studies of some of these methods that provide valuable information regarding their relative accuracies.

C.H. Thurber and N. Rabinowitz (eds.), Advances in Seismic Event Location, 71–99.

1. INTRODUCTION

Incorporating laterally heterogeneous Earth models into earthquake location procedures requires an efficient method for seismic wave travel-time calculations in three-dimensional (3-D) structures. Travel times are needed in order to calculate the arrival time residuals, and ray paths (or the travel time field in the vicinity of the source point) are needed for computation of the partial derivatives of travel times with respect to hypocenter parameters. 3-D travel time calculations are also required for nonlinear methods of seismic tomography.

There are as many techniques for determining ray paths and travel times as there are ways of representing the Earth's structure. It should be apparent that these two issues go hand in hand - the choice made for representing the 3-D structure often tends to define the most appropriate method to use for calculating ray paths and travel times, or vice-versa. The many methods can be categorized broadly as either exact or approximate, and the computational strategy involved usually can be classified as shooting, bending, perturbation, or grid-based. The latter two classes of methods are intended to circumvent problems intrinsic to shooting and bending. In the ray tracing approach (shooting and bending) to calculating travel times, the problem is treated mathematically as a two-point boundary value problem (2PBVP), involving the determination of the path of minimum travel time connecting two points (the source and receiver). A popular alternative approach is to evaluate the travel time field from a given point to all other points on a grid in a more brute-force manner, using finite-difference or graph theory.

The standard "Differential Equation Methods" (DEM) for solving the 3-D ray tracing problem are termed bending and shooting. The classic paper by Julian and Gubbins (1977) provides an excellent overview of these techniques. The bending method solves the 2PBVP directly by iteratively perturbing an initial path estimate with two fixed end-points until it satisfies the ray equations (Figure 1a). The shooting method solves the 2PBVP by treating it as an initial value problem (IVP), with a specified starting path point and trial propagation direction, and then iteratively adjusting the propagation direction until the target end point is reached (Figure 1b). The pitfalls of both techniques are widely recognized, including high computational cost, vulnerability to convergence failure, and potential for convergence to a local minimum (Julian and Gubbins, 1977; Thurber, 1986; Moser et al., 1992).

Several alternative strategies have been developed that have significant advantages over the standard DEM approaches. We will classify the main types of these alternative approaches as perturbation and grid methods. We present a review of the standard methods and alternative approaches for isotropic media, and discuss some recent variations and improvements to these techniques. We conclude by discussing some recent comparative studies of some of these methods, which provide valuable information regarding relative accuracy.

2. DIFFERENTIAL EQUATION METHODS

The differential equation methods (DEM's) of bending and shooting are distinguished by the "exact" nature of their solutions, due to their explicit use of the ray equations without approximations. The second-order ray equations can be written as

$$\frac{\partial}{\partial s}\left(\frac{1}{V(\mathbf{r})}\frac{\partial \mathbf{r}}{\partial s}\right) = \nabla\left(\frac{1}{V(\mathbf{r})}\right) \tag{1}$$

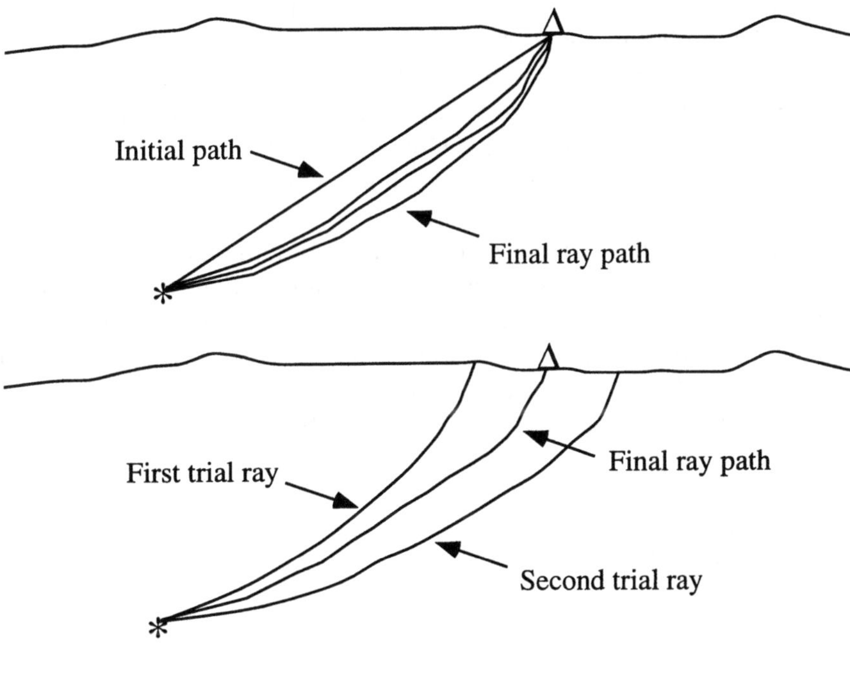

Figure 1. (a) Top - the bending method determines the ray path by iteratively perturbing an initial path estimate with two fixed end-points until it satisfies the ray equations. (b) Bottom - the shooting method determines the ray path by treating it as an initial value problem, with a specified starting path point and trial propagation direction, and then iteratively adjusting the propagation direction until the target end point is reached.

where s is the path length parameter, V is the 3-D velocity field, and **r** is the position vector along the ray. Most bending methods solve Equation 1 by discretizing the initial path estimate and iteratively solving a linearized finite-difference approximation to Equation 1 to determine perturbations to the current path estimate to bring it towards the true minimum-time ray path (Wesson, 1971; Julian and Gubbins, 1977; Pereyra et al., 1980).

Equation 1 can also be written as a set of 6 coupled first-order equations

$$\frac{\partial \mathbf{r}}{\partial s} = V(\mathbf{r})\,\mathbf{p}\ ;\ \frac{\partial \mathbf{p}}{\partial s} = \frac{\partial}{\partial \mathbf{r}}\left(\frac{1}{V(\mathbf{r})}\right) \tag{2}$$

where **p** is the "slowness vector" (i.e., the vector of the gradient of travel time along the ray at point **r**). Most shooting methods (Jacob, 1970; Engdahl and Lee, 1976; Julian and Gubbins, 1977; Sambridge and Kennett, 1990) solve Equation 2 (or the equivalent) by specifying an initial guess for **p** at one end point and obtaining a solution using numerical integration (e.g., Runge-Kutta or Predictor-Corrector methods). This approach solves what is known as the Initial Value Problem, so it will generally be necessary to modify the initial slowness vector so that the desired path end point is reached. Standard approaches are Newton's method and the method of False Position (extended to 2-D) (Julian and Gubbins, 1977).

If the target end point is defined on a horizontal plane at the appropriate depth, then the ending coordinates (Cartesian (x, y) or spherical(θ, ϕ)) can be related implicitly to the starting values of the azimuth and take-off angles. Newton's method involves carrying out a path integration of the equations evaluating geometric spreading along the ray in order to provide values for the partial derivatives of the end point coordinates with respect to the initial angles. Then the 2 x 2 system of equations relating initial angle change to end point coordinate change can be formed and solved:

$$\begin{bmatrix} \dfrac{\partial h}{\partial i_0} & \dfrac{\partial h}{\partial j_0} \\[2ex] \dfrac{\partial g}{\partial i_0} & \dfrac{\partial g}{\partial j_0} \end{bmatrix} \begin{bmatrix} \Delta i_0 \\ \Delta j_0 \end{bmatrix} = \begin{bmatrix} h_T - h_n \\ g_T - g_n \end{bmatrix} \tag{3}$$

where (h, g) represents (x, y) or (θ, ϕ), i_0 and j_0 are the initial azimuth and take-off angles, and subscripts n and T indicate the current and target values for the end point. Alternatively, using the method of False Position, three trial rays can be computed with slightly different initial angles, (i_0^1, j_0^1),

(i_0^2, j_0^2), (i_0^3, j_0^3). These initial angle values and their corresponding endpoint coordinates (h^i, g^i) on the target horizontal plane can be approximately related to the target coordinates and the estimate of the correct initial angles by the equations

$$\begin{vmatrix} i_0 - i_0^1 & i_0 - i_0^2 & i_0 - i_0^3 \\ h^1 - h_T & h^2 - h_T & h^3 - h_T \\ g^1 - h_T & g^2 - h_T & g^3 - h_T \end{vmatrix} = 0$$

$$\begin{vmatrix} j_0 - j_0^1 & j_0 - j_0^2 & j_0 - j_0^3 \\ h^1 - h_T & h^2 - h_T & h^3 - h_T \\ g^1 - h_T & g^2 - h_T & g^3 - h_T \end{vmatrix} = 0 \qquad (4)$$

(Julian and Gubbins, 1977). Both sets of equations (Equations 3 and 4) generally must be solved iteratively.

2.1 Problems and Alternatives

The bending and shooting methods yield results that are exact to the level of computational accuracy, in principle. However, the methods do suffer from some general problems. Both are subject to instability due to their nonlinear nature, so solutions may diverge. Convergence can be slow, or, more seriously, the solution may converge to a local minimum (a potential problem with essentially all non-grid methods). Discontinuities in velocity structure can be dealt with directly in shooting methods by the use of Snell's Law. For bending methods, discontinuities are generally handled by constraining one path point to lie along each discontinuity. Both methods require continuity of velocity derivatives (at least within model sub-regions), meaning that the velocity structure needs to be represented using relatively sophisticated methods (for example, 3-D cubic spline interpolation). This adds significantly to the computational burden.

Simulated annealing (SA) is an alternative strategy for determining the initial angle values that yield a ray reaching the target that increases the chance that the global minimum will be found. Velis and Ulrych (1996) have developed a 2-D method using Very Fast SA (Ingber, 1989). For a model with multiple local minima, they demonstrate their algorithm's ability to find the global minimum ray path. In addition, their method includes the ability to treat reflected and critically refracted rays.

One way to improve the computational efficiency of the DEM approach is to choose a velocity model representation that simplifies the calculations. Efficient 2-D and 3-D shooting approaches were presented by Aric et al. (1980) and Lin and Roecker (1996), who used velocity models composed of triangular and tetrahedral elements, respectively, with linear velocity gradient within each model element. As a result, ray paths are composed of piecewise circular arc segments that can be evaluated exactly using analytic expressions. An added advantage is that the triangular and tetrahedral elements can be arranged in extremely flexible manners, allowing for a more detailed velocity model in areas of greater heterogeneity (Figure 2).

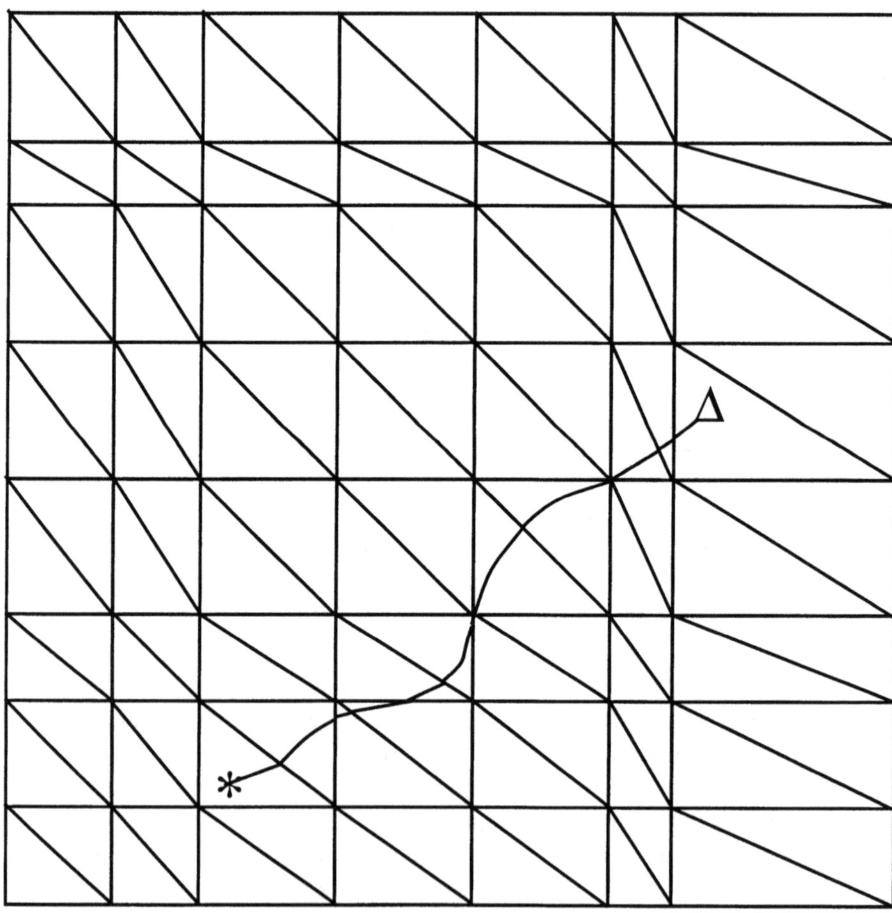

Figure 2. Representing a 2-D velocity model with triangular elements (or tetrahedra in 3-D) with a linear velocity gradient in each model element results in ray paths composed of piecewise circular arc segments that can be evaluated exactly using analytic expressions. This provides great flexibility in model parameterization, allowing for highly variable cell sizes.

Another model representation that leads to efficient solutions is the constant quadratic slowness gradient model (Cerveny, 1987). By defining the path variable (σ) along segments of the ray path as

$$\sigma(T) = \sigma(T_0) + \int_{T_0}^{T} V^2(t)\, dt \tag{5}$$

where T is travel time and T_0 is travel time to the start of the ray segment, the ray equations reduce to

$$\frac{d\mathbf{r}}{d\sigma} = \mathbf{p}\,;\ \frac{d\mathbf{p}}{d\sigma} = \frac{1}{2}\frac{\partial}{\partial \mathbf{r}}\left(\frac{1}{V^2}\right) \tag{6}$$

with the relation

$$\frac{dT}{d\sigma} = \frac{1}{V^2} \tag{7}$$

If the quadratic slowness has a constant gradient within model sub-volumes, that is, V^{-2} satisfies

$$V^{-2} = A_0 + A_1\, x_1 + A_2\, x_2 + A_3\, x_3 \tag{8}$$

(A_i are constants), then the solution to the ray equations are exact polynomials:

$$x_i(\sigma) = x_i(\sigma_0) + p_i(\sigma_0)\,(\sigma - \sigma_0) + \tfrac{1}{4} A_i\,(\sigma - \sigma_0)^2$$

$$p_i(\sigma) = p_i(\sigma_0) + \tfrac{1}{2} A_i\,(\sigma - \sigma_0)$$

$$T(\sigma) = T(\sigma_0) + \left(A_0 + A_i\, x_i(\sigma_0)\right)(\sigma - \sigma_0) + \tfrac{1}{2} A_i\, p_i(\sigma_0) \\ (\sigma - \sigma_0)^2 + \tfrac{1}{12} A_i^{\,2}\,(\sigma - \sigma_0)^3 \tag{9}$$

where $\sigma_0 = \sigma\,(T_0)$. Cerveny (1987) states that this is the most efficient model representation in terms of providing a simple analytic solution.

3. ALTERNATIVE STRATEGIES: PERTURBATION METHODS

Perturbation methods determine the ray path by starting with a path that is "near" the desired ray path and evaluating approximate perturbations to the path that bring it closer to the true ray path. By near, it is meant that either (case A) the starting path is a true ray path (e.g., one evaluated using a shooting method) and its end point is not too far from the desired end point, or (case B) that the starting path connects the start and end points but is not a true ray path. Thus one perturbation method strategy (case A) estimates the perturbation to the initial reference ray path that is required to reach the desired end point, and another strategy (case B) computes the improved path estimate by evaluating path perturbations that would bring the current path estimate closer to satisfying the ray equations. These strategies are thus comparable in nature to the shooting and bending methods, respectively.

Two common types of perturbation methods based on the ray equations have been developed, one using a Hamiltonian formulation approach (HFA) and the other a Langrangian formulation approach (LFA). A third type of approach is a hybrid perturbation method involving integration along the trial ray, called parameterized shooting, that combines elements of both the shooting and bending methods. Another strategy involves calculating a reference ray in a simple medium and determining the path perturbations due to perturbing the simple medium to the actual complex medium (the "Medium Perturbation Approach"). This method has not received wide application, so it will not be discussed further except in the context of ray tracing in anisotropic media. A fourth type of method evaluates path perturbations that would bring the current path closer to one of minimum travel time (the "Fermat's Principle Approach," FPA).

3.1 Lagrangian Formulation Approach (LFA)

A series of papers (Snieder and Sambridge, 1992, 1993; Snieder and Spencer, 1993; Pulliam and Snieder, 1996) explored ray perturbation using a Lagrangian formalism and the second-order equations of Equation 1. The approach involves expressing the desired ray as the sum of a reference ray or curve $\mathbf{r}_0$ and a perturbation

$$\mathbf{r}(s_0) = \mathbf{r}_0(s_0) + \varepsilon\, \mathbf{r}_1(s_0) \tag{10}$$

where s_0 is arc length along $\mathbf{r}_0$ and ε is a small constant. Note that the total arc length along the reference and perturbed curves will differ, so that when the path is parameterized in terms of arc length, it is necessary to introduce the concept of the "stretch factor" defined as

$$\text{stretch factor} = \dot{\mathbf{r}}_1 \cdot \dot{\mathbf{r}}_0 \tag{11}$$

where the dot superscript denotes differentiation with respect to arc length along the curve $\mathbf{r}_0$ (Snieder and Sambridge, 1993). Arc lengths along the unperturbed and perturbed curves are related by

$$\frac{\partial s}{\partial s_0} = 1 + \varepsilon\left(\dot{\mathbf{r}}_1 \cdot \dot{\mathbf{r}}_0\right) \tag{12}$$

Because the arc lengths of the reference and perturbed curves differ, there is ambiguity in how the reference curve is mapped into the perturbed curve. For example, points can be related by being positioned at equivalent fractional distances of the total arc length (Figure 3a), or the perturbed points can be displaced in a direction normal to the reference curve (Figure 3b).

LFA can readily treat three perturbation cases (Snieder and Sambridge, 1993):

1. perturbation of the slowness field ($u = 1/V$):

$$u(\mathbf{r}) = u_0(\mathbf{r}) + \varepsilon\, u_1(\mathbf{r}) \tag{13}$$

2. perturbation of an initial path estimate (the reference curve, not a true ray) to the desired true ray:

$$\nabla u_0 - \frac{d}{ds_0}\left(u_0 \frac{d\mathbf{r}_0}{ds_0}\right) = R_b \tag{14}$$

where R_b measures the degree to which the ray equations are violated;

3. perturbation of one or both ray end-points ($s_0 = 0$, S_0):

$$\mathbf{r}(0) = \mathbf{r}_0(0) + \varepsilon\, \mathbf{a}; \;\; \mathbf{r}(S_0) = \mathbf{r}_0(S_0) + \varepsilon\, \mathbf{b} \tag{15}$$

Case 1 is most relevant to seismic tomography, while cases 2 and 3 are directly applicable to two-point ray tracing and earthquake location problems. The corresponding equations have different forms depending on the choice of mapping from the reference curve to the perturbed ray. The

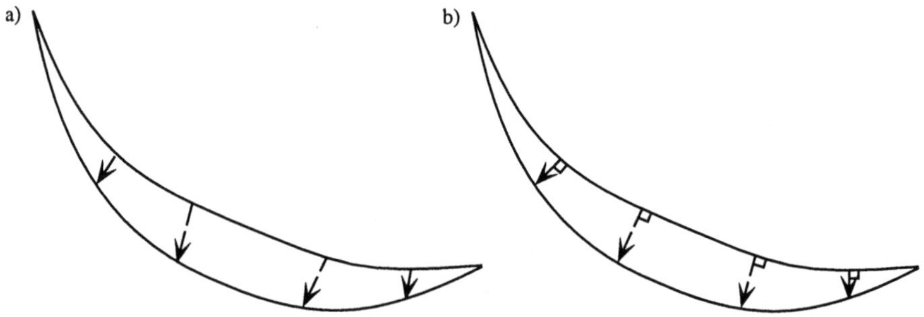

Figure 3. In the ray perturbation methods, points on the original and perturbed paths can be related by (a, left) being positioned at equivalent fractional distances of the total arc length (Snieder and Sambridge, 1992) or (b, right) being displaced in a direction normal to the reference curve (Snieder and Spencer, 1993).

method also provides the ability to evaluate travel time along the perturbed ray to second-order accuracy in ε even though the ray is only correct to first order, thanks to Fermat's theorem (Snieder and Sambridge, 1992).

Snieder and Spencer (1993) and Snieder and Sambridge (1993) demonstrate how the standard bending method (Julian and Gubbins, 1977) and the Hamiltonian approach (discussed below) are essentially equivalent to their Lagrangian approach with various mapping assumptions. Interestingly, one case of the Lagrangian approach of Pulliam and Snieder (1996) is equivalent to the Fermat's principle approach of Um and Thurber (1987). The case is one with straight reference curve segments and curve-normal perturbations. This will be discussed further in the "Fermat's Principal Approach" section. Recently, Bijwaard and Spakman (1999) have adapted this specific case to create an extremely versatile ray tracing algorithm. The true power of their algorithm lies in its ability to determine ray paths for direct, critically refracted, reflected, converted, and even some diffracted rays. Their method is likely to receive wide application in both event location and tomography studies.

3.2 Hamiltonian Formulation Approach (HFA)

HFA is based on the dynamic ray tracing method in ray-centered coordinates (Cerveny, 1987; Nowack and Lutter, 1988). The initial ray is called the central ray, denoted by $\mathbf{r}_c(\tau)$, with the corresponding slowness vector $\mathbf{p}_c(\tau)$, where τ is the sampling parameter along the ray. Paraxial rays are defined by perturbations to the central ray and its slowness, $\mathbf{r}(\tau) = \mathbf{r}_c(\tau) + \delta\mathbf{r}(\tau)$ and $\mathbf{p}(\tau) = \mathbf{p}_c(\tau) + \delta\mathbf{p}(\tau)$ (Figure 4). Expressions for the perturbations are derived via linear perturbation of the ray equations shown in Equation 2, yielding

$$\begin{bmatrix} \delta\dot{\mathbf{r}} \\ \delta\dot{\mathbf{p}} \end{bmatrix} = \begin{bmatrix} 0 & \mathbf{I} \\ \mathbf{W} & 0 \end{bmatrix} \begin{bmatrix} \delta\mathbf{r} \\ \delta\mathbf{p} \end{bmatrix} \tag{16}$$

where **I** is an identity matrix and **W** is a 2x2 matrix of second derivatives of velocity with respect to the path-normal coordinates evaluated along the central ray (Virieux et al., 1988). Because this system is linear, it is relatively straightforward to evaluate the path perturbations required to reach the desired target end point or to perturb a trial ray towards the true ray path.

A series of papers (Chapman, 1985; Virieux et al., 1988; Virieux, 1991; Virieux and Farra, 1991; Farra, 1992; Farra, 1993) explored in detail the Hamiltonian approach to ray tracing using paraxial rays. Farra et al. (1994) completely documented the equivalences between the LFA and HFA methods, and also presented a solution approach that does not require the use of a stretch factor (Equation 11). The HFA method can take full advantage of the efficiencies of constant quadratic slowness gradient models (section 2.1), allowing fast, exact ray tracing in complex 3-D structures (Virieux et al., 1988). The method can also be adapted to deal with anisotropic media and discontinuities (Farra, 1992).

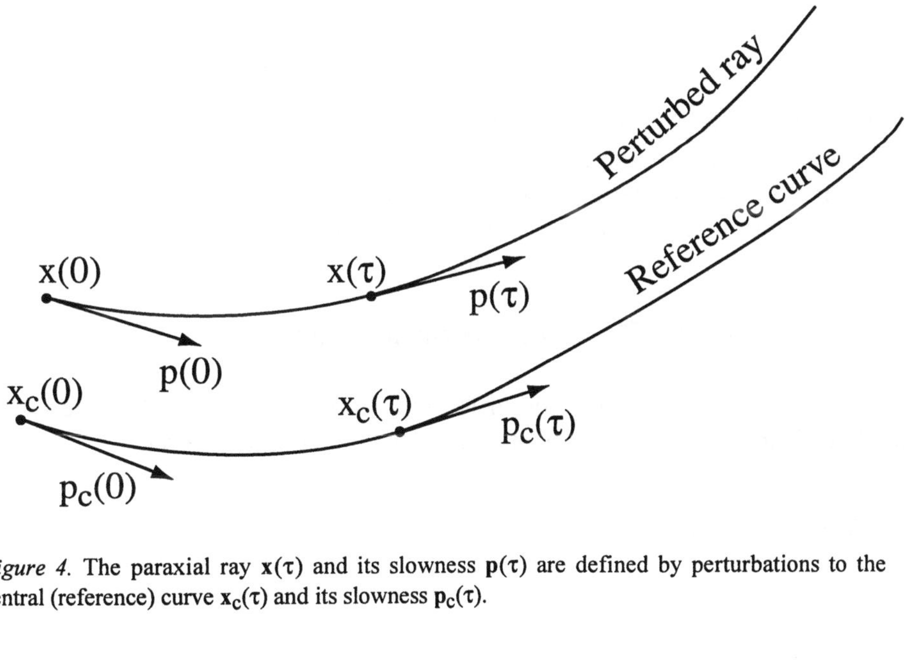

Figure 4. The paraxial ray $\mathbf{x}(\tau)$ and its slowness $\mathbf{p}(\tau)$ are defined by perturbations to the central (reference) curve $\mathbf{x}_c(\tau)$ and its slowness $\mathbf{p}_c(\tau)$.

3.3 Integral Method (Hybrid Perturbation)

An intriguing alternative to the HFA and LFA methods was presented by Sun (1993). Rather than solving additional systems of differential equations, Sun (1993) proposed a hybrid approach involving integrating the second-order ray equation (Equation 1) twice to yield a relation between the initial path **r**(s) (not necessarily a ray), the initial slowness vector $\mathbf{p}_0$ and a set of true rays $\mathbf{r}^{new}$

$$\mathbf{r}^{new}(s, \mathbf{p}_0) = \mathbf{r}(0) + \mathbf{p}_0 \, f(s) + \mathbf{q}(s) \tag{17}$$

where s is the path length parameter, **r**(0) is the starting point, and f(s) and **q**(s) are path integrals

$$f(S) = \int_0^S V(s)\, ds \; ; \; \mathbf{q}(S) = \int_0^S V(s) \int_0^s \nabla\left(\frac{1}{V(s')}\right) ds' \tag{18}$$

where S is the total length of the initial path. The initial slowness vector $\mathbf{p}^{new}$ that yields the path with the target end point $\mathbf{r}^T$ can be determined by solving

$$\mathbf{p}^{new} = \frac{\mathbf{r}^T - \mathbf{r}(0) - \mathbf{q}(S)}{f(S)} \tag{19}$$

Using $\mathbf{p}^{new}$ and Equations 17 and 18, the complete new path can be determined. The method parameterizes the rays in terms of the initial "shooting" direction, hence Sun (1993) termed it "parameterized shooting," but the solution technique involves integration, not the standard two-point shooting. In fact, it is closer to bending methods in its approach.

3.4 Fermat's Principle Approach (FPA)

An alternative to solving the differential equations of the HFA and LFA methods is to attempt to minimize directly the travel time integral

$$T = \int_{source}^{receiver} u \, ds \tag{20}$$

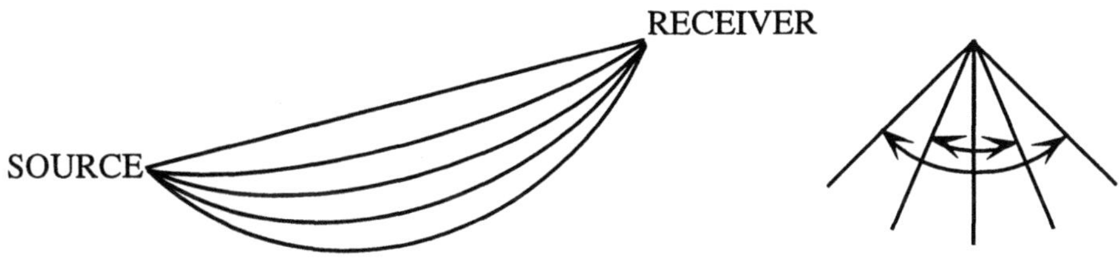

Figure 5. Approximate ray tracing via construction of arcuate ray paths connecting the source and receiver with a range of radii of curvature and oriented over a range of dip angles.

where u is slowness, by estimating perturbations to the path. Thurber (1981, 1983) introduced a simple brute-force search algorithm that involved rapid construction of arcuate ray paths connecting the source and receiver that had a range of radii of curvature and were oriented over a range of dip angles (Figure 5). Travel times along each arcuate path were computed, and the path of least travel time was adopted as the best estimated ray path. This method, known as Approximate Ray Tracing (ART), was shown to be reasonably accurate for complex velocity structures as long as path lengths were less than about 40 km (Eberhart-Phillips, 1986). The accuracy was limited mainly because paths were restricted to lie within planes.

The "pseudo-bending" approach of Um and Thurber (1987) involves discretizing the initial path into straight-line segments and using the following equation to estimate the direction of path perturbations in a piecewise manner:

$$\frac{d^2\mathbf{r}}{ds^2} = -\frac{1}{V}\left(\nabla V - \frac{dV}{ds}\frac{d\mathbf{r}}{ds}\right) \tag{21}$$

(Cerveny et al., 1977). The term on the left-hand side of Equation 21 is the ray path curvature, and the term in parentheses on the right-hand side is the component of the velocity gradient normal to the ray path (Figure 6). Um and Thurber (1987) created an algorithm that steps along a trial ray from the end points towards the middle, solving for the perturbation amplitude in the direction indicated by Equation 21 that locally minimizes travel time for each set of 3 path points. The algorithm starts with a path defined by 3 points (alternatively, the result of ART could be used as the starting path), and successively adds path points as convergence progresses. The trapezoidal rule is used for the numerical integration of travel time along the path.

This approach has two significant advantages over DEM, HFA, and LFA. One is the increased numerical stability afforded by the integral solution as opposed to the differential equation solution (Moser et al., 1992). The other

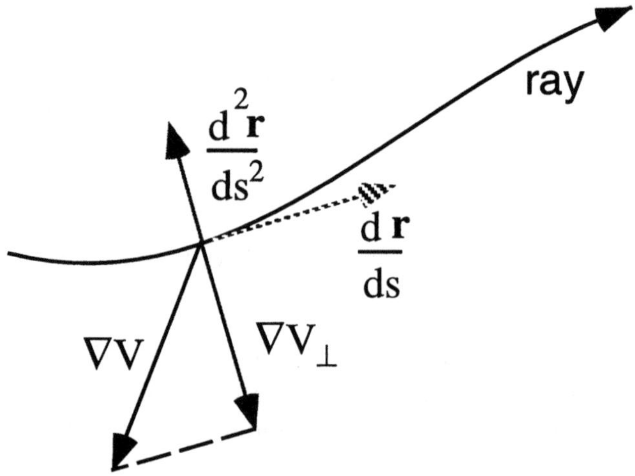

Figure 6. The pseudo-bending method perturbs the path using the relation between ray path curvature and the magnitude of the component of the velocity gradient normal to the ray path.

is the reduced computational burden due to the need for continuity of only the first spatial derivatives of velocity, not continuous second derivatives as required in other methods. The drawbacks of the Um and Thurber (1987) method are that it was not designed to handle discontinuities in velocity structure and that it has convergence problems in models dominated by nearly constant velocity due to the use of Equation 21 to define the perturbation direction. Moser et al. (1992) and Koketsu and Sekine (1998), however, show how discontinuities can be handled with the pseudo-bending method, and the method of Moser et al. (1992) successfully deals with constant-velocity regions (see below). Moser et al. (1992) and Koketsu and Sekine (1998) also extended the pseudo-bending method to the spherical-earth case.

There is an interesting parallel between the Um and Thurber (1987) pseudo-bending method and one case of the Pulliam and Snieder (1996) ray perturbation method - specifically, the case used by Bijwaard and Spakman (1999), discussed above. The latter define the measure of the difference between a trial path made up of straight line segments and the desired true ray by

$$\mathbf{R}_b = \nabla u_0 - \frac{d}{ds_0}\left(u_0 \frac{d\mathbf{r}_0}{ds_0}\right) \tag{22}$$

where the subscript 0 denotes values along the trial path. The term on the right-hand side of Equation 22 represents the component of the slowness right-hand side of Equation 21. The main differences between the Um and Thurber (1987) and Pulliam and Snieder (1996) solutions are the use of velocity versus slowness and the use of initially curved paths versus initially straight path segments, respectively.

Moser et al. (1992) improved on the pseudo-bending method of Um and Thurber (1987) in two other important ways. They parameterized the ray in terms of Beta-splines instead of linear segments. This provides greater flexibility in representing ray paths, along with improved efficiency and accuracy. In particular, they point out that it can be advantageous to use relatively few points for the perturbation computation but a greater number of points for the calculation of travel time. They also applied a global perturbation scheme to minimize travel times, rather than a local 3-point scheme. By making use of the conjugate gradient solution method, Moser et al. (1992) retained very high computational efficiency while improving the accuracy of the solution. Their method also has no difficulty with constant velocity regions due to their direct use of the velocity gradient vector in their perturbation calculation.

Prothero et al. (1988) developed an alternative FPA algorithm based on a simplex search algorithm and sinusoidal path perturbations. Starting with a path estimate from ART, horizontal and vertical perturbations (dh and dv) of the i'th point along the path were parameterized in terms of sums of sine wave terms up to order n:

$$dv(n, i) = A_v(n) \sin (n\pi \, d \, i/l) \, , \; dh(n, i) = A_h(n) \sin (n\pi \, d \, i/l) \qquad (23)$$

The simplex method (illustrated for one harmonic order and path point in Figure 7) is used to step towards the path of minimum travel time. The algorithm cannot diverge, but it can converge to a local minimum, as is the case for most methods. However a relatively large solution region is searched in this method, so the chances of finding the global minimum are improved compared to pseudo-bending.

4. GRID-BASED METHODS

Two totally different grid-based approaches to the travel time calculation problem are the wavefront-tracing method (Vidale, 1988, 1990; van Trier and Symes, 1991; Podvin and Lecomte, 1991; Coultrip, 1993; Klimes and Kvasnicka, 1994; Hole and Zelt, 1995) and the network (graph) theory method (Nakanishi and Yamaguchi, 1986; Moser, 1991; Fisher and Lees, 1993; Cheng and House, 1996; Bijwaard and Spakman, 1999). An exciting aspect of these techniques is the ability to determine the global minimum

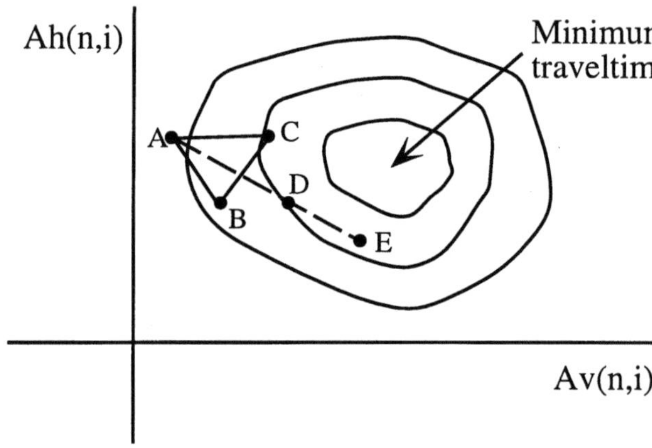

Figure 7. The simplex method illustrated for one harmonic order and path point. Starting from the initial simplex ABC, the worst point (A) is reflected to point D. The step is expanded to point E where the travel time is minimized along that direction and the next simplex step is taken using simplex BCE.

travel time. However, in most cases there is a significant trade-off between accuracy and computation speed – a dense grid is required for high accuracy, but computation times increase rapidly with increasing grid size.

Most wavefront-tracing algorithms use a finite-difference solution to the eikonal equation (Vidale, 1988, 1990; van Trier and Symes, 1991):

$$\left(\frac{\partial t}{\partial x}\right)^2 + \left(\frac{\partial t}{\partial y}\right)^2 + \left(\frac{\partial t}{\partial z}\right)^2 = u^2(x,y,z) \tag{24}$$

In the method of Vidale (1988, 1990), the finite-difference solution is carried out on a regular grid using circular or planar wavefront extrapolations. The method allows for the subsequent determination of ray paths by tracking normals to wavefronts (Vidale and Houston, 1990). The solution algorithm requires the sorting of grid points in order of increasing travel time before extrapolating times to subsequent grid points. This prevents the computation scheme from being vectorized for use on high-speed parallel computers. Van Trier and Symes (1991) improved on Vidale's approach by implementing a vectorizable "upwind" finite-difference scheme adapted from fluid mechanics. However, both methods can fail to find the global minimum travel time when the first arrival follows a relatively circuitous route (Coultrip, 1993). These methods can also have significant computational errors for sharp velocity changes (Podvin and Lecomte, 1991).

The finite-difference method of Podvin and Lecomte (1991) takes a Huygens' principle approach to travel time computation rather than using the eikonal equation. As shown in Figure 8a, the 3-D calculation of the arrival time at a new grid point is evaluated for a transmitted wave by assuming locally planar wavefronts emanating from all 4 possible sets of 3 adjacent gridpoints on the preceding grid layer, and using the earliest of the 4 computed times. For a given grid point, 24 different interface patches must be considered (Figure 8b). The method can also handle head waves, diffracted waves, and even reflected waves. Ray paths can be evaluated once the complete travel-time field has been evaluated.

Coultrip (1993) presented a 2-D "wavefront-tracing" method based on constant-velocity triangular model cells. Some factors that distinguish his approach are the use of circular or planar wavefronts and the inclusion of multiple nodes along cell boundaries (Figure 9). The use of triangular cells adds much flexibility to the model complexity. The algorithm readily handles critical refractions and diffractions, and could be extended to 3-D using tetrahedra. However, the number of nodes along the tetrahedral faces might need to be extremely large to provide adequate computational accuracy.

Another Huygens' principle approach is based on the use of network (graph) theory (Nakanishi and Yamaguchi, 1986; Moser, 1991; Fisher and Lees, 1993; Cheng and House, 1996; Bijwaard and Spakman, 1999). A set of nodes is established and connected by a network of paths (Figure 10). The "lengths" of the path segments are defined to be the corresponding travel times, and efficient sorting routines are used to identify the shortest time path (Moser, 1991). The nodes can be the corners of uniformly-sized cells (Nakanishi and Yamaguchi, 1986), or there can be multiple nodes along cell boundaries (Moser, 1991; Fischer and Lees, 1993) as in the triangular-cell approach of Coultrip (1993). An important issue is the sorting method used to find the minimum time path. Cheng and House (1996) recommend the "quick sort" over the "heap sort" due to the efficiency of the former method.

A drawback of the standard graph method is the need for huge numbers of nodes for large model volumes in order to retain acceptable computational accuracy. Nolet and Moser (1993) devised a solution to this problem by only discretizing a "cigar-shaped" volume about an initial path estimate, rather than extending the graph throughout the entire model volume. Equidistantly-spaced graph planes were oriented normal to the trial path except at discontinuities, where they were oriented parallel to the boundary. As convergence proceeded, the graph system was updated to the new path estimate at each iteration.

Bijwaard and Spakman (1999) modified Nolet and Moser's (1993) approach by using constant-sized graph planes with varying node spacing within the planes, rather than using variable-sized planes with constant node

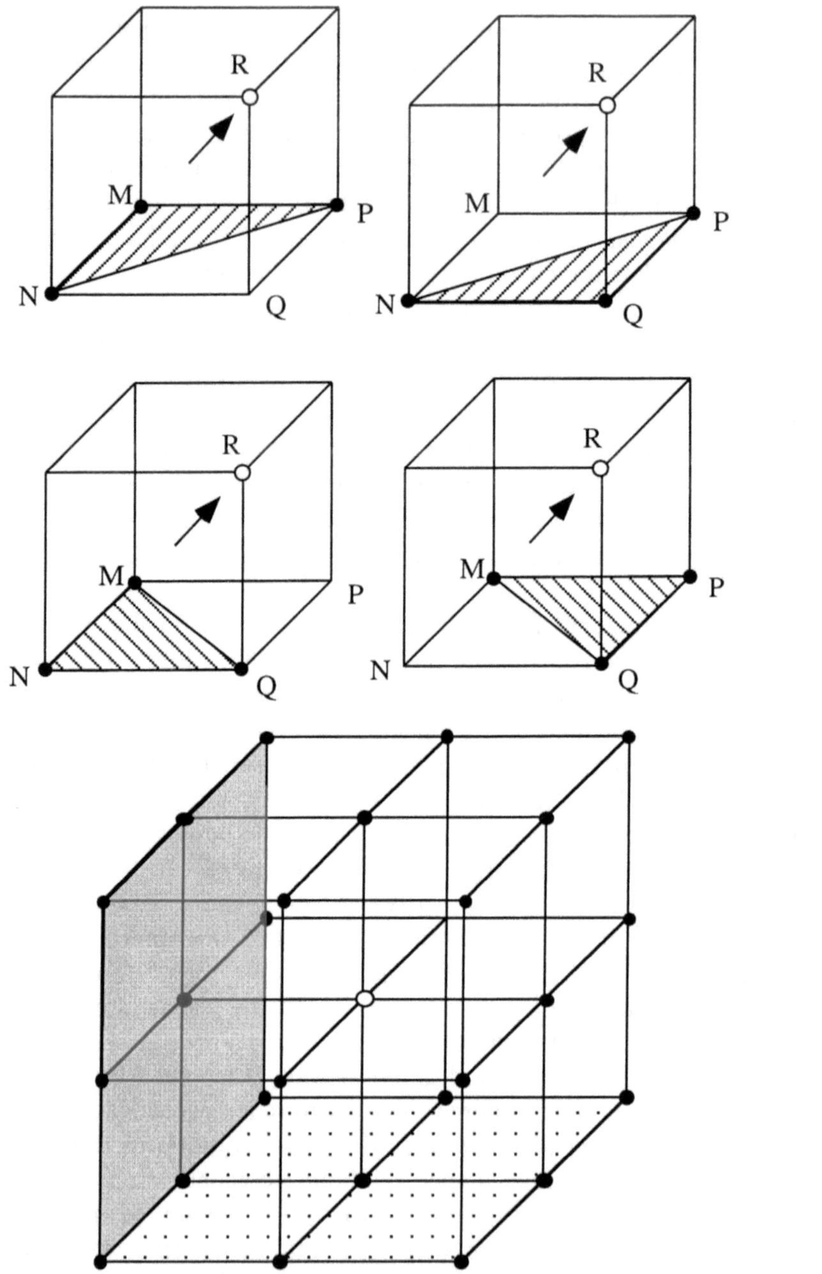

a)

b)

Figure 8. The finite-difference method of Podvin and Lecomte (1991) based on Huygens' principle. (a) The 3-D calculation of the arrival time at a new grid point is evaluated for a transmitted wave by assuming locally planar wavefronts emanating from all 4 possible sets of 3 adjacent gridpoints on the preceding grid layer. (b) For a given grid point, 24 different interface patches must be considered.

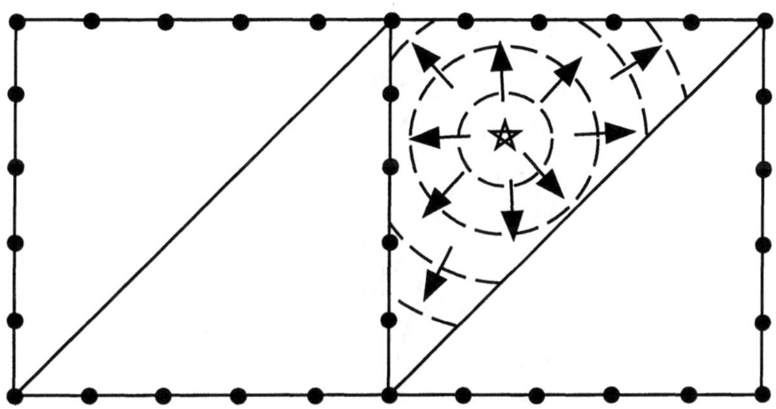

Figure 9. Coultrip's (1993) 2-D wavefront-tracing method, based on constant-velocity triangular model cells and the inclusion of multiple nodes along cell boundaries.

spacing. This simplifies the graph geometry and node indexing, and permits "zooming in" as a solution is approached. They also extended the method to include reflected, converted, and head waves.

5. 3-D RAY TRACING AND SEISMIC TOMOGRAPHY

One research area in which accurate and efficient algorithms for computing travel times and ray paths are essential is seismic tomography. Local earthquake studies have been incorporating some form of 3-D ray tracing for many years (Thurber, 1981, 1983), and currently is considered relatively routine. This required a merging of the use of 3-D ray tracing for local earthquake location (Engdahl and Lee, 1976) with the use of local earthquake arrival times for linearized seismic tomography (Aki and Lee, 1976). 3-D ray tracing is now being applied to global tomography as well (Bijwaard and Spakman, 1999).

With the availability of several accurate and fast ray tracing algorithms for complex velocity structures, one focus of our research attention shifts to the problem of finding the global travel time minimum (GTTM) and of identifying the local travel time minimum (LTTM) corresponding to an observed seismic phase. This demand is of equal importance to both parts of the coupled hypocenter-velocity model aspect of seismic tomography as it concerns non-linear dependencies of the solution on hypocentral and velocity parameters:

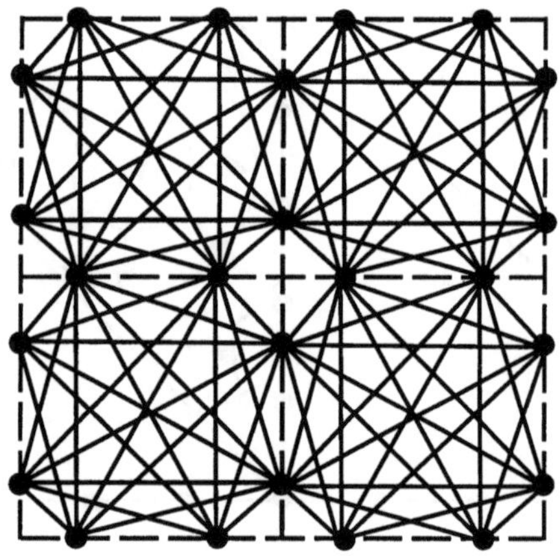

Figure 10. A set of nodes are established and connected by a network of paths in the graph theory approach. A search algorithm is then used to find the minimum travel time path.

- correlation of observed seismic phases with LTTM is necessary to correctly calculate take-off angles that are crucial for high-precision hypocenter locations in complex media.
- in seismic tomography, the correct correlation of seismic phases with travel time minima obviously has a strong influence on travel paths and model resolution.

Finding the GTTM in complex velocity structures is a difficult task for most ray tracing methods, and most grid-based methods are hampered by the significant trade-off between accuracy and computation speed (see above). This is one of the reasons that made SIMULPS (Thurber, 1993; Eberhart-Phillips, 1993; Evans et al., 1994) with its fast approximate ray tracer including an efficient search algorithm combined with subsequent pseudo-bending (ART-PB) so popular for local earthquake tomography applications.

With regard to the task of correct identification of LTTM corresponding to an observed seismic phase, strict ray-based methods even with sophisticated search algorithms for initial rays are clearly inferior to grid-based methods. Though full travel time fields theoretically contain the necessary information to identify most seismic phases in cases of complex velocity structure, the shear amount of information needed makes this route impractical. In addition, the calculation of travel paths of seismic waves for most grid-based methods is again based on ray theory only.

A practical alternative approach that relies on ray theory but includes, by first-order approximation, some elements of wave theory is based on Fresnel volume calculations or "fat rays" (Woodward, 1992; Husen, 1999) (Figure 11). Whereas in well-resolved regions, differences in seismic tomographic

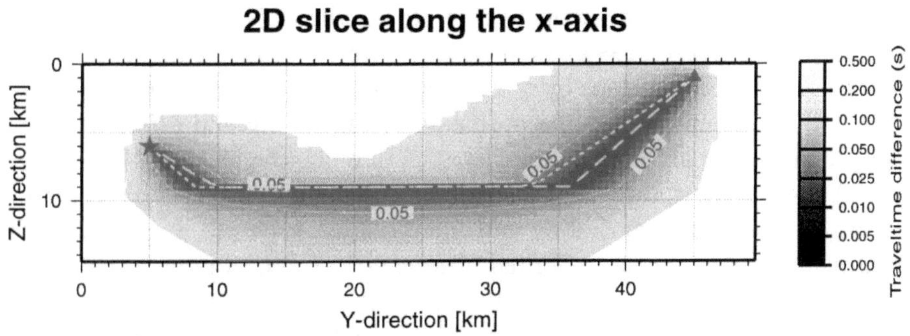

Figure 11. "Fat ray" example for a head wave (from Husen, 1999). The shading indicates the time difference between the source-receiver time and the sum of the source-to-point and point-to-receiver times. The contour denotes a fat ray width of 0.05 s. Dashed lines represent ray paths (source-receiver and receiver-source) found by locally following the steepest gradient in each travel time field.

results obtained by rays and fat rays are insignificant, in regions of only fair or poor resolution, application of strict ray theory may lead to distortions in the tomographic images (Husen and Kissling, 2000). Such artifacts are the combined result of model parameterization, ray path distortions, and problems in association of observed seismic phases with calculated ray paths. Ray tracing in complex structure by shooting or bending methods sometimes leads to very peculiar paths that denote a mathematical LTTM only (for waves of infinite frequency) whereas the more realistic LTTM - the fat ray - encompasses several such rays (Figure 11). In any case, the calculation of a specific fat ray (one LTTM) intrinsically carries the potential to map the neighboring LTTM's with little additional computational effort, thus significantly decreasing the potential for incorrect correlations of seismic phases with LTTM's.

6. TRAVEL TIME AND RAY PATH ACCURACY

Theoretically, high-precision forward solutions for ray paths and travel times for complex velocity structures may be obtained by several methods. Practical applications, however, are often limited by the representation of the velocity structure demanded by the forward method and perhaps *a priori* knowledge. For example, the ART-PB method (see above) implemented in the local earthquake tomography code SIMULPS (Thurber, 1993; Eberhart-Phillips, 1993; Evans et al., 1994) handles 3-D velocity models with uneven gridding very efficiently. Uneven gridding is a powerful model design tool easily allowing adaptation to *a priori* knowledge such as sedimentary basins or tectonic lineaments (Eberhart-Phillips, 1993). Most high-precision forward solvers, though, demand fine resampling of such a velocity grid. Recently, Haslinger (1998) incorporated the ray tracing algorithm RKP (Runga-Kutta Perturbation) by Virieux and Farra (1991) into the SIMULPS code and documented how the problems of regridding and different model representations of a complex velocity structure may be tackled. Haslinger's (1998) results show that the travel times calculated with two different ray tracers with optimal regridding and interpolation schemes might not differ significantly with respect to normal observation errors. Ray paths calculated with RKP and with ART mainly lie within the Fresnel volume and, hence, do not strongly affect the tomographic inverse solution in well-resolved regions, but they do affect resolution estimates (Haslinger and Kissling, 2000). Also, the differences in the take-off angles of the rays at the source might be significant for high-precision earthquake location (Figure 12).

Husen (1999) further compared travel times obtained by RKP, ART-PB, and a popular finite-difference (FD) algorithm (Podvin and Lecomte, 1991). In addition to evaluating travel time differences among the three algorithms, Husen (1999) used a reciprocity test to provide an additional measure of method accuracy. Analysis of travel times computed both from source to receiver and from receiver to source reveals that differences are due to numerical instabilities. The results of Haslinger (1998), Husen (1999), and Kissling et al. (2000) indicate that RKP and ART-PB perform extremely

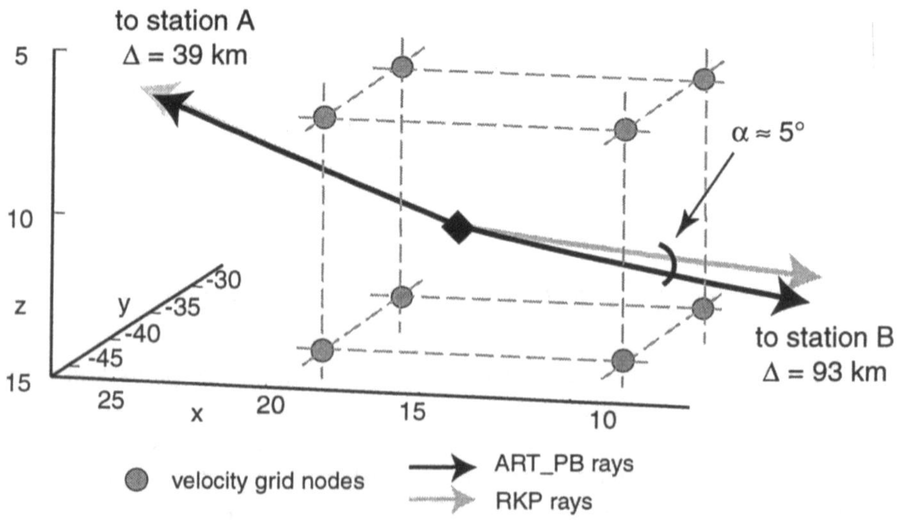

Figure 12. Differences in the take-off angles of the rays at the source for different ray tracing methods might be significant for high-precision earthquake location (from Haslinger and Kissling, 2000).

well up to epicentral distances of 60 km, both in terms of small time differences between them (10 ms or less, generally) and even smaller reciprocal time differences (Figure 13). The FD algorithm performed more poorly by both measures (Figure 14). These are encouraging results for the RKP and ART-PB methods, and suggest the need for some caution in using FD methods where numerical accuracy more strongly depends on grid size than with ray tracing methods (Husen and Kissling, 2000).

On the global scale, Bijwaard and Spakman (1999) have compared a modified graph method (Moser, 1991; Nolet and Moser, 1993) and a perturbation method (Pulliam and Snieder, 1996) against each other and against analytic solutions. In simple analytic models, both methods proved to be extremely accurate, with the perturbation methods about one order of magnitude better (errors of a few ms versus a few tens of ms for the graph method). For paths in a 3-D global model (Bijwaard et al., 1998), the time differences were still relatively minor, on the order of 0.1 to 0.15 seconds or less. Additional multi-method comparisons of these types would be extremely valuable for the user community.

7. OTHER CHALLENGES

Two other challenging issues that deserve further mention are the incorporation of model interfaces and the use of anisotropic velocity models. At high frequencies, Snell's Law can be applied to deal with the refraction of a ray across a curved interface with heterogeneous velocity structures on both sides (Cerveny, 1987). For shooting-type methods, then, the incorporation of interfaces is relatively straightforward, as the propagating ray simply alters direction when an interface is crossed. However, for bending-type methods, a more sophisticated approach is required. Examples can be found in Moser et al. (1992), Pereyra (1992), Zhao et al. (1992), Koketsu and Sekine (1998), and Bijwaard and Spakman (1999). Ray path and travel time calculation in anisotropic structures is a much more difficult subject. Possible approaches include medium perturbation methods (Cerveny and Firbas, 1984; Nowack and Psencik, 1991) for weakly anisotropic models and complete anisotropic ray tracing (Gajewski and Psencik, 1987; Cerveny and Firbas, 1984). The use of anisotropic models in body-wave tomography is generally limited to cases where extremely high quality, densely sampled data are available.

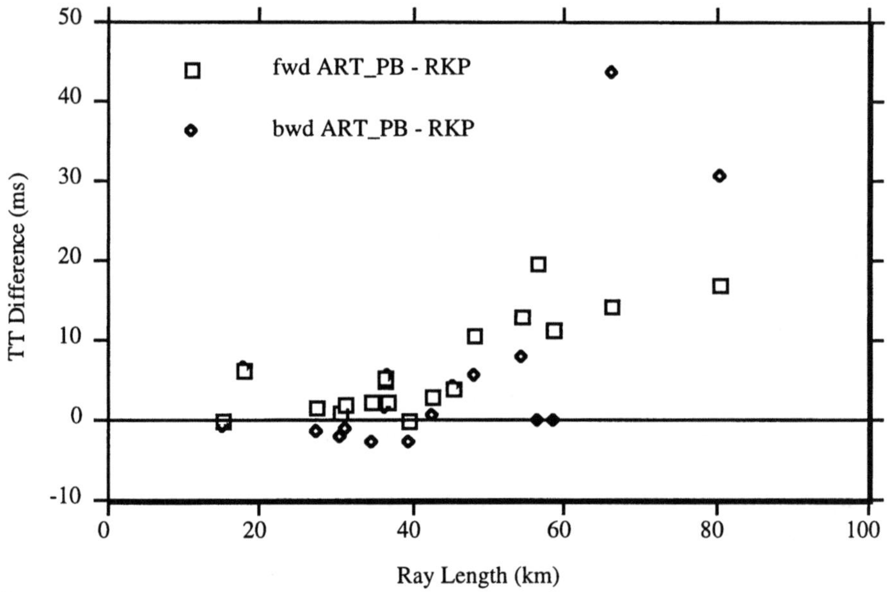

a)

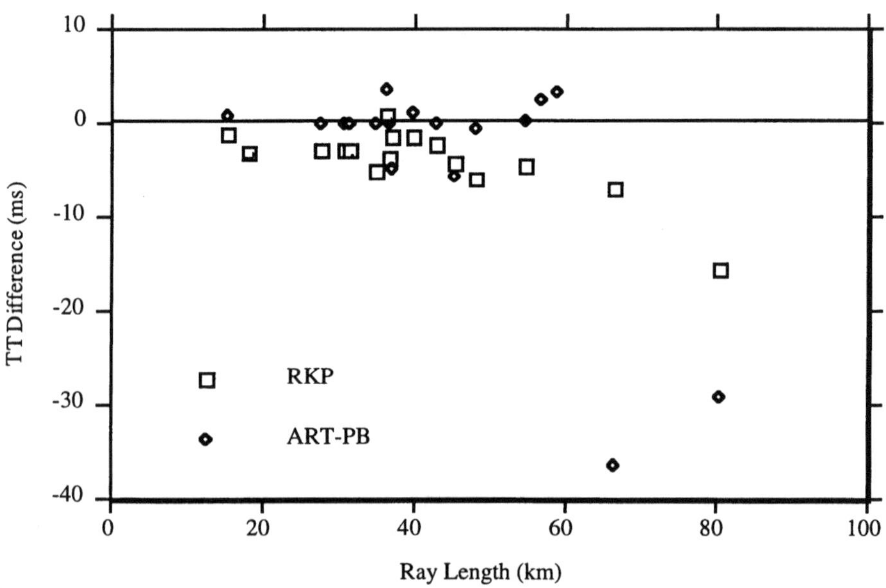

b)

Figure 13. (a) Differences between forward and reverse times for the RKP and ART_PB ray tracers. For rays less than 50 km, agreement is very good, but note that RKP times are faster beyond 50 km (Haslinger, 1998). (b) Comparison of forward (fwd) and reverse (bwd) travel times for RKP (squares) and ART_PB (triangles) ray tracers. RKP and ART-PB both perform extremely well up to ray lengths of 60 km, with errors on the order of 10 ms or less. Beyond 60 km, accuracy degrades significantly.

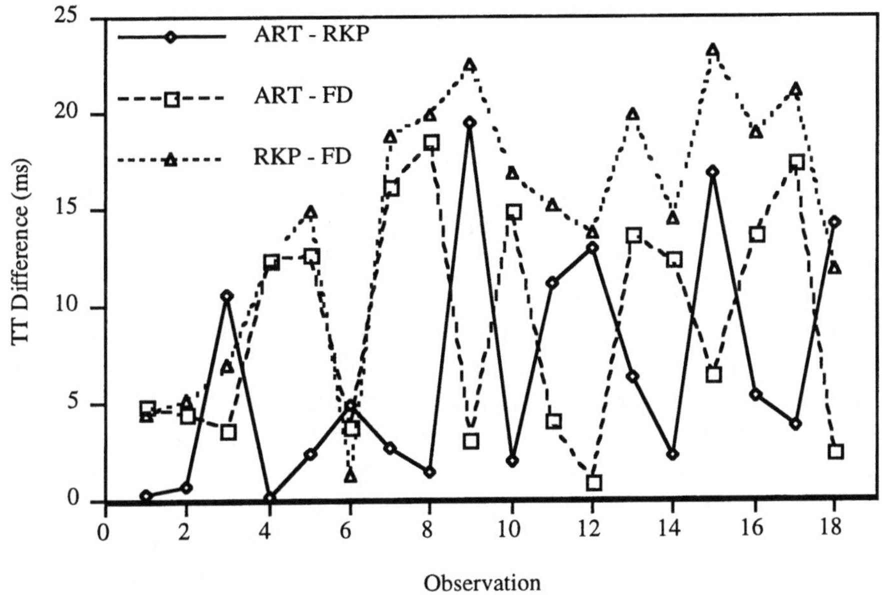

a)

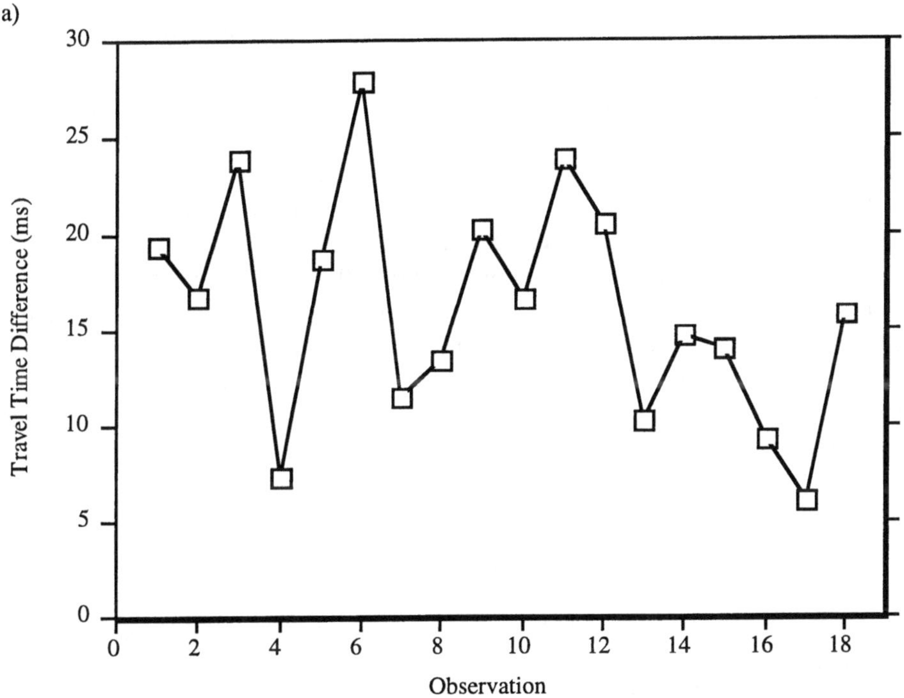

b)

Figure 14. (a) Travel time differences between the different forward solutions. Note the generally high difference between RKP and FD times. (b) Comparison of forward (fwd) and reverse (bwd) travel times for the FD solution. Compared to RKP and ART_PB (Fig. 13 b), the FD solutions generally show greater forward-reverse discrepancies.

8. DISCUSSION AND CONCLUSIONS

Enormous progress has been made in 3-D travel time calculation methods over the last 20 years. A wide variety of ray tracing and grid-based algorithms have been developed, some of which have been made openly available to the user community. Despite steady increases in computer speed, the computational burden of 3-D travel time calculations remains a significant issue, especially for global studies with millions of data (Bijwaard and Spakman, 1999).

Accuracy issues deserve additional examination. It is often presumed that "exact" ray tracing methods or finite-difference methods have superior accuracy to approximate or perturbation methods. However, recent studies (Haslinger, 1998; Husen, 1999) have shown that the supposedly more accurate methods have errors of the same order of magnitude as the approximate methods. This suggests the need for benchmark tests against which the various methods can be evaluated. Unfortunately, inherent differences in the ways in which 3-D models are represented (parameterized) in the various algorithms makes such benchmark testing very difficult (Kissling et al., 2000). This problem might be circumvented by using complex 3-D models defined by analytic functions for the testing.

Finally, 3-D travel time calculations will only be as good as the 3-D velocity models on which they are based, but of course the derivation of good 3-D models requires high-quality travel time and ray path calculations. At the start this requires high-quality data. We look forward to continued growth in numbers and improvement in quality of the instrumentation used in permanent and temporary networks around the world. We also strongly encourage an increased sharing of seismic data and software, allowing scientists with new ideas for analysis access to data for testing the new methods, and conversely allowing scientists with new data access to the best available methods for gaining the most insight into the Earth from their seismograms.

ACKNOWLEDGEMENTS

We thank Robert Nowack for valuable comments on an earlier draft of this chapter. S. Husen and S. Goes are thanked for constructive reviews. The first author (CHT) acknowledges the U.S. National Science Foundation and the Defense Threat Reduction Agency, U.S. Department of Defense, for support of research related to this chapter.

REFERENCES

Aki, K. and W.H.K. Lee (1976) Determination of three-dimensional anomalies under a seismic array using first P arrival times from local earthquakes, 1. A homogeneous initial model, *J. Geophys.Res.* **81**, 4381-4399.

Aric, K., B. Gutdeutsch, and A. Sailer (1980) Computation of travel times in a medium of two-dimensional velocity distribution, *Pure Applied Geophys.* **118**, 796-805.

Bijwaard, H. and W. Spakman (1999) Fast kinematic ray tracing of first- and later-arriving global seismic phases, *Geophys. J. Int.* **139**, 359-369.

Bijwaard, H., W. Spakman, and E. R. Engdahl (1998) Closing the gap between regional and global travel time tomography, *J. Geophys.Res.* **103**, 30,055-30,078.

Cerveny, V. (1987) Ray tracing alogorithms in three-dimensional laterally varying layered structures, in Nolet, G. (ed.), *Seismic Tomography with Applications in Global Seismology and Exploration Geophysics*, D. Reidel, pp. 99-133.

Cerveny, V. and P. Firbas (1984) Numerical modelling and inversion of travel of seismic body waves in inhomogeneous anisotropic media, *Geophys. J. R. Astr. Soc.*, **76**, 41-51.

Chapman, C. (1985) Ray theory and its extensions: WKBJ and Maslov seismograms, *J. Geophys. Res.* **58**, 27-43.

Cheng, N. and L. House (1996) Minimum traveltime calculation in 3-D graph theory, *Geophysics* **61**, 1895-1898.

Coultrip, R.L. (1993) High-accuracy wavefront tracing traveltime calculation, *Geophysics* **58**, 284-292.

Eberhart-Phillips, D. (1986) Three-dimensional velocity structure in Northern California Coast Ranges from inversion of local earthquake arrival times, *Bull. Seismol. Soc. Am.* **76**, 1025-1052.

Eberhart-Phillips (1993) Local earthquake tomography: earthquake source regions, in H. M. Iyer and K. Hirahara (eds.), *Seismic Tomography*, Chapman and Hall, pp. 613-643.

Engdahl, E.R. and W.H.K. Lee (1976) Relocation of local earthquakes by seismic ray tracing, *J. Geophys.Res.* **81**, 4400-4406.

Evans,J. R., D. Eberhart-Phillips, and C.H. Thurber (1994) User's manual for SIMULPS12 for imaging Vp and Vp/Vs: a derivative of the Thurber tomographic inversion SIMUL3 for local earthquakes and explosions, *U.S. Geol. Surv. Open File Rep. OFR* **94-431**, 101 pp.

Farra, V. (1992) Bending method revisited: a Hamiltonian approach, *Geophys. J. Int.* **109**, 138-150.

Farra, V. (1993) Ray tracing in complex media, *J. Applied Geophys.* **30**, 55-73.

Farra, V. and R. Madariaga (1987) Seismic waveform modeling in heterogeneous media by ray peturbation theory, *J. Geophys. Res.* **92**, 2697-2712.

Farra, F. and R. Madariaga (1994) Comments on "The ambiguity in ray perturbation theory" by Roel Snieder and Malcolm Sambridge, *J. Geophys. Res.* **99**, 21,963-21,968.

Fischer, R. and J.M. Lees (1993) Shortest path ray tracing with sparse graphs, Ge*ophysics* **58**, 987-996.

Gajewski, D. and I. Psencik (1987) Computation of high-frequency seismic wavefields in 3-D laterally inhomogeneous anisotropic media, *Geophys. J. R. Astr. Soc.* **91**, 383-411.

Haslinger, F. (1998) Velocity structure, seismicity and seismotectonics of Northwestern Greece between the Gulf of Arta and Zakynthos, *Diss. ETH No. 12966*, Swiss Federal Institute of Technology Zurich.

Haslinger, F. and E. Kissling (2000) Investigating effects of 3-D ray tracing methods in local earthquake tomography, submitted to *Phys. Earth Plan. Int.*

Husen, S. (1999) Local earthquake tomography of a convergent margin, North Chile, A combined on- and offshore study, *Dissertation*, Christian-Albrechts-Universität Kiel.

Husen, S. and E. Kissling (2000) Local earthquake tomography between rays and waves: Fat ray tomography, submitted to *Phys. Earth Plan. Int.*

Husen, S., E. Kissling, E. Flueh, and G. Asch (1999) Accurate hypocenter determination in the seismogenic zone of the subducting Nazca plate in northern Chile using a combined on-/offshore network, *Geophys. J. Int.* **138**, 687-701.

Ingber, L. (1989) Very fast simulated re-annealing, *Math. Comp. Mod.* **12**, 967-973.

Jacob, K.H. (1970) Three-dimensional seismic ray tracing in a laterally heterogeneous spherical earth, *J. Geophys. Res.* **75**, 6675-6689.

Julian, B.R. and D. Gubbins (1977) Three-dimensional seismic ray tracing, *J. Geophys.* **43**, 95-113.

Kissling, E., F. Haslinger, and S. Husen (2000) Model parametrization in seismic tomography: A choice of consequence for the solution quality, submitted to *Phys. Earth Plan. Int.*

Klimes, L. and M. Kvasnicka (1994) 3-D network ray tracing, *Geophys. J. Int.* 116, 726-738.

Koketsu, K. and S. Sekine (1998) Pseudo-bending method for three-dimensional seismic ray tracing in a spherical earth with discontinuities, *Geophys. J. Int.* **132**, 339-346.

Lin, C.H. and S.W. Roecker (1996) Three-dimensional P-wave velocity structure of the Bear Valley region of central California, *Pure Appl. Geophys.* **149**, 667-688.

Moser, T.J. (1991) Shortest path calculation of seismic rays, *Geophysics* **56**, 59-67.

Moser, T.J., G. Nolet, and R. Snieder (1992) Ray bending revisited, *Bull. Seismol. Soc. Am.* **82**, 259-288.

Nakanishi, I. and K. Yamaguchi (1986) A numerical experiment on nonlinear image construction from first arrival times for two-dimensional island arc structure, *J. Phys. Earth* **34**, 195-201.

Nolet, G. and T. Moser (1993) Teleseismic delay times in a 3-D Earth and a new look at the S discrepancy, *Geophys. J. Int.* **114**, 185-195.

Nowack, R.L. (1992) Wavefronts and solutions of the eikonal equation, *Geophys. J. Int.* **110**, 55-62.

Nowack, R.L. and W.J. Lutter (1988) Linearized rays, amplitude and inversion, *Pure Appl. Geophys.* **128**, 401-421.

Nowack, R.L. and I. Psencik (1991) Perturbation from isotropic to anisotropic heterogeneous media in the ray approximation, *Geophys. J. Int.* **106**, 1-10.

Pereyra, V. (1992) Two-point ray tracing in general 3D media, *Geophys.Prosp.* **40**, 267-287.

Pereyra, V., W.H.K. Lee, and H.B. Keller (1980) Solving two-point seismic ray-tracing problems in heterogeneous medium, Pt 1. A general adaptive finite difference method, *Bull. Seismol. Soc. Am.* **70**, 79-99.

Podvin, P. and I. Lecomte (1991) Finite difference computation of traveltimes in very contrasted velocity models: a massively parallel approach and its associated tools, *Geophys. J. Int.* **105**, 271-284.

Prothero, W.A., W.J., Taylor and J.A. Eickemeyer (1988) A fast, two-point, three-dimensional raytracing algorithm using a simple step search method, *Bull. Seismol. Soc. Am.* **78**, 1190-1198.

Pulliam, J. and R. Snieder (1996) Fast, efficient calculation of rays and travel times with ray perturbation theory, *J. Acoust. Soc. Am.* **99**, 383-389.

Sambridge, M.S. and B.L.N. Kennett (1990) Boundary value ray tracing in a heterogeneous medium: a simple and versatile algorithm, *Geophys. J. Int.* **101**, 157-168.

Snieder, R. and M. Sambridge (1992) Ray perturbation theory for traveltimes and ray paths in 3-D heterogeneous media, *Geophys. J. Int.* **109**, 294-322.

Snieder, R. and M. Sambridge (1993) The ambiguity in ray perturbation theory, *J. Geophys.Res.* **98**, 22,012-22,034.

Snieder, R. and C. Spencer (1993) A unified approach to ray bending, ray perturbation and paraxial ray theories, *Geophys. J. Int.* **115**, 456-470.

Sun, Y. (1993) Ray tracing in 3-D media by parameterized shooting, *Geophys. J. Int.* **114**, 145-155.

Thurber, C.H. (1981) Earth structure and earthquake locations in the Coyote Lake area, central California, *Ph.D. thesis*, M.I.T., 331 pp.

Thurber, C.H. (1983) Earthquake locations and three-dimensional crustal structure in the Coyote Lake area, central California, *J. Geophys. Res.* **88**, 8226-8236.

Thurber, C.H. (1986) Analysis methods for kinematic data from local earthquakes, *Rev. Geophys.* **24**, 793-805.

Thurber, C. H. (1993) Local earthquake tomography: Velocities and Vp/Vs - theory, in H. M. Iyer and K. Hirahara (eds.), *Seismic Tomography*, Chapman and Hall, pp. 563-583.

Um, J. and C.H. Thurber (1987) A fast algorithm for two-point seismic ray tracing, *Bull. Seismol. Soc. Am.* **77**, 972-986.

Van Trier, J. and W.W. Symes (1991) Upwind finite-difference calculation of traveltimes, *Geophysics* **56**, 812-821.

Velis, D.R. and T.J. Ulrych (1996) Simulated annealing two-point ray tracing, *Geophys. Res. Let.* **23**, 201-204.

Vidale, J.E. (1988) Finite-difference traveltime calculation, *Bull. Seis. Soc. Am.* **78**, 2062-2076.

Vidale, J.E. (1990) Finite-difference calculation of traveltimes in three dimensions, *Geophysics* **55**, 521-526.

Vidale, J.E., and H. Houston (1990) Rapid calculation of seismic amplitudes, *Geophyics* **55**, 1504-1507.

Virieux, J. (1991) Fast and accurate ray tracing by Hamiltonian perturbation, *J. Geophys. Res.* **96**, 579-594.

Virieux, J. and V. Farra (1991) Ray tracing in 3-D complex isotropic media: An analysis of the problem, *Geophysics* **56**, 2057-2069.

Virieux, J., V., Farra and R. Madariaga (1988) Ray tracing for earthquake location in laterally heterogeneous media, *J. Geophys. Res.* **93**, 6585-6599.

Wesson, R.L. (1971) Travel-time inversion for laterally inhomogeneous crustal velocity models, *Bull. Seismol. Soc. Am.* **61**, 729-746.

Woodward, M.J. (1992) Wave-equation tomography, *Geophysics* **57**, 15-26.

Zhao, D., Hasegawa, A., and Horiuchi, S. (1992) Tomographic image of P and S wave velocity structure beneath northeastern Japan, *J. Geophys. Res.* **97**, 19,909-19,928.

Chapter 5

PROBABILISTIC EARTHQUAKE LOCATION IN 3D AND LAYERED MODELS

Introduction of a Metropolis-Gibbs method and comparison with linear locations

Anthony Lomax and Jean Virieux
Géosciences-Azur, University of Nice - Sophia Antipolis, Valbonne, France

Philippe Volant and Catherine Berge-Thierry
Institut de Protection et de Sûreté Nucléaire, Fontenay-aux-Roses, Paris, France

Key words: 3D models, earthquake location, non-linear optimization, probability function

Abstract: Probabilistic earthquake location with non-linear, global search methods allows the use of 3D models and produces comprehensive uncertainty and resolution information represented by a probability density function over the unknown hypocentral parameters. We describe a probabilistic earthquake location methodology and introduce an efficient Metropolis-Gibbs, non-linear, global sampling algorithm to obtain such locations. Using synthetic travel times generated in a 3D model, we examine the locations and uncertainties given by an exhaustive grid-search and the Metropolis-Gibbs sampler using 3D and layered velocity models, and by a iterative, linear method in the layered model. We also investigate the relation of average station residuals to known static delays in the travel times, and the quality of the recovery of known focal mechanisms. With the 3D model and exact data, the location probability density functions obtained with the Metropolis-Gibbs method are nearly identical to those of the slower but exhaustive grid-search. The location PDFs can be large and irregular outside of a station network even for the case of exact data. With location in the 3D model and static shifts added to the data, there are systematic biases in the event locations. Locations using the layered model show that both linear and global methods give systematic biases in the event locations and that the error volumes do not include the "true" location – absolute event locations and errors are not recovered. The iterative,

C.H. Thurber and N. Rabinowitz (eds.), Advances in Seismic Event Location, 101–134.

linear location method can fail for locations near sharp contrasts in velocity and outside of a network. Metropolis-Gibbs is a practical method to obtain complete, probabilistic locations for large numbers of events and for location in 3D models. It is only about 10 times slower than linearized methods but is stable for cases where linearized methods fail. The exhaustive grid-search method is about 1000 times slower than linearized methods but is useful for location of smaller number of events and to obtain accurate images of location probability density functions that may be highly-irregular.

1. INTRODUCTION

The accurate location of earthquakes and a complete understanding of the location uncertainties are critical to seismotectonic and seismic hazard studies, to "real-time" seismic notification, and to nuclear test ban treaty verification. With the increasing availability of three-dimensional structural models from geologic and geophysical interpretations and seismic travel-time inversion, it is important to have location methods valid for 3D velocity models. Probabilistic earthquake location with non-linear, global-search methods allows the use of 3D models and produces comprehensive uncertainty and resolution information.

A complete, probabilistic earthquake location is represented by a probability density function over the unknown parameters, i.e. three spatial co-ordinates of the hypocenter and an origin time. This hyper-volume representation of a location may include multiple "optimal" solutions and may have a highly irregular form. Many studies of seismicity and seismotectonics make explicit use of a probabilistic representation of earthquake locations (i.e. Wittlinger et al., 1993; Vilardo et al., 1996; Calvert et al., 1997; Gresta et al., 1998; Jones and Stewart, 1997).

Commonly used iterative-linearized location programs produce a single, point solution, the preferred hypocenter, and uncertainty estimates based on Gaussian, or normal, statistics evaluated at this point. Such a solution will be a good representation of the complete, probabilistic location only for the case that the density function has a single optimum and a near-ellipsoidal form. Non-linear, global-search methods can be used to identify irregular volumetric, probability density functions required for complete, probabilistic locations. In addition, unlike linear approaches, these methods can be easily applied with 3D models because they do not require partial derivative information, which is difficult or impossible to obtain in complicated models. However, non-linear, global-search methods can be very time consuming, particularly exhaustive grid-search or pure Monte Carlo methods. For "near real-time" seismic notification and for CTBT monitoring, it is critical to locate events rapidly, while for seismotectonic and hazard studies, there is often a need to relocate many hundreds or thousands of earthquakes. It is thus of significant practical importance to investigate the performance and completeness of efficient, directed global search methods for probabilistic earthquake location in 3D models.

In this chapter we begin with a discussion of a probabilistic earthquake location and existing location methods. We then discuss the non-linear and linear location methods that we will use, and we introduce an efficient, non-linear, Metropolis-Gibbs, simulated annealing global search method. Using synthetic travel times generated in a 3D model and the ideal case of locating using the same 3D model, we compare the performance of this method for determining complete, probabilistic locations against a less efficient but exhaustive grid-search. We next examine a more useful and realistic case of earthquake location in a layered model using the 3D synthetic travel times. We compare the layered model locations and uncertainties obtained with linear and non-linear methods. For the various location methods in 3D and layered models, we investigate the spatial variation of uncertainties, the relation of average station residuals to know static delays in the travel times, and the quality of the recovery of known focal mechanisms.

2. PROBABILISTIC EARTHQUAKE LOCATION

2.1 Posterior density function (PDF)

The non-linear earthquake location algorithms described in this paper follow the probabilistic formulation of inversion presented in Tarantola and Valette (1982) and Tarantola (1987) and the equivalent methodology for earthquake location (*i.e.* Tarantola and Valette, 1982; Moser, van Eck and Nolet, 1992; Wittlinger, Herquel and Nakache, 1993). This formulation relies on the use of normalized and unnormalized probability density functions to express our knowledge about the values of parameters. Thus, given the normalized density function $f(x)$ for the value of a parameter x, the probability that x has a value between X and $X+\Delta X$ is

$$P(X \leq x \leq X+\Delta X) = \int_{X}^{X+\Delta X} f(x)dx. \tag{1}$$

In geophysical inversion we wish to constrain the values of a vector of unknown parameters **p**, given a vector of observed data **d** and a theoretical relationship $\theta(\mathbf{d},\mathbf{p})$ relating **d** and **p**. When the density functions giving the prior information on the model parameters $\rho_\mathbf{p}(\mathbf{p})$ and on the observations $\rho_\mathbf{d}(\mathbf{d})$ are independent, and the theoretical relationship can be expressed as a conditional density function $\theta(\mathbf{d}|\mathbf{p})\mu_\mathbf{p}(\mathbf{p})$, a complete, probabilistic solution can be expressed as a posterior density function (PDF) $\sigma_\mathbf{p}(\mathbf{p})$ (Tarantola and Valette, 1982; Tarantola,1987)

$$\sigma_\mathbf{p}(\mathbf{p}) = \rho_\mathbf{p}(\mathbf{p}) \int \frac{\rho_\mathbf{d}(\mathbf{d})\theta(\mathbf{d}\,|\,\mathbf{p})}{\mu_\mathbf{d}(\mathbf{d})} d\mathbf{d} \tag{2}$$

where $\mu_\mathbf{p}(\mathbf{p})$ and $\mu_\mathbf{d}(\mathbf{d})$ are null information density functions specifying the state of total ignorance.

For the case of earthquake location, the unknown parameters are the hypocentral coordinates $\mathbf{x} = (x, y, z)$ and the origin time T, the observed data is a set of arrival times **t**, and the theoretical relation gives predicted travel times **h**. Tarantola and Valette (1982) show that, if the theoretical relationship and the observed arrival times are assumed to have Gaussian uncertainties with covariance matrices $\mathbf{C}_T$ and $\mathbf{C}_t$, respectively, and if the prior information on T is taken as uniform, then it is possible to evaluate analytically the integral over **d** in (2) and an integral over origin time T to obtain the marginal PDF for the spatial location, $\sigma(\mathbf{x})$. This marginal PDF reduces to (Tarantola and Valette, 1982; Moser, et al., 1992)

$$\begin{aligned} \sigma(\mathbf{x}) &= K\,\rho(\mathbf{x}) \cdot \exp[-\tfrac{1}{2} g(\mathbf{x})] \\ g(\mathbf{x}) &= [\hat{\mathbf{t}}_0 - \hat{\mathbf{h}}(\mathbf{x})]^T (\mathbf{C}_t + \mathbf{C}_T)^{-1}][\hat{\mathbf{t}}_0 - \hat{\mathbf{h}}(\mathbf{x})]. \end{aligned} \tag{3}$$

In this expression K is a normalization factor, $\rho(\mathbf{x})$ is a density function of prior information on the model parameters, and $g(\mathbf{x})$ is an L2 misfit function. $\mathbf{t}_0$ is the vector of observed arrival times **t** minus their weighted mean, $\hat{\mathbf{h}}$ is the vector of theoretical travel times **h** minus their weighted mean, where the weights w_i are given by

$$w_i = \sum_j w_{ij}; \quad w_{ij} = [(\mathbf{C}_t + \mathbf{C}_T)^{-1}]_{ij}. \tag{4}$$

Furthermore, Moser et al. (1992) show that the maximum likelihood origin time corresponding to a hypocenter at (x, y, z) is given by

$$T_{ml}(\mathbf{x}) = \frac{\sum_i \sum_j w_{ij}[t_i - h_i(\mathbf{x})]}{\sum_i \sum_j w_{ij}}. \tag{5}$$

The posterior density function (PDF) $\sigma(\mathbf{x})$ given by Equation (3) represents a complete, probabilistic solution to the location problem, including information on uncertainty and resolution. This solution does not require a linearized theory, and the resulting PDF may be irregular and multi-modal because the forward calculation involves a non-linear relationship between hypocenter location and travel-times.

This solution includes location uncertainties due to the spatial relation between the network and the event, measurement uncertainty in the observed arrival times, and errors in the calculation of theoretical travel times. However, realistic estimates of uncertainties in the observed and theoretical times must be available and specified in a Gaussian form through $\mathbf{C}_t$ and $\mathbf{C}_T$, respectively. Absolute location errors due to incorrect velocity structure could be included through $\mathbf{C}_T$ if the resulting travel time errors can be estimated and described with a Gaussian structure. Estimating these travel time errors is difficult and often not attempted. When the model used for location is a poor approximation to the "true" structure (as is often the case with layered model approximations), the absolute location uncertainties can be very large.

2.2 Existing inversion methods and complete, probabilistic locations

In practice, a complete, probabilistic earthquake location is formed by some estimate, throughout the region of significant prior probability $\rho(\mathbf{x})$, of the posterior density function (PDF) $\sigma(\mathbf{x})$ given by Equation (3). Ideally, this solution will include multiple solutions and significant irregularities in the form of the PDF.

2.2.1 Linearized inversion

Direct and iterative linear inversion methods (see Aki and Richards (1980)) can be very rapid for optimization, and have been used frequently in recent years for this reason. These methods can be applied to non-linear problems when the theoretical relationship between the model and the data can be locally approximated by a linear expression and when there is a single, well-defined misfit "basin" containing the optimal solution (i.e., the problem is not ill-conditioned). If these conditions do not hold, then the results of the inversion, including information on uncertainty and resolution, may be incomplete or unstable, and they will not form a good estimate of the complete, probabilistic solution. In earthquake location, the combined effects of the depth-origin time trade-off, the network-event geometry, reading errors, and gradients or interfaces in the velocity model can result in a highly non-linear and ill-conditioned inverse problem that does not satisfy the above requirements. In this case, the inversion may fail or may produce a solution that is not a good representation of the complete PDF. However, we will see below that for the location of events within a network using a layered velocity model, an iterative linear method such as HYPOELLIPSE (Lahr, 1989) can produce a hypocenter and associated Gaussian uncertainties that are nearly identical to the PDF obtained with global search methods.

2.2.2 Exhaustive global search

Both an exhaustive grid-search and a "crude" Monte-Carlo search (Hammersley and Handscomb, 1967; Keilis-Book and Yanovskaya, 1967; Press, 1968; Wiggins, 1969; Sen and Stoffa, 1995) use global and well-distributed sampling of the model space. Thus the results obtained with these search methods can be used to estimate the PDF $\sigma(\mathbf{x})$ and hence to obtain complete, probabilistic locations. However, both methods are inefficient for problems with many unknown parameters, large parameter spaces, or time consuming forward calculations, because the number of models that must be tested can be very large. These methods have been successfully applied to earthquake hypocenter determination (Sambridge and Kennett, 1986; Kennett, 1992; Shearer, 1997; Dreger et al. 1998), and to probabilistic location (Moser, van Eck and Nolet, 1992; Wittlinger, Herquel and Nakache, 1993; Calvert et al., 1997), but their inefficiency may impose unacceptable limitations on the number of events or the size of the search volume.

Below we will use a nested grid-search as one of our non-linear, global location methods. This method gives excellent recovery of the location PDF $\sigma(\mathbf{x})$, but requires considerable calculation time and computing resources.

2.2.3 Directed, stochastic search

Directed, stochastic search techniques (Sen and Stoffa, 1995) include evolutionary, adaptive, global search methods such as the genetic algorithm (Goldberg, 1989; Holland, 1992; Stoffa and Sen, 1991; Sambridge and Drijkoningen, 1992) and simulated annealing (Kirkpatrick et al., 1983; Rothman, 1985; Tarantola, 1987). Most of these methods were developed for *optimization*, the identification of *some* very good solutions, which is equivalent to identifying global or local maxima of a solution PDF $\sigma(\mathbf{x})$. But in general these methods do not use a well-distributed sampling of the model space that can produce complete, probabilistic solutions to inverse problems. For example, the genetic algorithm obtains samples that cluster in small regions near locally optimum solutions and consequently cannot be used to obtain the global structure of a PDF (e.g. Goldberg and Richardson, 1987; Stoffa and Sen, 1991; Nolte and Frazer, 1994; Lomax and Snieder, 1995). Similarly, with simulated annealing, the interaction of the variable "temperature" parameter and step size with the local structure of the target function can lead to a slow gain of information about the global structure of this function (Scales et al., 1992). In practice, this leads to convergence and stalling near a locally optimum solution and a poorly distributed sample distribution.

Though not directly applicable to complete, probabilistic location, the genetic algorithm and simulated annealing are excellent for non-linear, earthquake hypocenter determination because of their efficiency (Sambridge and Gallagher, 1993; Billings, 1994).

2.2.4 Importance sampling – the Metropolis-Gibbs algorithm

The efficiency of a Monte Carlo algorithm used to estimate properties of a target function can be increased by choosing a sampling density which follows the function as closely as possible (Hammersley and Handscomb, 1967; Lepage, 1978; Press et al., 1992). Techniques that follow this rule are referred to as *importance sampling* methods, these were originally developed in physics for fast and accurate numerical integration of multi-dimensional functions.

The target function is unknown, however, and consequently the optimum importance sampling distribution cannot be determined a priori. But the efficiency can still be improved by adjusting (or adapting or evolving) the sampling by incorporating information gained from previous samples so that the sampling density tends towards the target function (Lepage, 1978; Press et al., 1992; Mosegaard and Tarantola, 1995; Sen and Stoffa, 1995). For example, importance sampling to determine the earthquake location PDF, Equation (3), can be obtained by beginning with a sampling that follows the prior density function $\rho(\mathbf{x})$ and then adjusting the sampling as the search progresses so that the sampling density approaches the location PDF $\sigma(\mathbf{x})$.

An importance sampling technique that is relevant to inversion for complete, probabilistic solutions is the Metropolis-Gibbs version of simulated annealing. The Metropolis-Gibbs sampling procedure is based on an algorithm (Metropolis et al., 1953) for the simulation of the distribution of atoms in a gas at a given temperature. Other importance sampling methods are discussed in Hammersley and Handscomb (1967), Lepage (1978) and in Press et al. (1992) in the context of numerical integration.

The Metropolis-Gibbs sampler (Sen and Stoffa, 1995; Mosegaard and Tarantola, 1995) is similar to simulated annealing with a constant temperature parameter. In the context of our earthquake location problem, the Metropolis-Gibbs algorithm performs a random walk in the solution space (x, y, z) with nearby trial moves which are accepted or rejected according to the PDF $\sigma(\mathbf{x})$ (after evaluation of the forward problem). Mosegaard and Tarantola (1995) show that this algorithm generates a set of "accepted" samples that are distributed according to the PDF $\sigma(\mathbf{x})$; it is therefore an importance sampling method. They also show that, in the limit of a very large number of trials, it will not become permanently "trapped" near a locally optimum solution and consequently can produce global sampling.

Because it is a random walk technique, the Metropolis-Gibbs sampler may perform well even if the volume of the significant regions of $\sigma(\mathbf{x})$ is small relative to the volume of the prior search space $\rho(\mathbf{x})$; this is usually the case in earthquake location. However, because it must use small steps in the model space between consecutive samples, and in practice the number of samples is limited, it may be subject to stalling near strong locally optimal solutions if $\sigma(\mathbf{x})$ is a complicated function. In addition, since samples are accepted or rejected *after* evaluation of the forward problem, the number of evaluations of the forward theory may be much larger that the set of "accepted" samples obtained.

Below we will introduce a new Metropolis-Gibbs algorithm for probabilistic earthquake location. We will show that it is much more computationally efficient than a grid search while producing nearly identical definition of the complete location PDF $\sigma(\mathbf{x})$ when this function is not highly irregular.

3. NON-LINEAR, PROBABILISTIC LOCATIONS IN 3D: NONLINLOC

For the non-linear, global search locations in this chapter, we use a set of location programs called NonLinLoc. This software can be obtained over the internet at the ORFEUS Seismological Software Library (http://orfeus.knmi.nl) under the "software links" page. The location program in NonLinLoc can be used with 3D velocity models; it produces an estimate of the posterior density function (PDF) for the spatial hypocenter location, $\sigma(\mathbf{x})$, using either a systematic grid-search or a stochastic,

Metropolis-Gibbs sampling approach. The location algorithm follows the probabilistic inversion approach of Tarantola and Valette (1982), described above, and the earthquake location methods of Tarantola and Valette (1982), Moser et al. (1992) and Wittlinger et al. (1993).

A grid of PDF values obtained by grid-search, samples drawn from this grid, or samples of the PDF obtained by the Metropolis-Gibbs sampler, represent the complete, probabilistic, spatial solution to the earthquake location problem. This solution indicates the uncertainty in the spatial location due to Gaussian-distributed picking and travel-time calculation errors, the network-event geometry, and the incompatibility of the picks. The location uncertainty will in general be non-ellipsoidal (non-Gaussian) because the forward calculation involves a non-linear relationship between hypocenter location and travel-times.

To make the location program efficient for complicated 3D models, the travel-times between each station and all nodes of an x,y,z spatial grid are calculated once using a 3D version (Le Meur, 1994; Le Meur, Virieux and Podvin, 1997) of the eikonal finite-difference scheme of Podvin and Lecomte (1991) and then stored on disk as travel-time grid files. This storage technique has been used by Wittlinger et al. (1993), and in related approaches by Nelson and Vidale (1990) and Shearer (1997). The forward calculation during location reduces to retrieving the travel-times from the grid files and forming the misfit function $g(\mathbf{x})$ in, Equation (3).

3.1 Travel-time and take-off angles calculations

The travel times between a station and all nodes of a 3D grid are calculated using the eikonal finite-difference scheme of Podvin and Lecomte (1991). This algorithm and related methods (Vidale, 1988) use a finite-differences approximation of Huygen's principle to find the first arriving, infinite frequency travel times at all nodes of the grid. The algorithm of Podvin and Lecomte (1991) gives stable recovery of diffracted waves near surfaces of strong velocity contrast and thus it accurately produces travel times for diffracted and head waves. A limitation of the current 3D version of the method is a restriction to cubic grids. This may lead to excessively large travel-time grids if a relatively fine cell spacing is required along one dimension since the same spacing must be used for the other dimensions. This can be a problem for regional studies where a fine node spacing in depth is necessary, but the horizontal extent of the study volume can be much greater than the depth extent. Thus a modification of the travel times calculation to allow use of an irregular grid would be very useful.

After the travel times are calculated throughout the grid, the NonLinLoc program uses the gradients of travel-time at the node to estimate the take-off angles at each node. Two gradients are estimated for each axis direction *x*, *y*, and *z*, one G_{low} between the node and its preceding neighbour along the axis, and a second G_{high} between the following neighbour and the node. The total gradient G_{axis} along an axis is the mean of these two gradients; the total gradient along the three axes determines the vector gradient of travel-time. The direction opposite to the vector gradient of travel-time gives the ray take-off angles for dip and azimuth. An estimate of the quality of the angle determination is given by a comparison of the magnitudes and signs of G_{low} and G_{high}. If these two values are not similar, then there may be two rays which arrive nearly simultaneously at the station, and the take-off angle determination at the node may be unstable.

3.2 Grid-Search Algorithm

The grid-search algorithm in NonLinLoc performs successively finer, nested grid searches within a spatial *x,y,z* volume to obtain an estimate of the location PDF, Equation (3). The grid-search performs a systematic, exhaustive coverage of the search region and thus can identify multiple optimal solutions and highly irregular confidence volumes. However, the grid-search is very time consuming relative to the Metropolis-Gibbs sampler and linear location techniques. It also requires a careful selection of the grid sizes and node spacing, otherwise, relative to the size of the most significant region of the PDF, the fine search grids may be too large (giving low resolution) or too small (leading to truncation of the PDF).

3.2.1 Procedure

An initial grid with a fixed size, number of nodes, and location defines the full search region. Subsequent, nested grids are centred automatically on the optimal hypocentral node of the containing grid in one or more of the *x,y,z* directions. The nested grids are typically smaller in size, but may have more nodes than the containing grid.

For every node of each location grid, the grid-search algorithm must obtain travel-times for every observation from the corresponding travel-time grid files. For 3D structures these files may be very large, and with many observations there will be many such files. To read these files efficiently without storing them entirely in memory, the search is performed systematically throughout each location grid with the *x* index varying last. Thus, it is only necessary to read into memory a 2D *y-z* plane or "sheet" corresponding to the current *x* index for each observation.

With these travel times, Equation (3) is evaluated to obtain a non-normalized location PDF value, which is stored at the appropriate node in the search grid in memory. When the grid-search for the final fine grid is complete, the gridded PDF values are normalized by assuming that the integral of the PDF over the search volume is unity. A set of samples of the PDF can be drawn from this normalized grid. The normalized fine grid, the PDF samples, and additional location results (see below) are saved to disk files.

3.3 Metropolis-Gibbs Sampling Algorithm

The Metropolis-Gibbs sampler introduced here performs a directed random walk within a spatial x,y,z volume to obtain a set of samples that follow the location PDF. This algorithm is similar to those described in Sen and Stoffa (1995) and Mosegaard and Tarantola (1995), with the addition of three distinct sampling stages to obtain adaptively an optimal step size for the walk.

Like the grid-search, the Metropolis-Gibbs sampler does not require partial derivatives and thus can be used with complicated 3D velocity structures. This method performs well with moderately irregular (non-ellipsoidal) PDFs with a single optimum solution, but because it is a stochastic search it may give inconsistent recovery of very irregular PDFs with multiple optimal solutions. The Metropolis-Gibbs sampler is only moderately slower (about 10 times) than linearized, iterative location techniques, and is much faster (about 100 times) than the grid-search. Because the Metropolis-Gibbs method samples the search space stochastically, it runs fastest with the full 3D travel-time grid files in memory. If memory limitations, the number of observations, or the size of the 3D travel-time grids prevents this, then the travel-time grids must be accessed on disk, and the search may run very slowly.

3.3.1 Procedure

The Metropolis-Gibbs sampler used in the program NonLinLoc for earthquake location consists of a directed walk in the solution space (x, y, z) which tends towards regions of high likelihood for the location PDF, $\sigma(\mathbf{x})$, given by Equation (3). At each step, the current walk location $\mathbf{x}_{curr}$ is perturbed by a vector $d\mathbf{x}$ of arbitrary direction and given length l to give a new location $\mathbf{x}_{new}$. The likelihood $\sigma(\mathbf{x}_{new})$ is calculated for the new location and compared to the likelihood $\sigma(\mathbf{x}_{curr})$ at the current location. If $\sigma(\mathbf{x}_{new}) \geq \sigma(\mathbf{x}_{curr})$, then the new location is accepted. If $\sigma(\mathbf{x}_{new}) < \sigma(\mathbf{x}_{curr})$, then the new location is accepted with probability $P = \sigma(\mathbf{x}_{new}) / \sigma(\mathbf{x}_{curr})$. When a new location is accepted it becomes the current location and may be saved as a sample of the location PDF.

In earthquake location, the dimensions of the significant regions of the location PDF can vary enormously and are not known a priori. It is important to choose an initial step size large enough to allow global exploration of the search volume, and to obtain a final step size that gives good coverage of the location PDF while resolving details and irregular structure of the PDF. The NonLinLoc Metropolis-Gibbs sampler uses three distinct sampling stages to determine adaptively an optimal step size l for the walk:

1. A *learning* stage where the step size is fixed and relatively large. The walk can explore globally the search volume and migrate towards regions of high likelihood. "Accepted" samples are not saved.
2. An *equilibration* stage where the step size l is adjusted in proportion to the standard deviations (s_x, s_y, s_z) of the spatial distribution of all previously "accepted" samples obtained after the middle of the learning stage. After each new accepted sample, the standard deviations are updated and the step size l is set equal to $f_s\,(s_x s_y s_z/N_s)^{1/3}$, where N_s is the total number of samples to be accepted during the saving stage, and f_s=8 is a step size scaling factor. This formula sets l in proportion to the cell size required to tile with N_s cells the rectangular volume with sides s_x, s_y and s_z. The walk can continue to migrate towards or may begin to explore regions of high likelihood. "Accepted" samples are not saved.
3. A *saving* stage where the step size l is fixed at its final value from the equilibration stage. The walk can continue to explore regions of high likelihood. "Accepted" samples are assumed to follow the location PDF and can be saved, but there may be a waiting time of several samples between saves to insure the independence of saved samples.

The NonLinLoc Metropolis-Gibbs sampling algorithm is initialized as follows:

1. The walk location is set at the x,y position of the station with the earliest arrival time and non-zero weight, at the mean depth of the search region.
2. If the initial step l size is not specified, it is set to the cell size required to tile with N_s cells the plane formed by the two longest sides of the initial search region. N_s is the total number of samples to be accepted during the saving stage, including samples that are skipped between saves.

The rejection by the algorithm of new walk locations for a large number of consecutive tries (the order of 1000 tries) may indicate that the last "accepted" sample falls on a sharp likelihood maxima that is narrower than the current step size. To allow the search to continue in this case, the new location is accepted unconditionally and the step size is reduced by a factor of two.

In the case that the size of the location PDF is very small relative to the search region, the algorithm may fail to locate the region of high likelihood or obtain an optimal step size. In this case the size of the search region must be reduced or the size of the initial step size adjusted. A more robust solution to this problem may be to add a temperature parameter to the likelihood function, as with simulated annealing. This variable parameter could be set to increase the effective size of the PDF during the learning and equilibration stages so that the region of high likelihood is located efficiently, and then set to 1 during the saving stage so that the true PDF is imaged.

3.4 Additional location results

In addition to an estimate of the location PDF, the NonLinLoc program determines an optimal hypocenter, traditional Gaussian statistics and other location parameters.

3.4.1 Maximum likelihood hypocenter

The maximum likelihood (or minimum misfit) point of the complete, non-linear location PDF is selected as an "optimal" hypocenter. The significance and uncertainty of this maximum likelihood hypocenter cannot be assessed independently of the complete solution PDF. The maximum likelihood hypocenter location is used for the determination of ray take-off angles, for the determination of phase residuals, and for magnitude calculation. The ray take-off angles can be used for a first-motion fault plane determination.

3.4.2 Gaussian estimators

"Traditional" Gaussian or normal estimators, such as the expectation $E(\mathbf{x})$ and covariance matrix $\mathbf{C}$ may be obtained from the gridded values of the normalized location PDF or from samples of this function (e.g. Tarantola and Valette, 1982; Sen and Stoffa, 1995). For the grid case with nodes at $\mathbf{x}_{i,j,k}$,

$$E(\mathbf{x}) = \Delta V \sum_{i,j,k} \mathbf{x}_{i,j,k}\sigma(\mathbf{x}_{i,j,k}), \tag{6}$$

where ΔV is the volume of a grid cell. For N samples drawn from the PDF with locations $\mathbf{x}_n$,

$$E(\mathbf{x}) = \frac{1}{N}\sum_n \mathbf{x}_n , \tag{7}$$

where the PDF values $\sigma(\mathbf{x}_n)$ are not required since the samples are assumed distributed according to the PDF. For both cases, the covariance matrix is then given by

$$\mathbf{C} = E[(\mathbf{x} - E(\mathbf{x}))\cdot(\mathbf{x} - E(\mathbf{x}))^T]. \tag{8}$$

The 68% confidence ellipsoid can be obtained from singular value decomposition (SVD) of the covariance matrix **C**, following Press *et al.* (1993; their sec. 15.6 and Equations 2.6.1 and 15.6.10). The SVD gives

$$\mathbf{C} = \mathbf{U}[diag\, w_i]\mathbf{V}^{\mathrm{T}}, \tag{9}$$

where **U** and **V** are square, symmetric matrices and w_i are singular values. The columns $\mathbf{V}_i$ of **V** give the principle axes of the confidence ellipsoid. The corresponding semi-diameters for a 68% confidence ellipsoid are $(3.53w_i)^{1/2}$, where 3.53 is the $\Delta\chi^2$ value for 68.3% confidence and 3 degrees of freedom. The Gaussian estimators and resulting confidence ellipsoid will be good indicators of the uncertainties in the location only in the case where the complete, non-linear PDF has a single maximum and has an ellipsoidal form.

4. LINEAR LOCATIONS: HYPOELLIPSE

The HYPOELLIPSE program (Lahr, 1989) performs local and near regional earthquake locations in layered or gradient-layered 1D models. The program uses Geiger's method (Geiger, 1912) to find a hypocenter that minimizes the root-man-square misfit between observed and calculated travel times by iterative update of trial locations. The updates are obtained from a damped least-squares inversion of the matrix of partial derivatives of travel time with respect to the three spatial hypocenter coordinates at the current trial location. Because the location problem is non-linear, this matrix represents a "linearization" of the problem, and the final solution must be obtained though iteration. Gaussian uncertainty information for the final hypocenter is given in the form of a 68% spatial confidence ellipsoid obtained from analysis of the matrix of partial derivatives at this solution. This ellipsoid forms a local approximation to the complete PDF for the location.

There are many programs to perform linear, iterative earthquake location following Geiger's method. We choose HYPOELLIPSE for this study because the 68% confidence ellipsoid can be compared directly to the complete PDF for the location obtained with the non-linear methods.

HYPOELLIPSE has options to perform a grid search in depth and to change individual observation weights during iteration based on residuals, distance or azimuth to hypocenter, and other quantities. For simplicity and compatibility we do not use these options in this study, but note that they may have significant effects on the hypocenter and confidence ellipsoid determinations, particularly for poorly constrained events.

5. A STUDY OF SYNTHETIC LOCATIONS

We examine various earthquake location scenarios using synthetic travel times for a 3D model. The purpose of this study is to 1) validate and compare the performance of the efficient Metropolis-Gibbs sampler algorithm against the slow but exhaustive grid-search, 2) examine the ideal case of location using the "correct" 3D velocity model, 3) study the realistic case of location in an "incorrect" layered model, and 4) compare the locations and uncertainties obtained by the linear, local method with those of the non-linear, global-search methods. Other synthetic location studies which address some of these issues have been performed by Pavlis (1986) and by Schwartz and Nelson (1991).

5.1 The synthetic problem

The station geometry, velocity structure and event locations are based on characteristics of the Institut de Protection et de Sûreté Nucléaire (IPSN) Durance micro-earthquake network and study region in southern France (Mohammadioun and Dervin, 1995; Volant et al., 2000). We select 11 IPSN stations to obtain a network of about 60 km by 30 km and typical station spacing of 10 to 20 km. For plotting we use a rotated, left-handed rectangular coordinate system where the Y axis is oriented along the true geographic azimuth N30°E, and Z is positive downwards.

The 3D velocity structure (Figure 1) is a simplified, 2.5D version of a preliminary 3D model for the Durance region. To reflect the transpressive tectonics of the region we include a strong velocity contrast of about 20% across a vertical plane through the network at Y=0 (representing the Middle-Durance fault) and a contrasts of 6 to 12% across a dipping plane that intersects this "fault" at depth. The S velocities are determined from the P velocity model using a V_P/V_S ratio of 1.75.

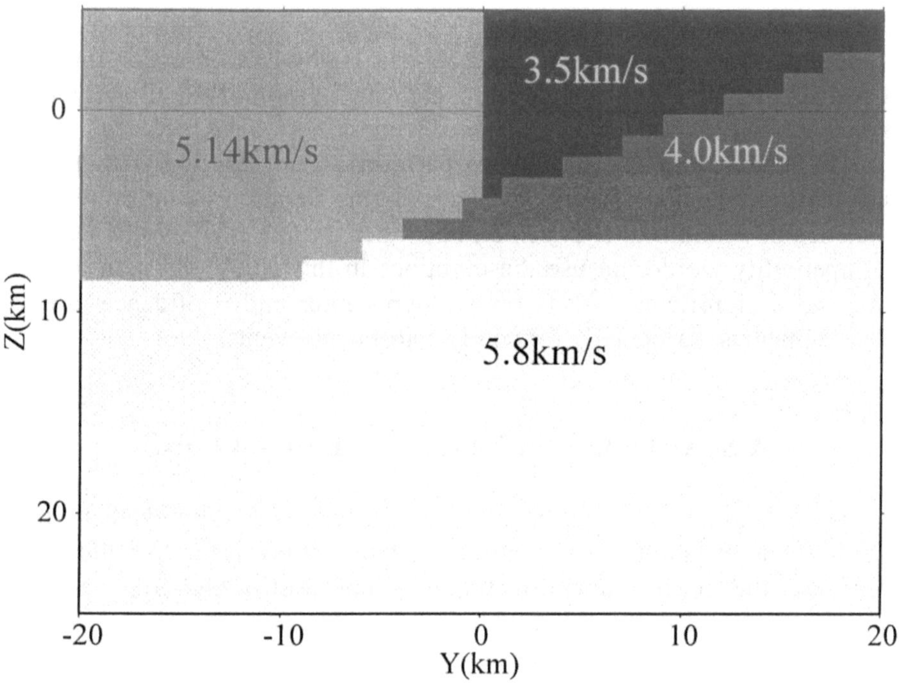

Figure 1. 3D velocity structure used for generating synthetic arrival times and for 3D model locations. The P velocity structure is shown in Y-Z cross section; the model is invariant in the X direction. S velocities are determined from the P velocity model using a V_P/V_S ratio of 1.75.

We consider 27 synthetic "events" regularly spaced on a vertical plane which is coincident with the "fault" plane (Y=0) and which passes through the network (Figure 2). The events are located up to 20 km outside of the network and between 0 and 20 km in depth.

The basic data set for the synthetic locations consists of 15 simulated P or S arrival times at the network stations for each of the 27 events. The times are calculated using the 3D velocity structure and the same eikonal finite-difference algorithm (Podvin and Lecomte, 1991) used for the NonLinLoc event locations. We generate a set of exact, "noise free" arrival times, and a set of arrival times with station "statics" by adding constant shifts to the times at each station of ±0.1 or ±0.2 second for P and S arrivals, respectively (Table 1). For an individual event, these statics can be considered as picking errors or as general velocity model errors; for the 27 events together these statics should be considered as errors in the velocity model near each station.

For both sets of arrival times and for all location methods the reading errors are set equal to the magnitude of the corresponding static shift (Table 1). The squares of the reading errors will form the diagonal terms of the diagonal covariance matrix $\mathbf{C}_t$ for the observed arrival times. For compatibility with HYPOELLIPSE, the covariance matrix $\mathbf{C}_T$ for theoretical relationship is set to the zero matrix $\mathbf{0}$; in practice the terms of this matrix should reflect the uncertainty in the theoretical travel times due to errors in the velocity model. We also generate synthetic first-motions for each P arrival corresponding to a pure left-lateral mechanism (strike=90°, dip=90°, rake=180°) in the unrotated, geographic coordinate frame.

Table 1. Station static shifts and reading errors

Station	phase	static shift (sec)	reading error (sec)
ART1	P	0.1	0.1
BLV1	P	-0.1	0.1
BST	P	0.1	0.1
BST	S	0.2	0.2
CAD	P	-0.1	0.1
CAD	S	-0.2	0.2
ESC	P	0.1	0.1
ESC	S	0.2	0.2
GRX	P	-0.1	0.1
JOU	P	0.1	0.1
JOU	S	-0.2	0.2
MEY	P	0.1	0.1
REV1	P	-0.1	0.1
OBS1	P	0.1	0.1
VAL1	P	-0.1	0.1

For the 1D model locations we use a two-layer over half-space velocity model (Table 2) where the layer depth differs from that of any horizontal interface in the 3D model, and with a different V_P/V_S ratio than the 3D model. The 1D model is constructed arbitrarily and not by averaging or transformation of the 3D model, nor by any analysis of the simulated arrival times. Thus the 1D model is not an optimized model for the problem, but represents a typical, "incorrect" reference model as might be used in a preliminary seismicity study for a region.

Table 2. 1D, layered velocity model ($V_P/V_S = 1.85$)

Depth to top (km)	P velocity (km/sec)	S velocity (km/sec)
-	4.5	2.43
4.5	5.8	3.14
30	8.0	4.32

Each grid-search location is run with an initial grid size of 100 by 100 km horizontally centred at X=0, Y=0 and a depth range of –5 to 30 km, with a node spacing of 2 km horizontally and 1 km vertically. Two further nested grids are each centred on the optimal hypocentral node of the preceding grid.

The final grid has a size of 10 by 10 km horizontally and 15 km in depth with a node spacing of 0.1 km, giving 1.5 million nodes. A total of about 1.7 million evaluations of the forward problem are performed.

Each Metropolis-Gibbs location is run with 1000 "accepted" samples in the learning stage, 4000 in the equilibration stage, and 5000 in the saving stage. This gives a total of 10000 "accepted" samples, which typically requires the testing (evaluation of the forward problem) at about 14000 locations. During the saving stage every fifth "accepted" sample is kept to give a total of 1000 samples of the PDF.

The HYPOELLIPSE locations use the same arrival times, arrival time uncertainties and layered model as the non-linear locations. HYPOELLIPSE is run for a maximum of 50 iterations, without any re-weighting of the observations during the iterations.

The relative calculation times for the 27 event locations for the three methods are HYPOELLIPSE 1 unit, Metropolis-Gibbs 10 units, and grid-search 1000 units, where a unit is the order of several second on a desktop workstation.

In the following we use the notation [*x*,*z*] to identify events using their approximate coordinates on the (X-Z) sections.

5.2 Location with 3D models

5.2.1 Exact data: verification of the algorithms and uncertainties due to station-event geometry

We first perform grid-search and Metropolis-Gibbs locations using the exact synthetic travel times and the correct 3D model. The location results are shown in Figure 2.

The Metropolis-Gibbs and grid-search PDFs and Gaussian estimators are very similar and the maximum likelihood hypocenters from both methods give excellent recovery of the true locations and origin times. Typical location errors for the grid-search are ~0km horizontally, <0.05km in depth and <0.01sec in origin time. For Metropolis-Gibbs the horizontal errors are <0.1km outside the network and <0.02km inside, <0.1km in depth (except for 2 events outside the network with errors of 0.25 and 0.15km) and < 0.01 in origin time. There is a very good agreement between the grid-search PDFs, which we take as near exact, and those for the Metropolis-Gibbs sampler (Figure 2). However, the Metropolis-Gibbs method does not recover the detailed form of the most irregular PDFs (i.e. events [–40,6] and [40,6]).

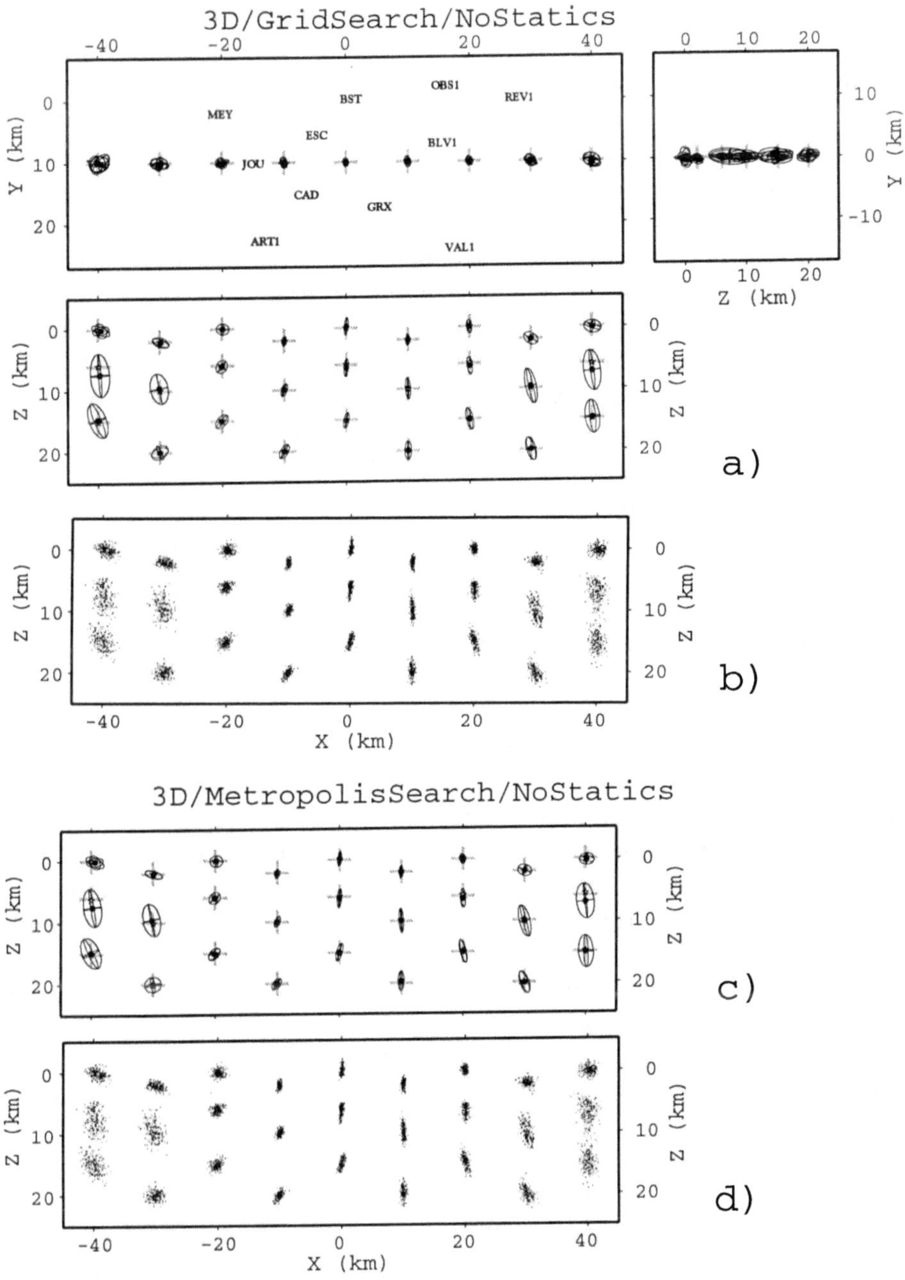

Figure 2. Location with 3D model and exact data. a) Grid-search locations in map view (XY) and perpendicular (ZY) and parallel (XZ) to the event plane. b) Grid-search PDF density plots in XZ section. c) Metropolis-Gibbs locations in XZ section. d) Metropolis-Gibbs PDF density plots in XZ section. Crosses: true event locations; stars: maximum likelihood hypocenters, dots: expectation hypocenters, ellipses: projections of 68% confidence ellipsoids.

For this "noise free" simulation with the correct velocity model, the shapes and sizes of the PDFs reflect the assumed reading errors, variations of the velocity model within the PDF, and the geometric relation between the events and stations (through the distribution of ray take-off angles). The location uncertainties are smallest and generally ellipsoidal within the network; they increase in size outside of the network, where some (i.e., events [–40,6] and [40,6]) become irregularly shaped and have off-center expectation hypocenter locations. The uncertainties are generally largest in depth, typically 2-4 times the horizontal uncertainty, even within the network. The depth uncertainty is smallest for the events within the network and with the least epicentral distance to the nearest station (e.g., events under station JOU). All of these variations reflect the improved distribution of ray take-off angles for events within the network and at shallower depths.

The effect of the horizontal interface in the velocity model at 5km depth is seen in the truncation of the tops of several PDFs (i.e., events [–20,6], [0,6] and [20,6]), and in a sharp bend in the PDFs for events [-40,6] and [40,6].

5.2.2 Data with station "statics": an ideal location scenario

We next examine Metropolis-Gibbs locations in the correct 3D model using the synthetic travel times with station statics. These results represent an ideal but realistic scenario where an accurate larger scale velocity model is available (perhaps determined independently by active source refraction experiments), but where there remain errors in the model near the stations. The locations obtained with the Metropolis-Gibbs sampler are shown in Figure 3. As above, these results are almost identical to those for the grid-search (not shown).

The maximum likelihood and expectation hypocenters in Figure 3 are shifted up to 5 km in depth and 2 km horizontally from the "true" event locations. In plan view (XY), the locations bend away from the plane of "true" locations with a larger shift for events outside of the network; in depth section (ZY), the distance of the locations from the plane of "true" locations increases in with increasing depth. Thus the added station statics do not cancel each other and are systematically biasing the locations.

In general, the size and orientation of the location uncertainties are similar to those from the previous simulation with exact data. One exception is the distorted and tilted PDF and confidence ellipsoid for the event near [40,5]. The station static errors have caused a shift in the PDF for this event such that it interacts more strongly with the model interface at 5 km depth. Another exception is the more elongated PDF and confidence ellipsoid for the event near [40,11]. Many of the confidence ellipsoids and PDF scatter samples do not contain the "true" location, even though the assigned reading errors (Table 1) are of the same magnitude as the static shifts. This result probably reflects a bias in the locations caused by the application of static shifts at a small number of irregularly distributed stations.

This example illustrates that formal location uncertainties are insensitive to the station static shifts and do not reflect the absolute location errors. Apparently the uncertainties depend primarily on the station geometry and assumed observational errors. Station statics or, equivalently, small errors in the model near the stations, lead mainly to a bias in the location.

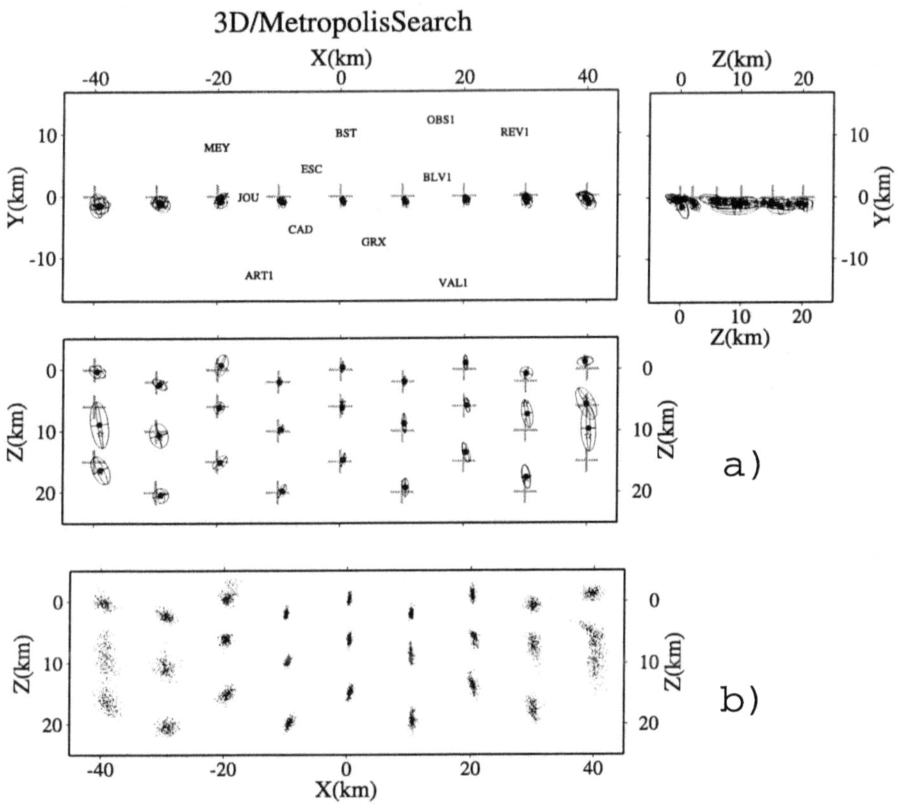

Figure 3. Location with 3D models and station static shifts. a) Metropolis-Gibbs locations in map view (XY), and views perpendicular (ZY) and parallel (XZ) to the event plane. b) Metropolis-Gibbs PDF density plots in XZ section. Crosses: true event locations; stars: maximum likelihood hypocenters, dots: expectation hypocenters, ellipses: projections of 68% confidence ellipsoids.

5.3 Location with layered models

We next perform grid-search, Metropolis-Gibbs, and linearized HYPOELLIPSE locations in the 1D layered model (Table 2) using the 3D model synthetic travel times with station statics. With this realistic and typical location scenario, we examine the location uncertainties and biases due to the use of a model that differs significantly from the true earth structure. We also compare the ellipsoidal approximations to the location PDFs given by HYPOELLIPSE to the complete PDFs and associated ellipsoids obtained with the non-linear, global search location algorithms.

5.3.1 Metropolis-Gibbs sampler

The maximum likelihood and expectation hypocenters and the PDF samples for the Metropolis-Gibbs location in the 1D model are shown in Figure 4. In depth section (ZY) the 1D model locations group near a vertical plane which is shifted significantly from the plane of "true" locations (the plane Y=0). In plan view (XY) the locations bend away from the plane of "true" locations with a shift of about 4 km horizontally towards higher velocity side of the 3D model for locations within network, increasing to about 10 km for events outside of the network. The (XZ) section shows that, relative to the "true" locations, most of the 1D locations are "pulled" towards the network centre and the relative locations are not preserved. There is a larger shift for deeper events and for events outside the network which are shifted up to 10 km shallower in depth and up to 8 km horizontally. The shifts in the (XZ) section are consistent with the generally lower velocities in the 1D layered model relative to the 3D model.

Overall, the location PDFs and confidence ellipsoids have the same or smaller size and a similar orientation to those for the 3D model. Two exceptions are the events near [–35,5] and [35,5] which have enlarged, elongated and bent PDFs and a maximum likelihood hypocenter located much shallower than the expectation hypocenter. The apparent decrease in uncertainty for some locations may be related to the shallower depth of the 1D locations and to the lower velocities in the 1D models. As with the 3D locations with station statics, the confidence ellipsoids and PDF samples do not contain the "true" locations and thus do not reflect the absolute location errors.

5.3.2 Grid search

Most of the grid-search locations and uncertainties are very similar to those obtained with Metropolis-Gibbs (Figure 4). Two exceptions are the

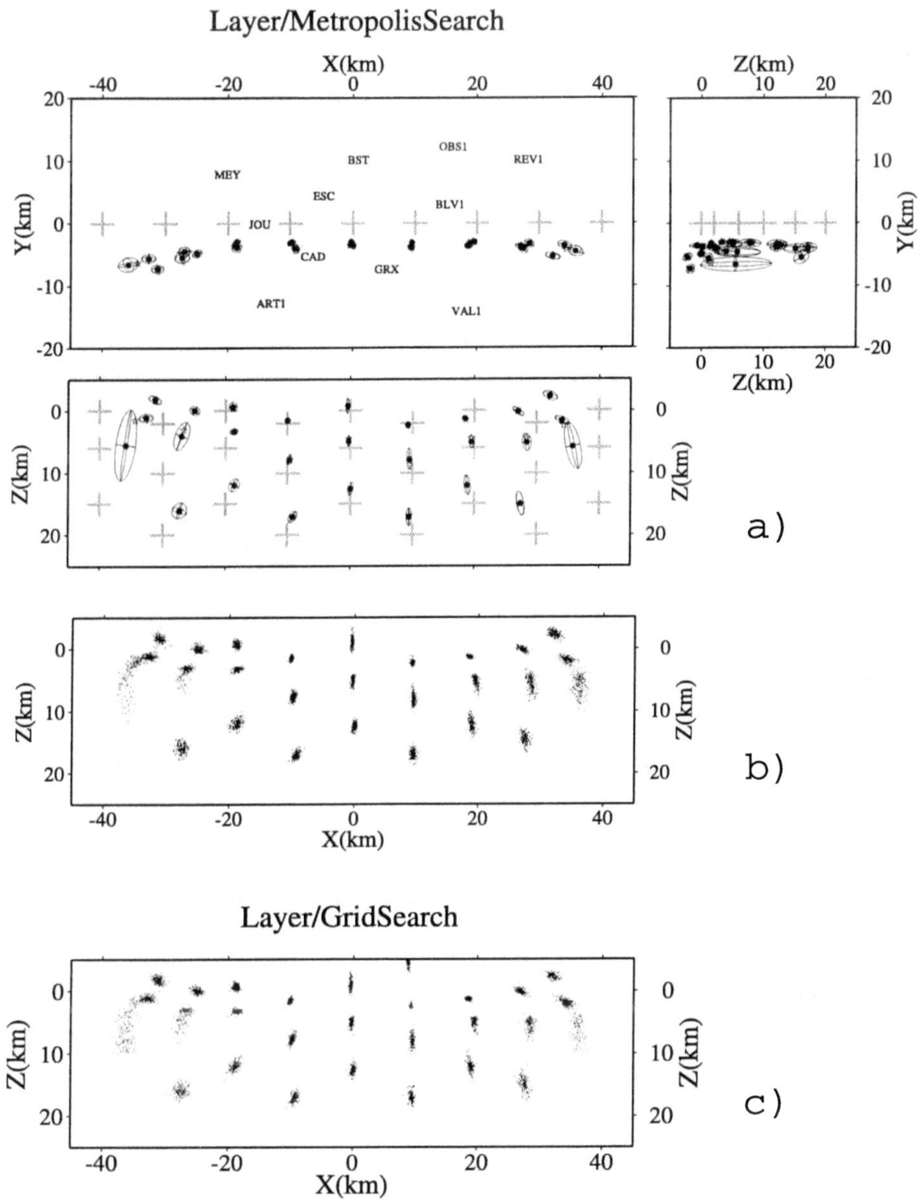

Figure 4. Location with the layered model. a) Metropolis-Gibbs locations and 68% confidence ellipsoids in map view (XY) and perpendicular (ZY) and parallel (XZ) to the event plane. b) Metropolis-Gibbs PDF density plots in XZ section. c) Grid-search PDF density plots in XZ section. Crosses: true event locations; stars: maximum likelihood hypocenters, dots: expectation hypocenters, ellipses: projections of 68% confidence ellipsoids.

events at [-37,5] and [36,7] where the fine grid-search volume is not large enough to contain the elongated PDF which is clipped at depth. This problem can be avoided by extending the fine grid in depth, at the expense of increased calculation time. Other exceptions are the events at [-27,4], [36,7] and [10,2] where the Metropolis-Gibbs PDFs differ notably from those of the grid-search. Apparently for these events the PDF is of sufficient complexity that it cannot be fully recovered by the Metropolis-Gibbs sampler. This is most clear for the event at [10,2] where the grid-search PDF shows two completely separate maxima in the PDF, one at a depth of about 2 km and one at the top of the search region at –5 km. The Metropolis-Gibbs sampler recovers only the deeper of these two solutions.

5.3.3 HYPOELLIPSE

The HYPOELLIPSE hypocenters and confidence ellipsoids for locations in the layered model are shown in Figure 5. For events within the network (-20km ≤ X ≤ 20km) most of the HYPOELLIPSE hypocenters and confidence ellipsoids are nearly identical to the corresponding Metropolis-Gibbs and grid-search results (Figure 4). Though, like the Metropolis-Gibbs sampler, HYPOELLIPSE only recovers the deeper of the two solutions for the event at [10,2]. Two other exceptions are the events at [0,5] and at [20,5] both of which have a much larger depth uncertainty with HYPOELLIPSE than Metropolis-Gibbs. This discrepancy can be explained for both events by examining the form of the grid-search PDF for the event near [20,5] (Figure 6). The PDF is strongly distorted at the model layer boundary at 4.5km depth and the maximum likelihood hypocenter is located on this boundary. At this boundary the gradients in the misfit function (i.e., $g(\mathbf{x})$ in Equation (3)) will be discontinuous which can lead to convergence problems for a linear method like HYPOELLIPSE. For this event, HYPOELLIPSE converges to a hypocenter just below the boundary. In this region the misfit function and PDF change very slowly with depth (Figure 6) because the first arriving rays from points just below the boundary are "head waves" with horizontal take-off directions. Thus, for the linear approach, the partial derivatives of travel-time with respect to depth will be near zero in this region, and the corresponding error estimates very large, leading to the elongated Z axis of the confidence ellipsoid. Note that for these 2 events the difference in expectation and maximum likelihood hypocenters obtained by the global search methods is an indicator of complexity in the PDFs.

For events outside of the network, there are significant differences between the HYPOELLIPSE and non-linear, global search locations. For many events with highly irregular PDFs, the HYPOELLIPSE error ellipsoids are greatly elongated in the Z direction, indicating poor depth constraint, and the HYPOELLIPSE hypocenters are far from the maximum likelihood point of the PDFs. Such events include [-35,5], [28,5], [36,7], and [39,6], which correspond to true event locations [-40,6], [30,10], [40,15], and [40,0], respectively. Because of the presence of the layer boundary at 4.5km, and their location outside the network, the PDFs for these events are highly irregular (Figure 4). For the event with true location [40,0], it is important to note that the global optimum solution, as indicated by the grid-search or Metropolis-Gibbs maximum likelihood point, is located near [32,-2], more than 10km from the HYPOELLIPSE hypocenter near [39,6].

A comparison with the un-clipped grid-search PDF for the event near [36,7] is shown in Figure 6. The HYPOELLIPSE hypocenter is located near a secondary maximum of the grid-search PDF, probably because the global maximum is in a basin with a much smaller volume than that of the secondary maximum and because the starting depth of 10km used for HYPOELLIPSE is closer to the secondary maximum. The hypocenter is in a relatively flat region of the PDF, which explains the large size of the confidence ellipsoid in the depth direction since uncertainty is inversely proportional to the second derivative of the misfit surface. The very small gradients of the PDF in the Z direction near the hypocenter make the linear

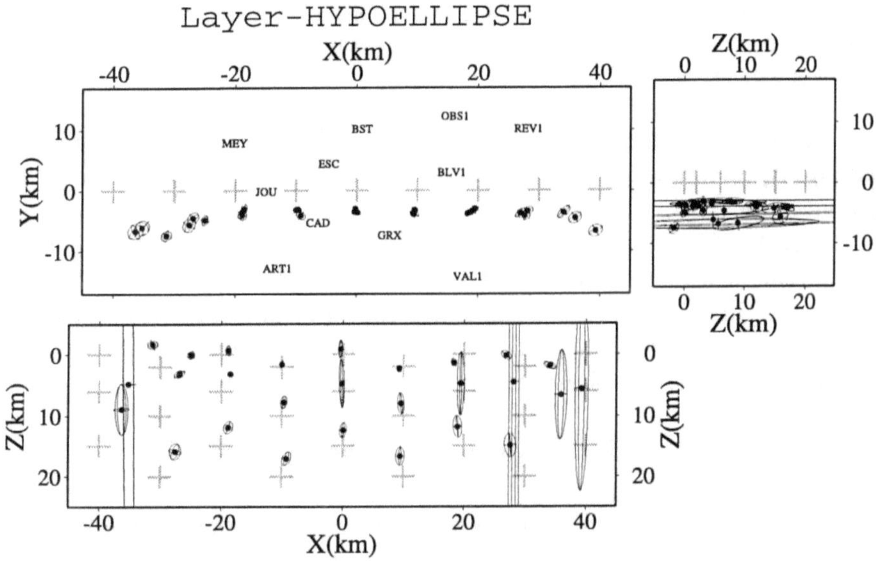

Figure 5. HYPOELLIPSE location with the layered model. Locations and 68% confidence ellipsoids in map view (XY), and views perpendicular (ZY) and parallel (XZ) to the event plane. Crosses: true event locations; stars: maximum likelihood hypocenters, dots: expectation hypocenters, ellipses: projections of 68% confidence ellipsoids.

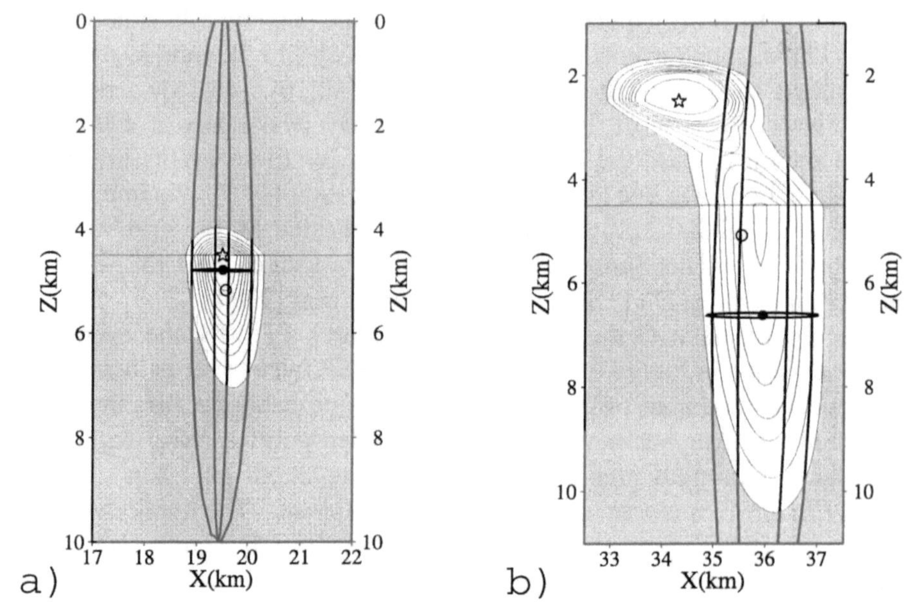

Figure 6. Grid-search and HYPOELLIPSE locations for (a) event [20,5] and (b) event [36,7]. Grid-search PDF: (white contoured region) - XZ slice of grid-search 90% confidence volume with a contour interval of 10%. Star: maximum likelihood hypocenter, open circle: expectation hypocenter. HYPOELLISE location - ellipses: projections of 68% confidence ellipsoid, large dot: expectation hypocenter.

system ill-conditioned and cause instability in the hypocenter depth perturbations; this may explain why the location of the hypocenter is not at a maximum of the PDF.

HYPOELLIPSE has convergence problems or produces a confidence ellipsoid that is a poor estimate of the complete PDF when the PDF is strongly non-ellipsoidal. This is expected since a local linearization gives a poor representation of the global properties of an irregular function. In effect, an irregular PDF implies a strongly non-linear problem.

5.4 Average station residuals and station "statics"

We next investigate the relation of average station residuals to the known static shifts in the arrival times. Figure 7 shows the applied 3D model static shifts, theoretical layered model static shifts, and average residuals for each phase at each station for different location scenarios. The theoretical layered model statics are given by the sum of the 3D model statics and the travel time difference between the 3D model times and the layered model times. For the goal of obtaining station adjustments to give improved absolute event locations, the average station residuals should match the corresponding applied static shifts.

As expected, the average station residuals are effectively zero for locations using the exact synthetic travel times and the correct 3D model with both Metropolis-Gibbs (dots in Figure 7) and the grid-search (not shown).

The average station residuals for Metropolis-Gibbs locations in the correct 3D model using the synthetic travel times with station statics (stars in Figure 7) generally follow the true station statics in both polarity and amplitude. However, there are some significant mismatches (i.e. BST P and S, MEY P, VAL1 P) which can be attributed to the fact that the average station residuals are calculated for incorrect, though optimal, locations, and so reflect location errors as well as model errors. It may be expected that an iterative simultaneous inversion for true statics and event locations beginning with these average residuals will converge to near the correct solutions.

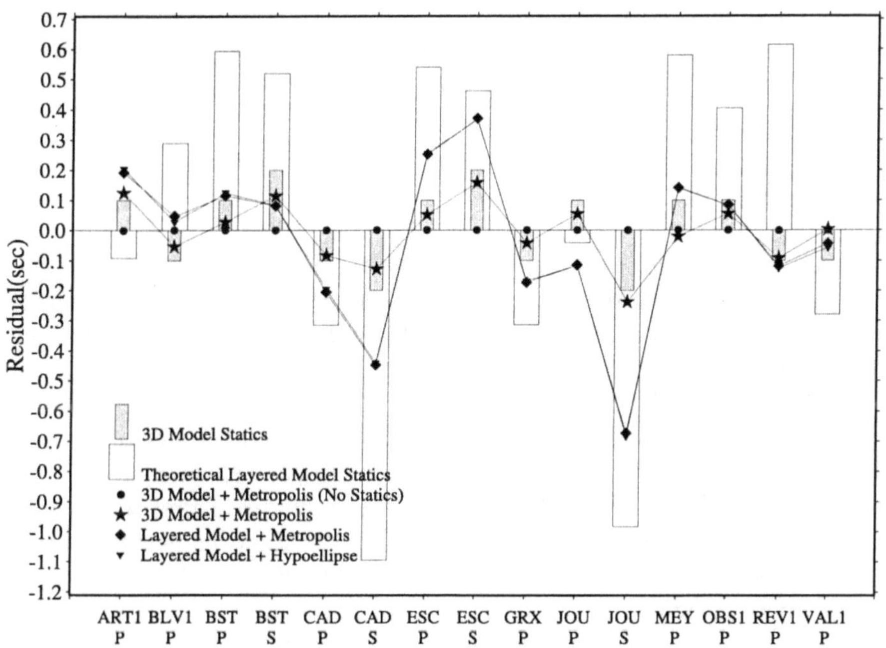

Figure 7. Comparison of station statics and average station residuals. (grey bars) True 3D model station statics (static shift in Table 1). (white bars) True layered model station statics (added static shift in Table 1 plus travel time difference between layered and 3D models for the "true" event location). Average station residuals (observed – calculated arrival times) for the maximum likelihood hypocenters are shown by symbols and lines for several location scenarios.

The average station residuals for the Metropolis-Gibbs and HYPOELLIPSE locations in the layered model (diamonds and triangles in Figure 7) are nearly identical. This agreement reflects the similarity in optimal hypocenters from both methods for most events. However, these residuals do not match and generally have much lower amplitude than the theoretical layered model statics. As above in the 3D case, this mismatch can be attributed to the fact that the average station residuals are calculated for incorrect locations, but with the layered model these locations are much further from the "true" locations. In this case it does not seem likely that a simultaneous inversion for true statics and event relocations beginning with these average residuals will necessarily converge towards the correct solutions. In general there is a better match between average residuals and theoretical statics for stations in the middle of the network (*i.e.* CAD, ESC, GRX and JOU) than for stations on the edge of the network (*i.e.* ART1, BST, OBS1, REV1 and VAL1). This pattern is consistent with the greater variety of paths and coverage of the model for rays between the internal stations and the events than for the external stations.

We have checked the sensitivity of this analysis to the particular set of static shifts used (Table 1) by repeating the locations using a reversed polarity for all of the static shifts. We found that all of the conclusions in this subsection were effectively unchanged.

5.5 Mechanisms

In this section we examine the quality of the recovery of known focal mechanisms in the various location scenarios. Synthetic P first-motions (up or down) corresponding to a pure left-lateral mechanism (strike=90°, dip=90°, rake=180°) were calculated along with the travel time for each synthetic arrival in the 3D velocity model. For each event location, the ray take-off angles are calculated at the optimal hypocenter (maximum likelihood point of PDF for the non-linear, global search methods, final hypocenter for HYPOELLIPSE). Using these synthetic first motions and location dependant take-off angles, composite focal mechanisms (Figure 8) are determined with the grid-search algorithm FPFIT (Reasenberg and Oppenheimer, 1985).

As is expected, the composite focal mechanism for the exact synthetic travel times and the correct 3D model (Figure 8a) is nearly identical to the synthetic left-lateral mechanism. The distribution of first-motion polarities on the focal sphere allows the exact synthetic mechanism as a possible solution with no discrepant observations out of 297 total. Figure 8b shows the composite focal mechanism for Metropolis-Gibbs locations in the 3D model using the synthetic travel times with station statics. For this case there are now 10 discrepant observations, but the preferred focal mechanism remains very similar to the synthetic mechanism.

The composite focal mechanisms for Metropolis-Gibbs and HYPOELLIPSE locations in the layered model using the 3D model synthetic travel times with station statics are shown in Figures 8c and 8d. The Metropolis-Gibbs and HYPOELLIPSE results have, respectively, up to 6 and 10° errors in rake direction, and 43 and 48 discrepant observations. These differences from the synthetic mechanism are significant, but not extreme. This indicate that it is possible to have good recovery of a true composite solution using an incorrect velocity model, if a large and well distributed set of events is available and it can be assumed that they all have the same mechanism.

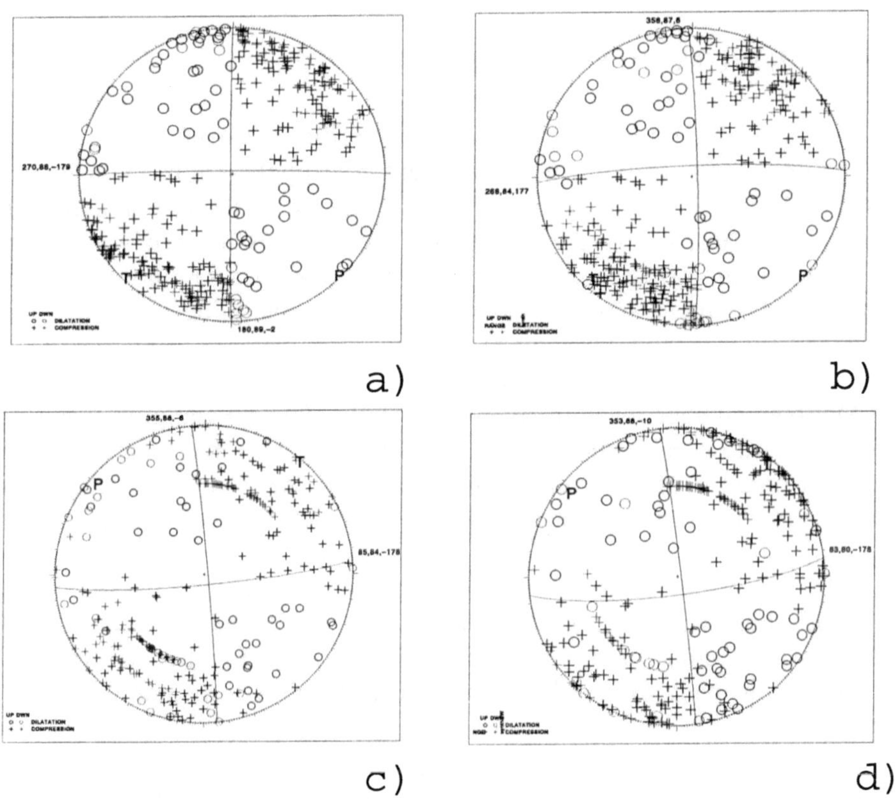

Figure 8. Composite focal mechanisms for Metropolis-Gibbs locations in the 3D model with (a) exact travel times and (b) station statics, and layered model locations using (c) Metropolis-Gibbs and (d) HYPOELLIPSE. Because of coordinate rotation, the positive Y axis of the location plots falls along the azimuth N30°E on these mechanism plots.

We do not investigate individual event mechanisms in detail because the number of observations and coverage of the focal sphere is not adequate to constrain well these solutions. Thus, for all location scenarios, including the exact, 3D case, the single preferred solution given by FPFIT differs significantly from the synthetic left-lateral mechanism for most events outside the network and for many events inside. However, for locations in the 3D model the distribution and polarity of first motions on the focal sphere remain compatible with the synthetic solution. That is, there are no discrepant data for the exact, 3D case and few discrepant data for the 3D case with station statics. This is not the case for the layered model locations where the ray take-off angles often have large errors due to strong velocity model errors and large absolute location errors.

6. DISCUSSION

We have described a complete, probabilistic earthquake location methodology and introduced an efficient Metropolis-Gibbs, non-linear, global sampling algorithm to obtain such locations. With several synthetic location scenarios, we have examined the locations and uncertainties given by an exhaustive grid-search and the Metropolis-Gibbs sampler using 3D and layered velocity models, and by the iterative, linearized method HYPOELLIPSE in the layered model.

With an exact 3D model and no added station static shifts, the grid-search and Metropolis-Gibbs location results are nearly identical and give excellent recovery of the true event location. However, because the location PDFs vary as a function of the event-station geometry, the location uncertainties can be large and irregular outside of the station network even for this exact case.

With an exact 3D model and added station static shifts there are systematic biases in the event locations. The PDFs and 68% confidence ellipsoids are similar in size and orientation to those for the case without statics, but for many events these error indicators do not include the "true" location.

The locations using the incorrect, layered model show that both linear and global methods give systematic biases in the event locations and that the 68% confidence volumes do not include the "true" location – absolute event locations and errors are not recovered. The results indicate that HYPOELLIPSE and related linearized, local methods can give stable and complete location and uncertainty results within a network, except for hypocenters near strong gradients in velocity. Outside a network these methods may have problems with depth determination, depth uncertainty estimation, and even epicentral location. In contrast, the grid-search and Metropolis-Gibbs sampler global methods are always stable and give complete solution information, even near velocity gradients and outside of a network.

Our examination of average station residuals indicate that theoretical station statics can be recovered approximately if a model similar to the true velocity structure is available. However, if a poor model is used for location, then the average station residuals can be affected strongly by event location errors and will not be representative of the true station statics.

Our analysis of focal mechanisms shows that composite mechanisms can be recovered very well if a good velocity model is available, and fairly well even with a poor model. Individual event mechanisms, however, can be erroneous if there are few readings, or if a poor velocity model is used when there are strong lateral variations in the true structure.

HYPOELLIPSE and related linearized approaches are very fast and do not require significant computer memory. However, these methods are difficult or impossible to apply in 3D velocity structures, and they can fail if the location PDF is irregular or multi-modal. The systematic grid-search algorithm we use here is inefficient, about 1000 times slower than HYPOELLIPSE, but will generally give accurate recovery of the complete, probabilistic location PDF. However, if not selected carefully, the grid-search volume can clip parts of the location PDF; this problem can be avoided with an adaptive algorithm. Our implementation of the more efficient Metropolis-Gibbs sampler is only about 10 times slower than HYPOELLIPSE but may require considerable memory with 3D models. The Metropolis-Gibbs sampler gives good recovery of complete, probabilistic location PDFs, except for those that are highly irregular or multi-modal with widely separated maxima.

Thus the Metropolis-Gibbs sampler is a practical method to obtain complete, probabilistic locations for large numbers of events and for location in 3D models. With layered models, the Metropolis-Gibbs sampler is stable for cases where linearized methods fail, including locations near sharp contrasts in velocity and outside of a network. The grid-search method is useful for location of smaller number of events and to obtain an accurate images of location PDFs that may be highly-irregular.

ACKNOWLEDGEMENTS

We thank Pascal Podvin for providing the computer code for the calculation of finite differences travel times. This work was funded in part by the Institut de Protection et de Sûreté Nucléaire and in part by the European Commission "TOMOVES" project ENV4 4980698.

REFERENCES

Aki, K., and P.G. Richards (1980) *Quantitative seismology*, W.H. Freeman, San Francisco.

Billings, S.D. (1994) Simulated annealing for earthquake location, *Geophys. J. Int.*, **118**, 680-692.

Calvert, A., F. Gomez, D. Seber, M. Barazangi, N. Jabour, A. Ibenbrahim, A. and Demnati (1997) An integrated geophysical investigation of recent seismicity in the Al-Hoceima region of North Morocco, *Bull. Seism. Soc. Am.* **87**, 637-651.

Dreger, D., R. Uhrhammer, M. Pasyanos, J. Frank, and B. Romanowicz (1998) Regional and far-regional earthquake locations and source parameters using sparse broadband networks: A test on the Ridgecrest sequence, *Bull. Seism. Soc. Am.* **88**, 1353-1362.

Geiger, L. (1912) Probability method for the determination of earthquake epicenters from the arrival time only (translated from German), *Bull. St. Louis Univ.* **8** (1), 56-71.

Goldberg, D.E. (1989) *Genetic Algorithms in Search, Optimization and Machine Learning*, Addison-Wesley, Reading, MA.

Goldberg, D.E., and J. Richardson (1987) Genetic algorithms with sharing for multimodal function optimization, in J.J. Grefenstette (Ed.), *Genetic Algorithms and their Applications, Proceedings of the Second International Conference on Genetic Algorithms and their applications*, Lawrence Erlbaum Associates, Hillsdale, NJ, 41-49.

Gresta, S., L. Peruzza, D. Slejko and G. Distefano (1998) Inferences on the main volcano-tectonic structures at Mt. Etna (Sicily) from a probabilistic seismological approach, *J. Seis.* **2**, 105-116.

Hammersley, J.M., and D.C. Handscomb (1967) *Monte Carlo Methods*, Methuen, London.

Holland, J.H. (1992) *Adaptation in natural and artificial systems*, Bradford Books/MIT Press, Cambridge, MA, 211 pp.

Jones, R.H., and R.C. Stewart (1997) A method for determining significant structures in a cloud of earthquakes, *J. Geophys. Res.* **102**, 8245-8254.

Keilis-Book, V.I., and T.B. Yanovskaya (1967) Inverse problems in seismology (structural review), *Geophys. J. R. Astr. Soc.* **13**, 223-234.

Kennett, B.L.N. (1992) Locating oceanic earthquakes – the influence of regional models and location criteria, *Geophys. J. Int.* **108**, 848-854.

Kirkpatrick, S., C.D. Gelatt, and M.P. Vecchi (1983) Optimization by simulated annealing, *Science* **220**, 671-680.

Lahr, J.C. (1989) HYPOELLIPSE/Version 2.0: A computer program for determining local earthquake hypocentral parameters, magnitude and first motion pattern, *U.S. Geol. Surv. Open-File Rep.* **89-116**, 92p.

Lepage, G.P. (1978) A new algorithm for adaptive multidimensional integration, *J. Comp. Phys.* **27**, 192-203.

Le Meur, H. (1994) Tomographie tridimensionelle a partor des temps des premieres arrivées des ondes P et S, application a la région de Patras (Grece), *These de Doctorate*, Université Paris VII, France.

Le Meur, H., J. Virieux, and P. Podvin (1997) Seismic tomogrphy of the Gulf of Corinth: a comparison of methods, *Ann. Geofis.* **40**, 1-24.

Lomax, A., and R. Snieder (1995) Identifying sets of acceptable solutions to non-linear, geophysical inverse problems which have complicated misfit functions, *Nonlinear Processes in Geophys.* **2**, 222-227.

Mohammadioun G., and P. Dervin (1995) A full scale laboratory for seismic studies in Southeastern France: The Middle Durance Fault, in *Proc. 5th International Conference on Seismic Zonation, Ouest Editions*, **2**, 1635-1642

Metropolis, N., A.W. Rosenbluth, M.N. Rosenbluth, A.H. Teller, and E. Teller (1953) Equation of state calculations by fast computing machines, *J. Chem. Phys.* **1**, 1087-1092.

Mosegaard, K., and A. Tarantola (1995) Monte Carlo sampling of solutions to inverse problems, *J. Geophys. Res.* **100**, 12431-12447.

Moser, T.J., T. van Eck, and G. Nolet (1992) Hypocenter determination in strongly heterogeneous earth models using the shortest path method, *J. Geophys. Res.* **97**, 6563-6572.

Nelson, G.D., and J.E. Vidale (1990) Earthquake locations by 3-D finite-difference travel times, *Bull. Seism. Soc. Am.* **80**, 395-410.

Nolte, B., and L.N. Frazer (1994) Vertical seismic profile inversion with genetic algorithms, *Geophys. J. Int.* **117**, 162-178.

Pavlis, G.L. (1986) Appraising earthquake hypocenter location errors: a complete practical approach for single event locations, *Bull. Seism. Soc. Am.* **76**, 1699-1717.

Podvin, P. and I. Lecomte (1991) Finite difference computations of traveltimes in very contrasted velocity models: a massively parallel approach and its associated tools, *Geophys. J. Int.* **105**, 271-284.

Press, F. (1968) Earth models obtained by Monte Carlo inversions, *J. Geophys. Res.* **73**, 5223-5234.

Press, W.H., S.A. Teukolosky, W.T. Vetterling, and B.P. Flannery (1993) *Numerical recipies in C: the art of scientific computing*, Cambridge Univ. Press, Cambridge, 994 pp.

Reasenberg, P. and D. Oppenheimer (1985) FPFIT, FPPLOT and FPPAGE: FORTRAN computer programs for calculating and plotting earthquake fault-plane solutions, *U.S. Geol. Surv. Open-File Rep.* **85-739**, 109p.

Rothman, D.H. (1985) Nonlinear inversion, statistical mechanics, and residual statics estimation, *Geophysics* **50**, 2784-2796.

Sambridge, M. and G. Drijkoningen (1992) Genetic algorithms in seismic waveform inversion, *Geophys. J. Int.* **109**, 323-342.

Sambridge, M. and K. Gallagher (1993) Earthquake hypocenter location using genetic algorithms, *Bull. Seism. Soc. Am.* **83** 1467-1491.

Sambridge, M.S., and B.L.N. Kennett (1986) A novel method of hypocenter location, *Geophys. J. R. Astron. Soc.* **87**, 313-331.

Scales, J. A., M. L. Smith, and T.L. Fischer (1992) Global optimization methods for multimodal inverse problems, *J. Comp. Phys.* **103**, 258-268.

Schwartz, S.Y., and G.D. Nelson (1991) Loma Prieta aftershock relocation with S-P traveltimes: effects of 3D structure and true error estimates, *Bull. Seism. Soc. Am.* **81**, 1705-1725.

Shearer, P.M. (1997) Improving local earthquake locations using the L1 norm and waveform cross correlation: Application to the Whittier Narrows, California, aftershock sequence., *J. Geophys. Res.* **102**, 8269-8283.

Sen, M.K., and P.L. Stoffa (1995) *Global optimization methods in geophysical inversion*, Elsevier, Amsterdam, 281p.

Stoffa, P.L., and M.K. Sen (1991) Nonlinear multiparameter optimization using genetic algorithms: Inversion of plane-wave seismograms, *Geophysics* **56**, 1794-1810.

Tarantola, A. (1987) *Inverse problem theory: Methods for data fitting and model parameter estimation*, Elsevier, Amsterdam, 613p.

Tarantola, A. and B. Valette (1982) Inverse problems = quest for information, *J. Geophys.*, **50**, 159-170.

Vidale, J.E. (1988) Finite-difference calculation of travel times, *Bull. Seism. Soc. Am.*, **78**, 2062-2078.

Vilardo, G., G. De Natale, G. Milano, and U. Coppa (1996) The seismicity of Mt. Vesuvius, *Tectonophys.*, **261**, 127-138.

Volant P., C. Berge, P. Dervin, M. Cushing., G. Mohammadiou and F. Mathieu (2000) The Southeastern Durance fault permanent network: preliminary results, *J. Seism.*, in press.

Wiggins, R. A. (1969) Monte Carlo inversion of body wave observations, *J. Geophys. Res.* **74**, 3171-3181.

Wittlinger, G., G. Herquel, and T. Nakache (1993) Earthquake location in strongly heterogeneous media, *Geophys. J. Int.* **115**, 759-777.

Chapter 6

LOCATION CALIBRATION BASED ON 3-D MODELLING

Steps towards the calibration of the Global CTBT Network

Petr Firbas
CTBTO, International Data Centre, Vienna Int'l Centre, P.O.Box 1250, A-1400 Vienna, Austria; email: Petr.Firbas@ctbto.org

Key words: earthquake location, location accuracy, velocity model, Earth's mantle, CTBT, 3-D modelling, Source-Specific Station Corrections

Abstract: This paper provides a comprehensive analysis of 3-D velocity modelling for the purpose of the Comprehensive Nuclear Test Ban Treaty (CTBT) monitoring. The method is based on the usage of ground-truth data for network calibration and combines all available information into one 3-D global velocity model. The 3-D modelling allows predicting travel-time Source-Specific Station Corrections (SSSCs) for any (even currently non-existent) station in the calibrated region and for events with arbitrarily deep hypocenters. The calibration, based on the method described, can be extended to global coverage. Based on the 3-D modelling, travel-time SSSCs can be predicted for all regional and teleseismic phases and this can be done in a way that guarantees the mutual consistency of the SSSCs. A method for calculation of the SSSCs is described and results of test calculations and relocation experiments are discussed. The travel-time calculation method employed is based on several steps including a perturbation approach and ray morphing. In all documented cases the obtained event mislocations of a range of 20 or more kilometres, using the globally averaged 1-D global iasp91 velocity model, can be reduced by calibration based on 3-D modelling. Mislocations of the order of just a few kilometres were obtained, using a very low number of well-calibrated stations. Based on the summarised results, the presented 3-D modelling approach appears to be a feasible way to calibrate the global network of the CTBT International Monitoring System (IMS).

C.H. Thurber and N. Rabinowitz (eds.), Advances in Seismic Event Location, 135–161.

1. INTRODUCTION

The Comprehensive Nuclear Test Ban Treaty (CTBT) verification system includes an International Monitoring System (IMS) composed of 321 stations of four different monitoring technologies: seismic, hydroacoustic, infrasound, and radionuclide monitoring. The first three mentioned technologies are also known collectively as seismo-acoustic monitoring. It is expected that all four technologies will be used in a synergetic manner for event detection, location, and identification.

All the three corresponding media (i.e. solid Earth, oceans, and atmosphere) are known to be inhomogeneous and in some cases they are even anisotropic. Further, the velocity of wave propagation in oceans is seasonally dependent. The same is true for the atmosphere where the changes in wave propagation can be observed on an hourly basis and they depend, among other factors, on the winds in stratosphere and thermosphere. In the synergetic approach (i.e. applying all three waveform technologies jointly) events of interest can be located using observed times of arrival, azimuths, and slownesses. Examples of types of events for which the synergetic approach to event location and event characterization is valuable include underground explosions, mine explosions, underwater explosions, atmospheric explosions (including bolides), volcanic eruptions (including underwater ones), etc..

In this paper the high-accuracy event location using seismic travel-time data only will be discussed. The synergetic approach using more types of observed quantities and multiple technologies will be discussed elsewhere.

Most commonly the location of events within the solid earth is performed using laterally homogeneous (radially symmetric) globally averaged velocity models. Event locations based on these one-dimensional (1-D) velocity models show substantial bias in many cases. The process of improving the ability to locate events more accurately (based on information on events with precisely known origin parameters) is known as "calibration".

In this paper, calibration for the purpose of seismic monitoring is considered. The IMS system is to be composed of 50 "primary" and 120 "auxiliary" seismic stations distributed around the globe with positions defined in the CTBT text and its Protocol. The primary stations are used for the detection and initial location of events. The data from auxiliary seismic stations are to be used for the location refinement. The IMS is configured to record not only large seismic events at teleseismic ranges, but also smaller events recorded only by a few stations within regional distances (in ranges less than 20 degrees). Location based on a combination of regional and teleseismic phases, using a relatively sparse network, dictates the scenario in which the IMS network has to be calibrated.

The so-called "On-Site-Inspection" (OSI) provisions of CTBT require specifying a search area of not more than 1000 km^2. To achieve this event location accuracy on a global scale using a quite sparse network of IMS stations is not an easy task, but it was demonstrated that this is a realistic task (e.g. Firbas, 1997a, Firbas et al., 1998, and Ryaboy et al., 1999).

The generally accepted method for estimation of event location precision is calculation of the so-called 90% uncertainty error ellipse. Such uncertainty error ellipses can be generally defined by the criteria that for 90% of events with known epicenters, the known epicenters have to fall inside of the calculated uncertainty ellipses.

To be able to state that the true event location is (with 90% probability) within the determined 1000 km^2 uncertainty ellipse area, the travel-times must be known with high accuracy. For events of magnitude 4 and larger detected by a subset of IMS stations with reasonable azimuthal coverage, the travel-times of regional P-phases must be known with an accuracy of about 1.0 second (in the Root-Mean-Square (RMS) sense). This is quite a strict requirement, but compared to the requirements used in the interpretation of Deep Seismic Sounding (DSS) profiles or long-range profiles, the conditions for achieving "acceptable" fit to travel-times are actually relaxed for IMS calibration.

It is a well-known fact that not only globally averaged travel-time curves but also regional 1-D velocity models are not ideal for accurate locations of events recorded at regional distances (Ryaboy et al., 1999). This is due to large variations of travel-times of regional phases caused by lateral heterogeneities in the crust and in the uppermost mantle.

The travel-time curves derived from explosions and earthquakes recorded in the regional distance zone (up to the range of 20 degrees) reveal major differences between the waves propagating in tectonically active regions and the waves where the propagation is confined to platforms. Differences can amount up to 10 seconds or more for Pn. Besides these differences, there is also a large "scatter" of about 3 – 5 seconds for Pn in tectonic provinces and somehow smaller "scatter" for pre-Cambrian platforms. This "scatter" is due to smaller-scale lateral heterogeneities in any of these provinces. The differences in travel-times for Sn waves and Lg waves due to lateral heterogeneities in the upper mantle are even larger than the differences for Pn waves and can easily reach 20 seconds for epicentral distances of 1500 – 2000 km.

The effects of the lateral heterogeneity of the Earth were recognized decades ago. To accommodate the lateral heterogeneities, 1-D velocity models were abandoned in favour of 2-D velocity models in analyzing the long-range profile data. This resulted in very successful travel-time predictions along profiles down to an accuracy of several tenths of a second.

To get a high-accuracy and consistent interpretation of crossing long-range profiles, anisotropic velocity models had to be used in some cases. The effects of lateral heterogeneity and anisotropy are not so easy to distinguish as both these effects can be modelled with limited accuracy by laterally heterogeneous velocity models with curved interfaces (Firbas, 1984, 1988). In global and regional seismology 2-D velocity models are of limited use. 3-D velocity models have to be used to model observed travel-times. This paper will concentrate on 3-D isotropic velocity models.

Any so-called “event-definition-criteria”, which determine the events to be included in the bulletins, are and will be always debatable. A balance has to be struck to reach the best compromise between bulletin completeness, its reliability, and usability for a given purpose. Limiting ourselves to the set of IMS stations we can expect, however, that the number of IMS stations used in the location of an event of interest for the CTBT monitoring may be quite low in many cases. The sparse IMS cannot often provide data from more than 5 to 7 IMS stations for an event and still its location accuracy should be acceptable. The current event-definition criteria used at the CTBT International Data Center do permit situations when as low as three IMS stations are used to build an event to be included in the so-called Reviewed Event Bulletin (REB). The fact that events have to be located using phase readings from only a very low number of IMS stations highlights the need for predicted (modelled) travel-times to be as accurate as possible.

This paper deals with various aspects of the calibration process based on 3-D modelling and utilization of reliable travel-time information provided by “ground-truth” events. Ground-truth events are those events for which the origin (hypocenter) parameters are known with a very high accuracy. Ground-truth events present the most important input data on which the calibration process is based. In an ideal case, all the source parameters (origin coordinates, origin depth and origin time) are known with such accuracy that the predicted travel-times are not substantially influenced by uncertainty of the event origin parameters.

Earthquake seismologists consider the exact knowledge of origin time to be of the least importance from all of the source parameters. For the calibration effort it is very important to know the origin time with high accuracy. This is because the error of the travel-time (which we want to “calibrate”) does not depend only on the accurate knowledge of the onset time, but it depends to the same extent on the accurate knowledge of the origin time.

Besides the uncertainty of the origin time, the travel-time accuracy depends on errors called measurement error and modelling error. There will always be a measurement error as the onset-times of the phases are being detected within the background noise. The onset-time measurement error is dependent in part on the signal-to-noise ratio (SNR). We can try to lower the onset-time measurement error by various means, e.g. by signal filtration or other signal-to-noise enhancement techniques. This type of error cannot be, however, lowered by the calibration.

The other component of the error budget is the modelling error. The modelling error includes the effects due to our incomplete understanding of the Earth structure. We try to minimize the modelling error by calibration using travel-time information provided by events with well-known origin parameters (known sources).

Let us assume that the knowledge of the event origin parameters (origin time, latitude, longitude, and depth) can be cumulatively characterized by an uncertainty of the order of a few tenths of a second or less. In such a situation the accuracy of determination for P travel-times of 1.0 second (RMS) or better can be achieved for signal detections with a reasonable signal-to-noise ratio. The events of magnitude 4 recorded within the regional zone (up to the distance of 20 degrees) mostly have the reasonable signal-to-noise ratio needed to achieve the above stated accuracy.

The goal of the calibration process is to decrease the modelling errors of the travel-times for all phases used in the event location. For the group of P-phases, both the regional phases (Pn, Pg) as well as the teleseismic phases (P, PKP, PcP, etc.) have to be calibrated in a mutually consistent way. The calibration should also encompass the S-type phases, Lg and Rg phases and various multiply reflected and converted phases, including depth phases, if they are used in the event location process.

It is important to emphasize that for achieving stable and accurate event location (in particular event depth), it is very important to guarantee that not only the predicted travel-times for all used phases are mutually consistent but also travel-time differences, e.g.: Sn-Pn, PcP-P, PP-P, pP-P, sP-P, etc. are mutually consistent. This is true for the currently most used travel-time tables, the IASPEI tables (Kennett, 1991). In the case of IASPEI travel-time tables the mutual consistency of travel-times for all phases is achieved by deriving them from one velocity model.

Once we embark on calibration of travel-times we need to preserve the mutual consistency of travel-times for all phases. This cannot be achieved if the travel-times for individual phases are calibrated "independently". This mutual consistency can be, however, well guaranteed if the travel-times are predicted using one single 3-D velocity model.

For 1-D velocity models "travel-time curves" are defined. When working with 3-D velocity models, "travel-time surfaces" are used. As travel-time curves can be multivalued (e.g. can have triplications), the travel-time surfaces can also be multivalued. It seems to be quite an art to define suitable parameterization of the travel-time curves and travel-time surfaces to avoid, or at least minimize, the impact of the fact that travel-time curves / surfaces are multivalued. It is critical to use the predicted travel-time derived from the appropriate travel-time curve segment/surface sheet when comparing it with the measured arrival (propagation) time. It has to be emphasized that the smoothness of the travel-time surfaces has direct implication on the stability of the location process.

The travel-time surfaces for individual phases are defined by existence of ray endpoints and travel-time values in some grid covering the area of existence of a particular phase. This area will be typically of a quite irregular shape as the presence of lateral heterogeneities results in the fact that rays for a particular phase can reach the Earth's surface from a selected source over a range of epicentral distances that are azimuth-dependent. This is different from the situation obtained for a 1-D velocity structure (model). In case of a 1-D velocity model, the area of existence of a particular phase is always radially symmetric.

The difference in the size and shape of the area of phase existence for a 1-D and 3-D structure (model) poses a challenge for any travel-time calculation algorithm. Using a perturbation approach, the full travel-time value for a 3-D velocity model can be defined as a sum of a full travel-time value for a background 1-D velocity model (e.g. iasp91 velocity model) and the appropriate Source-Specific Station Correction (SSSC) value. For any perturbation approach used to calculate the full travel-time values for a 3-D velocity model, SSSCs can be "defined" only on the radially symmetric area of phase existence described in the previous paragraph. When we employ a 3-D structure (model) then along some azimuths from the source, the true range of phase existence for a particular phase can be larger than the azimuth-independent range, described in the previous paragraph, and some extrapolation of the travel-time surface may be needed.

Using the iasp91 velocity model as the 1-D background velocity model for defining the "areas of phase existence" can be considered helpful when making a step from event location based on 1-D globally averaged velocity model to event location employing a 3-D velocity model. This can be seen as useful by analysts who are used to searching for phases in the ranges of phase existence defined by the popular iasp91 velocity model. It can be also useful for the automatic phase association programs.

2. GEOLOGICAL AND GEOPHYSICAL FACTS IMPORTANT FOR CALIBRATION

The Earth is radially symmetric to a first approximation. The effect of ellipticity of the Earth is typically corrected by the ellipticity corrections

(Dziewonski and Gilbert, 1976, Dornboos, 1988, Kennett, 1991). We will expect that these ellipticity corrections are applied before we get engaged in the calibration on the global scale. In future, the elliptical, not radially symmetric Earth can be considered as the starting point. The corrections for topography are also expected to be applied before we get engaged in the calibration.

Earth's crust and the uppermost mantle present quite a challenge, as they are strongly laterally heterogeneous. The Moho discontinuity is curved and the Moho depth can vary from less than 10 km characteristically for the oceanic areas to 70-80 km under some parts of the continents. Uppermost mantle heterogeneities result in Pn velocities (just under the Moho discontinuity) between 7.6 km/s and 8.4 km/s. It has to be kept in mind that even a 0.05 km/s change in Pn velocity makes a very important difference for Pn travel-times at regional distances.

To get reliable travel-times both for regional and teleseismic phases, 3-D velocity structure of the mantle has to be taken into account. In particular, it has to be understood that subducting slabs have a significant horizontal extent and thus do influence the travel-times of regional phases substantially. The existence of subducting slabs makes using any regional 1-D velocity model for an extensive region even more difficult. Moreover, it has to be kept in mind that a high percentage of earthquakes originate in the provinces where subduction zones are present and thus a lot of regional phases from these sources are strongly influenced by these lateral heterogeneities.

The DSS and long-range profiles have revealed some correlation between the Earth's crust and the upper mantle velocity structure on one hand and types of tectonic provinces on the other hand. Also, the correlation between the geology of the basement and/or basement rock age and the deep Earth's structure was considered in the past decades. Furthermore, the correlation between the Earth's crust and uppermost mantle velocity structure was studied. It has to be taken into account that all these types of correlations are of limited value when deriving an accurate 3-D velocity model of the Earth.

In some cases, for example for the inner parts of the large platform/shield provinces, the correlation between the type of the Earth's crust and the velocity structure of the uppermost mantle seems to be quite well developed down to depths of several hundreds of kilometres. This is true unless there is a significant laterally heterogeneous structure involved (e.g. a subducting slab). For tectonic provinces, which include major seismically active zones, the correlation between crustal and upper mantle structures may be valid for shallower depths only, if at all. DSS studies have shown that cases do exist, when crustal and uppermost mantle velocity structures are actually in "anti-correlation", examples being some very deep sedimentary basins with elevated Moho and high-velocity material in the uppermost mantle (e.g. the North Caspian depression (Vinnik and Ryaboy, 1981)).

It is no surprise that the location of regional events, i.e. the events whose location is based predominantly on regional phases, depends greatly on our knowledge of velocities in the uppermost mantle and, last but not least, on the shape of the Moho boundary. Thus, we can take any correlation between the crustal and upper mantle structures only as the first gross guess and we must proceed with modelling of the upper mantle structures using full-fledged 3-D velocity models. In such a modelling, crustal and upper mantle velocity structures are to be varied independently without any limiting assumptions on correlation. At the same time we can benefit from the fact that the Moho shape is quite well known from DSS results and the Moho depth can be fixed during the 3-D modelling in many cases.

3. 1-D VELOCITY MODELS AND STATION CORRECTIONS

Before discussing the 3-D modelling and its results, the 1-D velocity models and station corrections should be mentioned.

It is well known that the globally averaged 1-D velocity models (e.g. iasp91) represent travel-times of the teleseismic phases relatively well. Nevertheless, some important deviations still do exist and limit the accuracy of event location. The accuracy of travel-times of teleseismic phases was attempted to be increased, using the so-called "station corrections". These corrections were represented either as a single number per station (bulk corrections) or by a function which includes an azimuthal dependency (but not dependency on range) for each station (Cleary and Hales, 1966, Herrin et al., 1968, Herrin and Taggart, 1968, Lilwall and Douglas, 1970, Sengupta and Julian, 1976, Dziewonski and Anderson, 1981, 1983) (see Chapter 1).

Although the general distribution of station correction values reflects the major distribution of tectonic zones and platform areas, their contribution to event locations based on IMS data did not result in such a location accuracy improvement that would be sufficient for the CTBT monitoring. Also the station corrections, as presented in studies of various authors over the last 3 decades, show a substantial scatter. The scatter can be at least partially explained by the selection of a data set used in deriving the corrections.

The globally averaged 1-D velocity models do not represent the uppermost mantle well enough to be used for event location for CTBT purposes. Thus, usage of 1-D globally averaged velocity models for prediction of travel-times of regional phases is not possible for the CTBT monitoring.

The usage of 1-D regional velocity models does not seem to be satisfactory for this purpose either. The rays of regional phases often cross several geologically different regions with different velocity distributions. Applying a 1-D regional velocity model may improve event location accuracy inside of the region where the 1-D regional velocity model is applicable if the stations from this region only are used for the event location. Once the stations from outside the region of applicability of the 1-D regional velocity model are used, the accuracy of travel-time prediction for regional phases deteriorates. Application of 1-D regional velocity models may, in fact, lead to substantial decrease of the location accuracy for events outside the region for which the 1-D regional velocity model is applicable.

If regional 1-D velocity models are to be used to predict the travel-times for regional phases (i.e. up to 20 degrees range), it has to be taken into account that there are very few continental regions on the Earth which have a radius of 20 degrees or more and that can be at least approximately represented by one 1-D velocity model. Furthermore, even in the largest regions, there are substantial smaller-scale lateral heterogeneities that do not allow achieving the required regional P travel-time prediction accuracy of 1.0 second (RMS) or better, as the "scatter" of the travel-time values can reach several seconds.

In trying to assess the contribution of some smaller tectonic provinces to the overall travel-time value, it has to be understood that in some of these regions rays of regional phases will be crossing the upper mantle only (and not the crust), and such contributions to the travel-times can never be directly measured.

Knowledge of the Moho boundary shape and of the Pn velocity variations just below the Moho boundary is critical for the accurate location of events based on regional phases. Any regional 1-D velocity model ignoring the Moho depth variations and/or Pn velocity variations will not be sufficiently accurate for achieving the 1.0 second (RMS) travel-time accuracy requested for CTBT monitoring.

The conclusion of all these arguments is that not 1-D, but only 3-D velocity models allow representing the real structure of the Earth with the required precision and as such have the potential to reach the travel-time prediction accuracy required for the CTBT monitoring. The calibration examples included in this paper are to demonstrate that the 3-D modelling for IMS calibration is feasible and practical.

4. GROUND-TRUTH DATA AND 3-D VELOCITY MODELS

High-quality ground-truth data are the primary resource for any calibration approach. Any velocity models are just means of representing the available information from ground-truth data in a consistent manner. 1-D velocity models are based on a fit to ground-truth data. Fitting a 3-D model to the data set is the same, but just a more sophisticated approach. So there is no principal "mathematical" difference between fitting a 1-D and fitting a 3-D velocity model.

3-D velocity models have, however, a lot of advantages compared to 1-D velocity models. A 3-D velocity model can have a resolution varying for different parts of the Earth based on the quality and amount of ground-truth data used to constrain the model. Contrary to 1-D velocity models, the 3-D modelling can potentially cover all structural features observed (including anisotropy if necessary).

As an added benefit, the 3-D velocity models can also be extrapolated into the areas that are not covered by ground-truth data at all or are not covered by ground-truth data with sufficient density. The velocity-depth sections for latitude-longitude points outside of the area covered by ground-truth data can be obtained by smooth extrapolation reaching a "fallback" (or default) velocity-depth section far away from the area well covered by the ground-truth data. The simplest fallback velocity-depth section can be a 1-D velocity model generally accepted for a particular type of region. When there is no information on the velocity structure of the province available, the 1-D globally averaged velocity model may be used as the fallback velocity-depth section for this part of the 3-D velocity model.

There is enough information on the detailed structure of the crust and on the Moho shape for many regions of the Earth and this information can be directly implemented in the 3-D velocity model to increase the accuracy of the travel-time modelling. The impact of the crust and Moho represent, in many cases, travel-time deviations of the order of 1-2 seconds. Without the proper use of this information, the overall accuracy goal of 1.0 s (RMS) or better for the travel-time prediction of P-phases cannot be reached. For a well designed 3-D velocity model parameterization, the information on the Earth's crust and on the Moho shape can be directly used in the 3-D velocity model even when the deeper parts of the Earth will still have to be independently modelled.

Every velocity model will have a finite resolution. The current realistic threshold on global 3-D velocity models is around 0.5 – 1.0 degree in any horizontal direction. The approach used here allows variable resolution. In such a way, more detailed regional velocity models with higher resolution can be incorporated into the global velocity model. In case of a very precise calibration of some small areas (e.g. test sites) additional residual corrections may be needed. In such a case, a bulk correction for a station and/or travel-time correction surfaces (e.g. fitted by local splines or kriging) representing the very local phenomena can be used to supersede the regional averaging in the global 3-D velocity model. It should be noted that even a slight anisotropy could be rather easily included in the 3-D modelling, using the perturbation approach (Cerveny and Firbas, 1984; Firbas, 1984, 1988).

Once the global 3-D velocity model allows incorporating regional velocity models provided by national or regional authorities, it is possible to use the framework of 3-D modelling for verification of location accuracy improvement based on these regional velocity models. The 3-D modelling approach documented in this paper makes it possible.

5. TRAVEL-TIME CURVES VERSUS VELOCITY MODELS

Before going into the discussion on "travel-time curves versus velocity models", it is appropriate to stress again in this context that ground-truth data and nothing else are primary to any calibration approach. Let us compare, however, what are the inherent similarities and differences between "fitting travel-time curves" and "fitting velocity models" to the ground-truth data and what are the advantages and disadvantages of these two approaches.

Fitting a travel-time curve or a velocity model of any type (1-D, 2-D, or 3-D; isotropic or anisotropic) is a similar type of operation from the mathematical point of view. In both cases some unknown parameters are attempted to be determined, either describing the shape of the travel-time curve or the structure of the velocity model. So in both cases some optimization is involved.

The data can be fitted either independently for individual phases or simultaneously for all phases (or a group of phases):

- *In the former case, the fit does not guarantee consistency between individual phases (it is an inconsistent fit). This approach is more typically employed when a single travel-time curve is being fitted. The inconsistencies between travel-time curves / surfaces for individual phases fitted independently to ground-truth data can result in substantial degradation of the location quality and in unpredictable event location biases.*

- *In the latter case, the fitting operation can guarantee consistency between predicted travel-times of various phases (or all phases). This type of fit can be considered a partially (or fully) consistent fit. If both P and S velocity models are developed in a consistent way, then all P, S, and transformed phase travel-times are consistently predictable. It is worth mentioning that the generally used IASPEI travel-time tables are based on one single 1-D globally averaged velocity model (P and S velocities included) to guarantee mutual consistency of all phases (including phases like sP, PKS, etc.).*

Although fitting a travel-time curve or a velocity model are mathematically similar types of operations, they are different types of operations from the geophysical point of view. Contrary to the fitting of a travel-time curve, the fitting of a velocity model allows inclusion of a priori information on geology, on the Earth structure (e.g. Moho depths) or information from other geophysical methods (e.g. gravity surveys).

Even more important, however, is the "prediction potential" of the velocity models. Based on a fitted velocity model, it is possible to predict travel-times for any (even currently non-existent) station and also predict the travel-times for any event hypocenter depth. Further, it is also possible to predict consistently travel-times even for those phases for which the travel-time data are difficult to collect or have not be been collected so far (e.g. pP, PP, etc.). As mentioned above, velocity models also allow a smooth and geophysically sound extrapolation into regions for which travel-times have not been collected yet (no or insufficient coverage by ground truth events).

Further, it is worth stressing that the velocity models can impose some "bounds" on the fitted parameters. The parameters of the velocity models (i.e. velocity values, boundary depth values, or related parameters) have some physical meaning and, as such, they are constrained to intervals of permissible values. These bounds on the fitted parameters impose in turn some realistic bounds on the overall fit and on the resulting predicted travel-time curves / surfaces. This is not the case when just travel-time curves / surfaces are fitted directly to the ground-truth data without relation to any velocity models.

When a 3-D velocity model of the crust and uppermost mantle is able to predict all regional phases successfully (i.e., when the parameters of the crust, Moho boundary, and upper mantle are fitted to ground truth data), then the derived model can be utilized outside of the domain of calibration. It can also be viewed as a part of a step-by-step approach which will additionally address calibration of teleseismic phases. This advocated step-by-step approach allows including either parts of 1-D velocity models (e.g. iasp91 velocity model) or parts of 3-D global tomography velocity models (e.g. the

Harvard model SP12WM13) for the lower parts of the mantle and/or for the core. This capability is included in the program package for which results are presented in this paper.

Based on these arguments, it is quite obvious that using velocity models is superior to using independent travel-times for individual phases.

It is quite clear that besides model sophistication there is no conceptual difference between deriving a 1-D or 3-D velocity model based on ground truth data. In both cases the quality of the obtained velocity model is assessed based on the quality of fit of the predicted travel-times to the ground-truth data.

Generally, one set of ground-truth data is used to develop a 1-D or 3-D velocity model. It would be beneficial to have a second independent set of ground-truth data (which was not used in deriving the velocity model) and use it to verify the derived velocity model. We can be satisfied when the velocity model derived from the first set of ground-truth data predicts all travel times for the second set of ground-truth events. If it is found that the second set of ground-truth data contains arrivals whose travel times cannot be modelled with sufficient accuracy, then all the ground-truth data together are to be used to develop the next iteration of the velocity model. This iterative sequence is to be continued until the velocity model adequately supports all the information obtained from the available ground-truth data. In the case of a 3-D velocity model, it means that both the lateral and depth velocity distribution reach the required resolution to explain all the important effects observed in the ground-truth data.

As the resolution of any model is finite, there will always be a need for combining the 3-D velocity model with some additional mechanism to represent "short-wavelength" phenomena. As mentioned already above, residual corrections for "test sites" or other spatially rather limited areas (where the extremely high location accuracy is needed) may be consistently added to the travel-times predicted by the velocity model. Local splines or other approaches, e.g. kriging, can be used to fit residual correction surfaces for areas of high interest. Areas where such residual corrections will be applied will be typically limited in size and can be seen as "patches" in relation to the travel-times predicted by the global 3-D velocity model. Also, additional residual corrections may need to be included for stations (bulk corrections or azimuth dependent station corrections). These bulk station corrections can "absorb" the very local effects at the station, e.g. sediment layer under the station, some problems with station equipment timing or station specific signal delays in the station equipment, etc.

6. OBSERVED TRAVEL-TIMES

As the first example of travel-time observations, data on Pn travel-times for North America are shown. Figure 1 shows the observed Pn travel-time deviations from the globally averaged Pn travel-times based on the iasp91 velocity model (Firbas et al., 1998; Ryaboy et al., 1999) for four ground truth events. Figure 2 shows two other data sets of Pn travel-time data for Europe (Firbas, 1997a; Firbas et al., 1997; Firbas et al., 1998). These figures confirm that the travel-time anomalies can reach ± 5 s or more relative to IASPEI travel-time curves for Pn. In some areas the large scale anomalies (with spatial dimensions even over several thousands km) dominate. It is clearly seen, however, that also the smaller scale anomalies (with spatial dimensions from about 100 km to several hundreds of kilometres) are important once the travel-times of Pn-waves are to be predicted with accuracy of 1.0 s (RMS) or better.

7. 3-D MODELLING APPROACH

The modelling approach depends on two major steps. The first step is the representation of the 3-D velocity model and the second step is the modelling algorithm employed. In the modelling approach it is critical to keep good balance between the model complexity (number of parameters in the model) and ease with which the model parameters can be changed, observing any known constraints (a priori information).

In the presented case, the parameterization of any 3-D velocity model used is based on independent representation of the major "layers" inside the Earth. The first "layer" (laterally heterogeneous) currently used is the Earth's crust. Further, the curved Moho boundary is included. The mantle can be subdivided into several laterally heterogeneous "layers" and each of them can be represented independently in the 3-D velocity model and varied independently during the modelling process. This type of 3-D velocity model parameterization gives the chance to model the uppermost mantle in detail and, at the same time, to use either iasp91 global velocity model or a 3-D tomography model for any other part of the mantle.

The current best global representation of the crust is based on 5 x 5 degrees velocity model "Crust 5.1" (Mooney et al., 1998). This velocity model also includes the 5 x 5 degrees averaged information on the Moho depths. While the 5 x 5 degrees global velocity model for the crust was found to be a good start for the 3-D modelling (and, in fact, until now it was kept fixed in the modelling), the 5 degrees resolution for the Moho depths was found to be too rough. As the used parameterization of the 3-D velocity

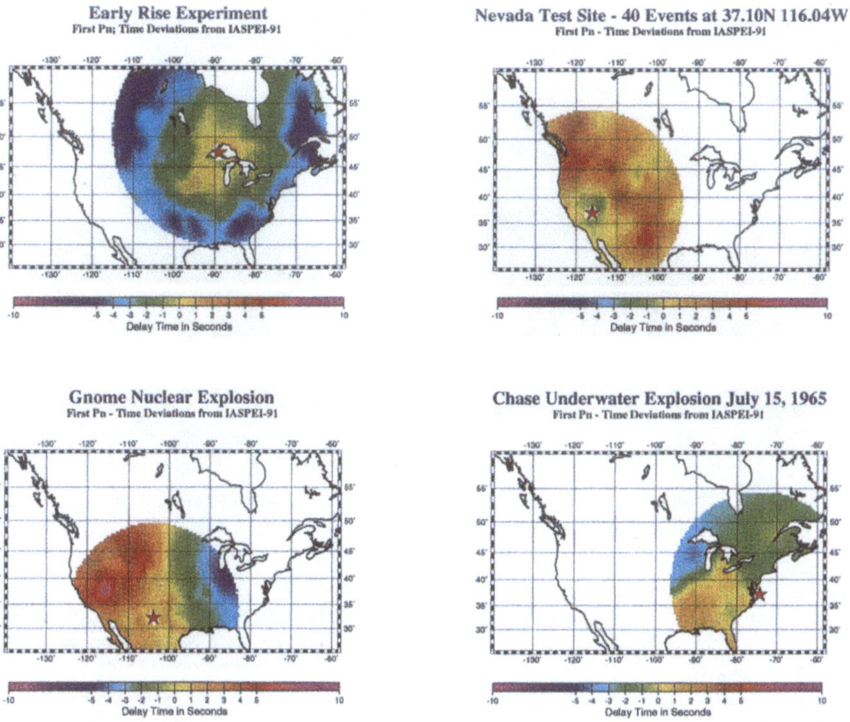

Figure 1. Pn-wave travel-time anomalies for four "calibration events" in North America.

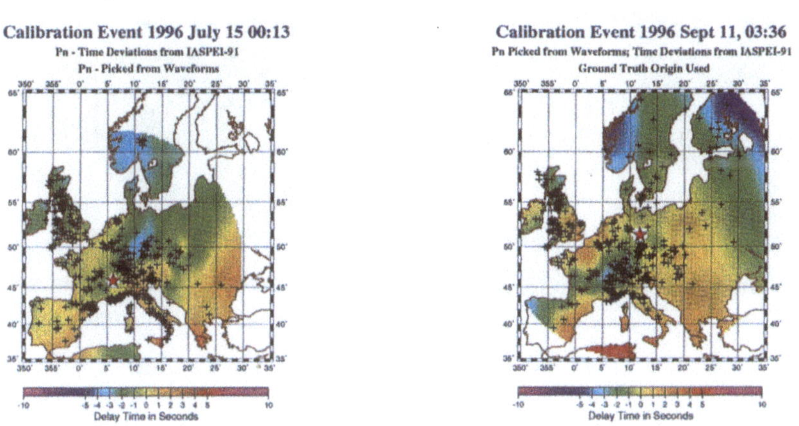

Figure 2. Pn-wave travel-time anomalies for two "calibration events" in Europe. The crosses are stations for which Pn onset time data were measured.

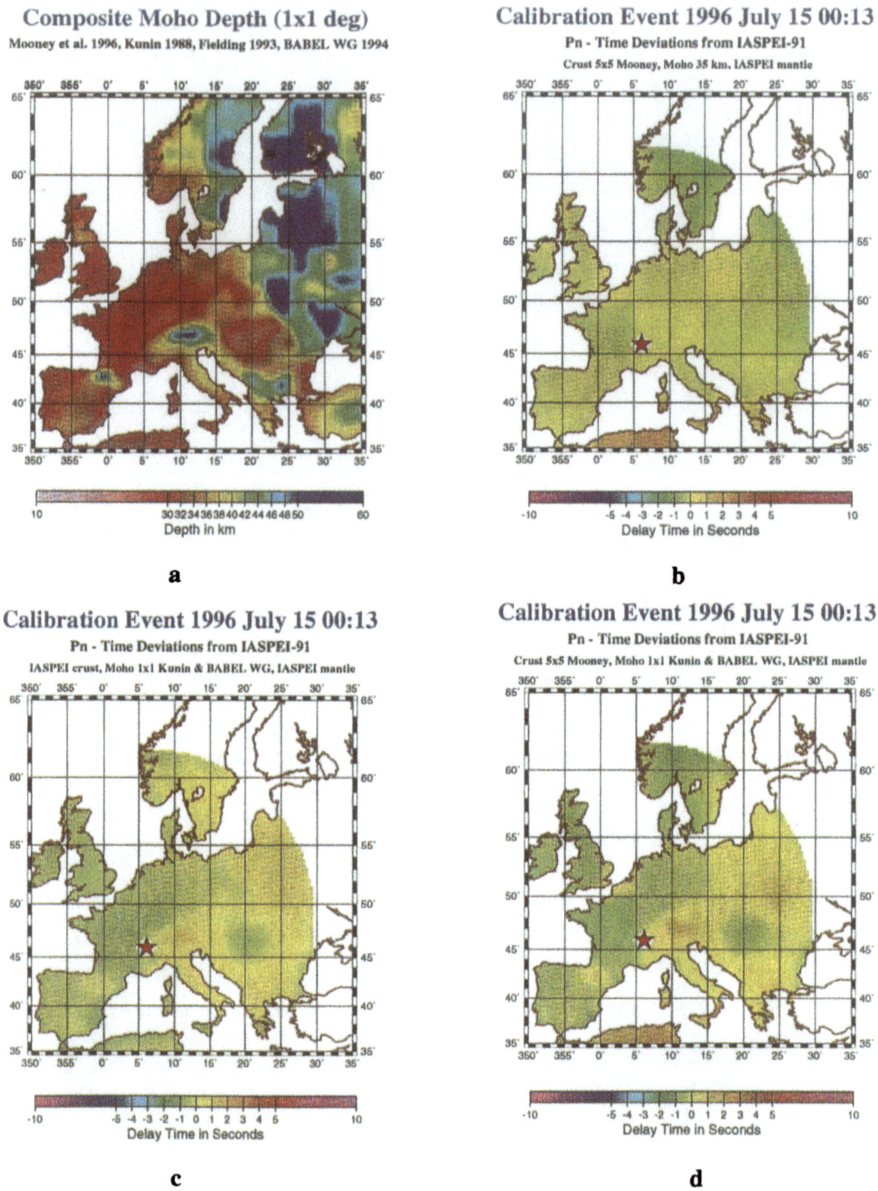

Figure 3. a) shows the shape of the Moho used; b) shows the contribution of crust (resolution 5 degrees) to SSSC; c) shows the contribution of Moho to SSSC and d) shows the combined effect of crust and Moho.

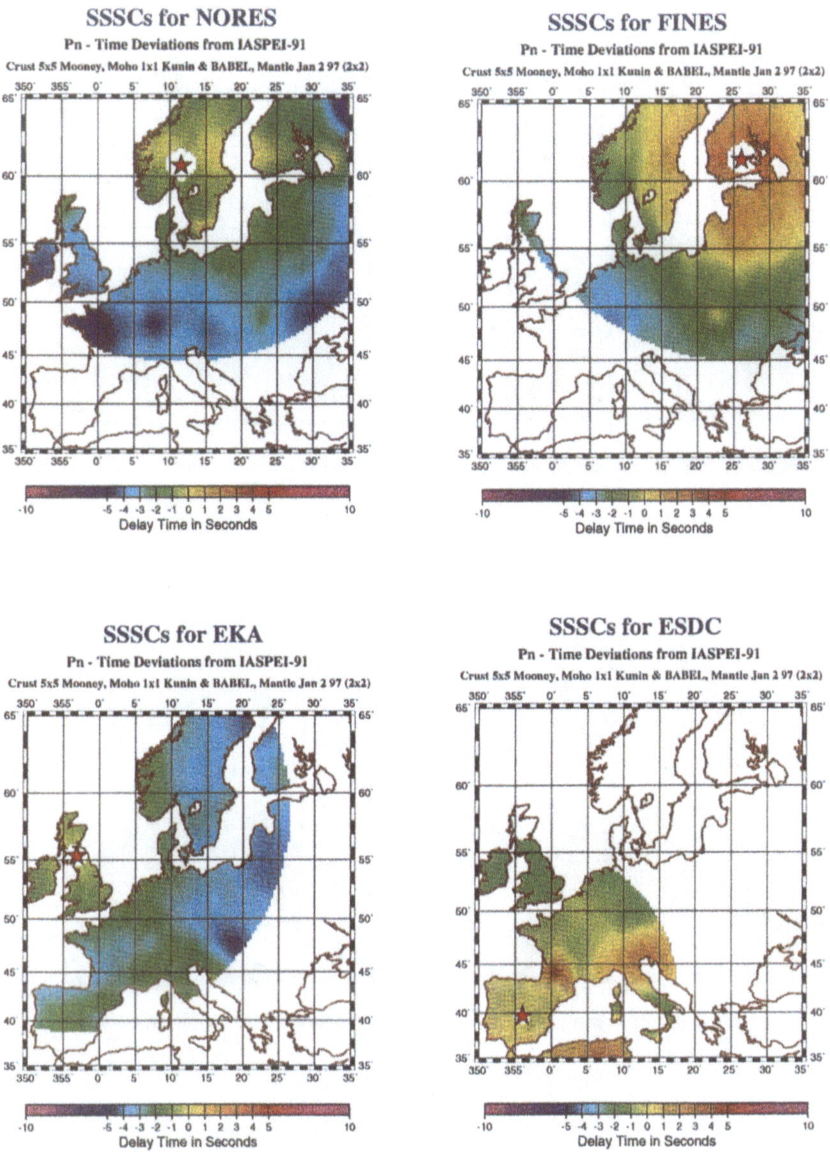

Figure 4. Source-Station Specific Corrections (SSSC), i.e. travel time anomalies, computed based on the used (preliminary) 3-D model for selected stations of the IMS.

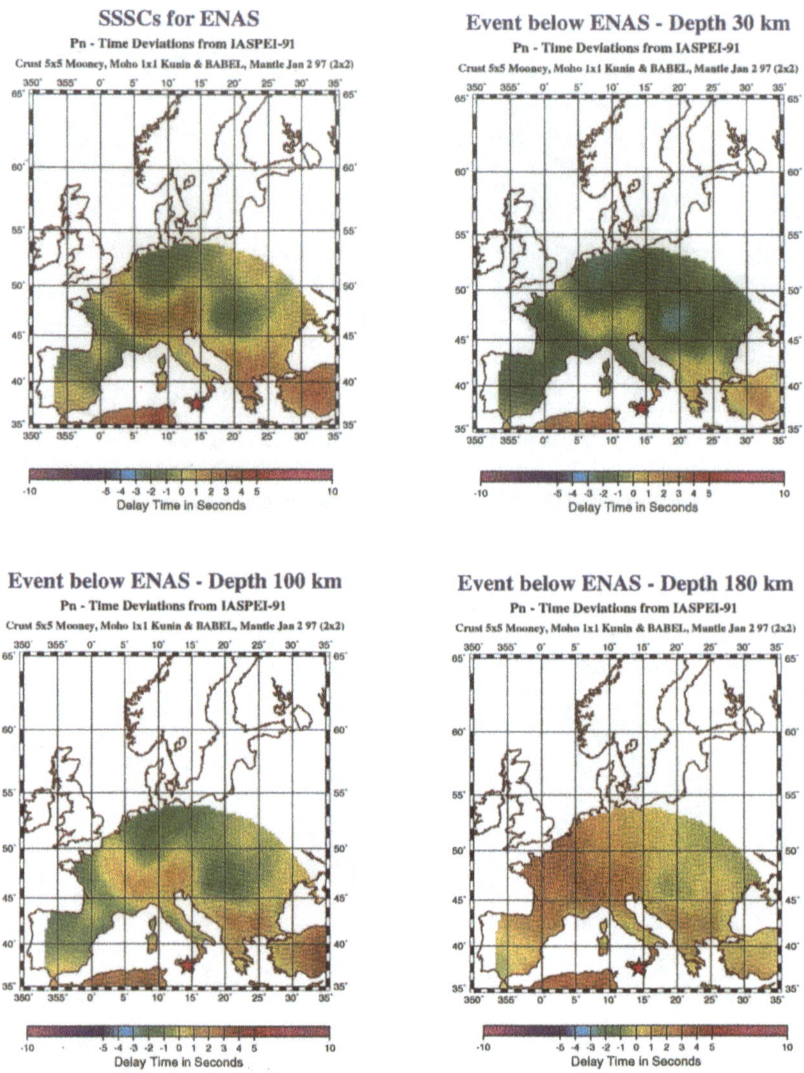

Figure 5. Source-Station Specific Corrections (SSSC), i.e. travel time anomalies, computed based on the used (preliminary) 3-D model for a planned IMS station ENAS (depth = 0 km) and for hypothetical events at the 30, 100, and 180 kms.

model permits the Moho shape to be modeled with a resolution independent of the resolution of the Earth crust and mantle, it was decided to create a Moho depth map with a higher resolution. Currently 1 x 1 degree resolution of the Moho depth map was used for any region for which 3-D modeling was attempted.

The Moho used for the test case of Western Europe is shown in Figure 3a. The Figures 3b-d show the contribution of the crust (with 5 x 5 degrees resolution) and Moho (with 1 x 1 degree resolution) and a combination of both of these contributions. These figures confirm that without proper representation of the crust and Moho, it would not be possible to represent the Pn travel-times with 1.0 second (RMS) accuracy.

The parameterization of the 3-D velocity model used permits to implement easily the patching of the global Moho depth surface by regional Moho maps known with higher reliability. In all the 3-D modelling cases attempted so far, using the approach documented below, the crust and Moho were kept fixed during the modelling and only the parameters describing the upper mantle structure were varied.

The mantle is represented independently of the representation of the crust. In all of the 3-D modelling cases attempted so far, the upper mantle was modelled from the Moho down to 440 or 670 km. The 3-D velocity model of the upper mantle was up to now always represented by velocity-depth sections at corners of a 2 x 2 degrees regular longitude-latitude grid. The velocity at any arbitrary position and depth is computed by interpolation of the 4 nearest velocity-depth sections forming the 3-D velocity model representation. The 2 x 2 degrees resolution for the mantle was found quite adequate in most cases, but, in principle, it is possible to use a higher resolution, e.g. 1 x 1 degree or 0.5 x 0.5 degree.

The approach of representing the upper mantle in the described manner has proven to be successful and practical. It was found rather easy to start with a small number of typical velocity-depth sections representing large-scale provinces. Populating individual grid points by these typical velocity-depth sections was used to create the starting 3-D velocity model for the upper mantle. Thus "irregular" blocks were effectively defined in the upper mantle. Step-by-step modifying the shape of these "irregular" blocks and step-by-step (but very carefully) increasing the number of used velocity-depth sections and modifying them allowed fitting the available ground-truth data in a reasonable number of trial-and-error steps. This approach was applied for Western Europe (Firbas, 1997a; Firbas et al., 1997; Firbas et al., 1998) and for North America (Ryaboy et al., 1999).

The computer program "sssc" (Firbas, 1997b) was used for the computation of SSSCs for Pn travel-times. The computation of the travel-times in 3-D velocity models is quite a time-consuming task. Direct application of either the shooting or the bending ray-tracing method does not seem to be practical in this case, especially when the travel-time accuracy is not required to be better than 0.1 second and no more than an overall fit with 1.0 second (RMS) accuracy is required. Both the shooting and bending ray-tracing methods have their pitfalls and are quite time-consuming (see Chapter 4). These facts led to a development of a robust approach that is based on a combination of the ray shooting and ray bending (morphing).

The individual steps of the approach used are the following ones:

a) First, any travel-time branch is assigned a unique code that describes all the segments of the ray, type of wave (P or S) in each segment and behaviour of the ray at all interfaces. This code helps to guarantee that during the bending (morphing) process the ray will not change its properties. This approach, in principle, allows computing the rays for all the branches of the individual travel-time curves.

b) The SSSC calculation can be considered to be, to some extent, a perturbation approach. There is a background velocity model used to start with. In this case the background velocity model can be, for example, an arbitrary 1-D global velocity model. To "guarantee" the existence of standard ranges of teleseismic phases, the standard iasp91 velocity model is currently used. Other 1-D global models were tested and can be easily incorporated.

c) Rays are computed in the 1-D background velocity model, using a very fast ray-tracing scheme. A paraxial approximation helps to find the ray for the selected receiver range. The tests have shown that the computed travel-times for the iasp91 velocity model are precise to at least several hundredths of a second.

d) The linearization scheme used in tomography (see e.g. Firbas, 1981, 1987) is applied to include the impact of the 3-D velocity model. This generally applied linearization step has, however, been proven not to be sufficiently precise for prediction of travel-times of all phases. To further increase the accuracy of the travel-time prediction, in particular for regional phases like Pn or Sn, a ray morphing (ray bending) approach is further applied (guided by Fermat's principle) to get the best estimate for the full travel-time for a particular phase and range.

Using this approach, the full predicted travel-time for a selected 3-D velocity model is calculated. This is sufficient for any location procedure. For those who are interested in the SSSCs, the IASPEI travel-time can be subtracted from the full travel-time in the 3-D velocity model and the travel-time anomaly (or SSSC) can be obtained.

It is worth mentioning that the program "sssc" is able to employ (for the depths not covered by 3-D modelling) any globally averaged 1-D velocity model or results of global 3-D tomography or combination thereof. Currently it is capable of using the Harvard 3-D velocity model SP12WM13 (Su and Dziewonski, 1993; Ekstrom et al., 1994).

8. CALCULATED SSSC

Fitting a tentative 3-D velocity model using ground-truth data allows predicting SSSCs for any (even currently non-existent) seismic station. For the case of North America some examples of SSSCs were given in Ryaboy et al. (1999).

In Figure 4 some examples of SSSCs are given for several European IMS stations based on a feasibility study (Firbas, 1997a; Firbas et al., 1997; Firbas et al., 1998). These calculated SSSCs do show the expected features based on common knowledge of geology and tectonics. In particular the SSSCs for the Spanish seismic array station ESDC nicely depict the effects of Pyrenees and of the Alps. The SSSCs both for the Norwegian seismic array station NORES and for the UK seismic array station EKA show the expected faster propagation of Pn waves in the consolidated areas of Fennoscandia. The Finnish seismic array station FINES is also sited above the high-velocity mantle in Fennoscandia. The SSSCs for this station show a different pattern compared to the SSSCs for NORES. This can be explained by the fact that NORES is sited in area where the crust is rather shallow whereas FINES is sited close to the area where the Moho is assumed to be very deep. This comparisons documents the importance of proper representation of undulated Moho when calculating the SSSCs. This can be accommodated only in 3-D models.

The 3-D velocity model also allows predicting SSSCs for events with any hypocenter depth. The Figure 5 shows the estimated SSSCs for various hypocenter depths for a case of hypothetical events located at latitude and longitude of the planned IMS station in the Mediterranean. These 4 calculated examples document the importance of proper representation of depth when using the SSSCs. If the effect of depth is neglected and the SSSCs for zero depth would be used only then, based on the provided comparison, it is obvious that the applied zero-depth SSSCs can be substantially erroneous for events with deep hypocenter. There does not seem to be any hope to collect enough data to derive the depth-dependent SSSCs based on very precise knowledge of hypocenters and origin times of deep events. The solution here is in usage of a 3-D velocity model which can consistently predict, if properly validated, SSSCs for any hypocenter depth. In the same way the SSSCs for phases as pP, sP, etc. can be derived consistently from a validated 3-D velocity model.

9. LOCATION ACCURACY IMPROVEMENT

The ultimate measure of success of any IMS calibration can be assessed based on location accuracy improvement. As an example of the real-life situation, Figure 6 shows various estimated epicenters for the July 15, 1996 earthquake in France. The scatter of the computed epicenters is substantial. This scatter occurred despite the fact that in many event locations undertaken by the data centers there were several tens of regional stations used in locating this earthquake. In case of the CTBT monitoring, there will be only 5 IMS stations in the regional zone for which reliable travel-times could be expected for this earthquake. It is a big challenge to locate the event reliably and with high accuracy, using just a very limited number of stations.

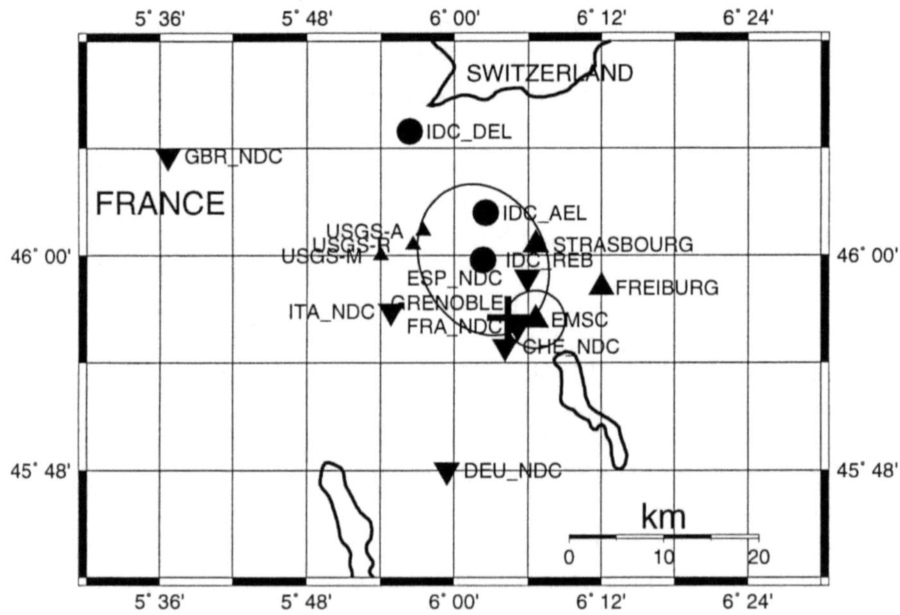

Figure 6. Epicenter solutions provided by national and international data centers and by local networks. The symbols are centered on individual epicenter solutions. Error ellipses for the ISC and the European Mediterranean Seismological Centre (EMSC) solutions are shown. The Grenoble University solution is shown by a cross (at the point where the two ellipses intersect).

For assessment of accuracy of a calculated epicenter, ground-truth information is needed. Contrary to a controlled explosion, in case of an earthquake, the true hypocenter and true origin time will hardly ever be known with such a high accuracy. Nevertheless, in some favorable situations the hypocenter and origin time for shallow earthquakes can be determined with a sufficiently high accuracy for these types of comparisons. This was the case of the July 15, 1996 earthquake. The earthquake happened beneath quite a dense local seismic network and that allowed locating the earthquake with very high accuracy (Observatoire de Grenoble, 1996). This location information is labelled as "Grenoble University" in Table 1. Subsequent aftershock studies and on-site geological studies confirmed the location results based on local seismic network. The latitude, longitude, and depth can be considered to be known within ± 1 km or better and the origin time known within a few tenths of a second.

The second line in Table 1 shows the location from the Reviewed Event Bulletin (REB) of the Prototype International Data Center. The REB location was based on several tens of stations. The distance of the computed REB epicenter from the Grenoble University "ground-truth" solution is 6.5 km. When only the 5 IMS European stations only were used (and no SSSCs applied) then the distance of computed epicenter increased to 18.8 km. This effect of accuracy decrease, when fewer stations are used, is quite typical.

Calibration for this area was needed. To calibrate this area, data from many hundred European stations were used to construct the map of travel-time Pn anomalies (against iasp91 travel-times) measured on waveforms obtained for this earthquake (see Figure 2a). Due to some averaging, some measurement errors in the used onset readings, and due the above-described limited quality of the "ground-truth" solution, one cannot expect that the travel-time corrections, derived from such a composite map of travel-time anomalies, can be absolutely exact. Nevertheless, when these "smoothed" travel-time corrections were applied to Pn travel-times for 5 IMS European stations, the computed epicenter was found very accurately, just 2.5 km away from the "ground-truth" epicenter. This relocation test can be considered an end-to-end consistency check, but this is not an independent location attempt.

To test the calibration approach, a preliminary 3-D velocity model for Europe was derived. For this 3-D velocity model the SSSCs were calculated for the 5 IMS stations, using the program "sssc". The predicted travel-time anomalies (SSSCs) for Pn were applied to the measured travel-times and the event was relocated. Table 1 shows that the computed epicenter was only 4.6 km from the "ground-truth" epicenter. This improvement from 18.8 km mislocation to 4.6 km mislocation measures the quality of the achieved calibration.

Table 3. Comparison of hypocenters calculated for the July 15, 1996 seismic event.

Hypocenter Solution	**Latitude [deg]**	**Longitude [deg]**	**Depth [km]**	**Distance from the Grenoble University Solution [km]**
Grenoble University	45.942	6.073	4	-
PIDC REB	45.996	6.039	0	6.5
5 IMS Stations only Pn only No Correction	**46.086**	**5.946**	**4 (fixed)**	**18.8**
5 IMS Stations only Pn only Correction from one SSSC map only (constructed for the event)	45.950	6.042	4 (fixed)	2.5
5 IMS Stations only Pn only Correction from five SSSC maps (derived from the 3-D velocity model)	**45.930**	**6.016**	**4 (fixed)**	**4.6**

It can be mentioned that this case seems to be quite typical and that improvements on this level can be found realistic also in other cases. In calibration of North America (Ryaboy et al., 1999) the average mislocation of large nuclear and chemical explosions when using IASPEI travel-times was found to be 19.4 km. The average mislocation when SSSCs based on a 3-D velocity model for North America were used was found to be just 5.7 km. This means lowering of the mislocation by more than 70% when the calibration information is used.

The study of Ryaboy et al. (1999) further showed that not only the mislocation was lower, but also the size of the calculated "90%-error ellipse" is consistently decreasing and, at the same time, the expected 90% of coverage is either kept or even increased. In case of North America test events studied, the size of the calculated 90%-error ellipse decreased from 4260 km^2 to 640 km^2. It was an interesting finding that 100% of the test events had their epicenters inside the computed respective 90%-error ellipses. This means that the estimates of the measurement and modelling errors used in 90%-error ellipse calculation may have been too conservative. To get just 90% event epicenters inside the calculated 90%-error ellipses, the average 90%-error ellipse could be further decreased to 410 km^2 in this case. Both the improvement of the location accuracy and the consistency of the solutions (confirmed by the dramatic decrease of the 90%-error ellipse size) measure the successful improvement achieved by the calibration based on the 3-D modelling.

10. CONCLUSIONS

The method presented was specifically designed to strike a good balance between the amount of the available ground-truth data on one hand and the accuracy and consistency of the travel-time prediction required to calibrate the IMS global network on the other hand.

This paper demonstrates that the calibration based on 3-D modelling is feasible. Furthermore, it documents that for the selected test events (recorded by a very limited number of IMS stations) the location accuracy necessary to comply with the On-Site Inspection requirements of the CTBT monitoring can be achieved. After careful calibration, the search area for many events could be 1000 km^2 or smaller. The approach described in this paper will be further applied to other areas of the globe and tested on larger sets of events from all around the world.

The paper was focused on the calibration for IMS and CTBT monitoring. However, the method described here has a much broader application. In principle, the 3-D modelling can be applied on any scale: local, regional, and global.

ACKNOWLEDGEMENTS

Substantial work on the feasibility study on calibration of Western Europe was carried out during a long-term visiting stay of the author at the prototype International Data Center in Arlington, USA. During this period also the above mentioned 3-D modelling package "sssc" was developed. The author is grateful for fruitful discussions and advice from his colleagues: R. Cook, J. Coyne, H. Israelsson, R. North, V. Ryaboy, and C. Romney. Data for the feasibility calibration study for Western Europe were provided by more than 50 institutes and data centers in 25 European countries and several International Data Centers (for a full list see please Firbas et al., 1997). GMT software (Wessel and Smith, 1995) was used for making maps. The crustal velocity models collected by USGS (Mooney et al., 1998) and co-operating institutions from the whole world were an important input to this calibration effort. The co-operation with Harvard University (G. Ekstrom and G. Smith) and with Australian National University (O. Gudmundsson and M. Sambridge) and Cornell University (M. Barzangi, D. Seber, and D. Steer) was sincerely appreciated. The author wishes also to thank C. Thurber for helpful comments.

REFERENCES

Cerveny V., and P. Firbas (1984) Numerical modelling and inversion of travel times of seismic body waves in inhomogeneous anisotropic media, *Geophys. J. R. Astr. Soc.* **76**, 41-56.

Cleary J., and A.L. Hales (1966) An analysis of travel times of P waves to North American stations, in the distance range 30° to 100°, *Bull. Seismol. Soc. Am.* **56**, 467-489.

Doornbos D.J. (1988) Asphericity and ellipticity corrections, in: D.J. Doornbos (Ed.), *Seismological algorithms* Academic Press, London, pp. 75-85.

Dziewonski, A.M. and D.L. Anderson (1981) Preliminary reference Earth model, *Phys. Earth Planet. Inter.* **25**, 297-356.

Dziewonski, A.M. and D.L. Anderson (1983) Travel times and station corrections for P waves at teleseismic distances, *J. Geophys. Res.* **88**, 3295-3314.

Dziewonski A.M., and F. Gilbert (1976) The effect of small, aspherical perturbations on travel times and re-examination of the correction for ellipticity, *Geophys. J. R. Astr. Soc* **44**, 7-17.

Ekstrom G., A.M. Dziewonski, W. Su, and G.P. Smith (1994) Elastic and anelastic structure beneath Eurasia, *Proceedings of the 16th Annual Seismic Research Symposium*, **PL-TR-94-2217**, 78-84.

Fielding, E., M. Barazangi, and B. Isacks (1993) A network-accessible geological and geophysical database for Eurasia, North Africa, and the Middle East, *Proc. 15th Annual Seismic Research Symposium*, **PL-TR-93-2160**, 93-106.

Fielding E., M. Barazangi, B. Isacks, and D. Seber (1994) A network-accessible geological and geophysical database for Eurasia, North Africa, and the Middle East: Digital database development for the Middle East and North Africa, *Proc. 16th Annual Seismic Research Symposium*, **PL-TR-94-2217**, 85-91.

Firbas P. (1981) Inversion of travel-time data for laterally heterogeneous structure - linearization approach, *Geophys. J. R. Astr. Soc.* **64**, 189-198.

Firbas P. (1984) Travel-time curves for complex inhomogeneous slightly anisotropic media, *Studia Geoph. Geod.* **28**, 393-406.

Firbas P. (1987) Seismic Tomography for profiles, in: G. Nolet (Ed.) *Seismic tomography with applications in global seismology and exploration geophysics*, Reidel.

Firbas P. (1988) Inhomogeneity versus anisotropy in the interpretation of seismic dat,. *Phys. Earth Planet. Inter.* **51**, 36–41.

Firbas P. (1997a) IMS Calibration for Europe, *Technical Report* **CMR-97/02**, Center for Monitoring Research, Arlington, USA, 44 pp.

Firbas, P (1997b) Program sssc for computation of station-source specific time corrections for 3-D velocity models, *Technical Report* **CMR-97/08**, Center for Monitoring Research, Arlington, USA, 31 pp.

Firbas P., K. Fuchs, and W.D. Mooney (1998) Calibration of seismograph network may meet test ban treaty's monitoring needs, *EOS, Trans. Am. Geophys. Union* **79**, 413-421.

Firbas P., A. B. Peshkov, and V. Ryaboy (1997) From IASP-91 global model to a 3-d model for a CTBT monitoring, in: K. Fuchs (Ed.) *Upper Mantle Heterogeneities from Active and Passive Seismology*, Proceedings of the NATO ARW Moscow April 1997, NATO ASI Series, Kluwer Academic Publishers, Dordrecht, Netherlands, pp. 199-214.

Herrin E., and J. Taggart (1968) Regional variations in P travel times, *Bull. Seismol. Soc. Am.* **58**, 1325-1337.

Herrin E., W. Tucker, J. Taggart, D.W. Gordon, and J.L. Lobdell (1968) Estimation of surface focus P travel times, *Bull. Seismol. Soc. Am.* **58**, 1273-1291.

Kennett, B.L.N. (1991) IASPEI 1991 Seismological Tables, Res. School of Earth Sci., Austr. Natl. Univ., Canberra, Australia, 167 pp.

Kunin, N.Ya., N.V. Goncharova, G.I. Semenova, S.V. Usenko, E.R. Sheikh-Zade, V.V. Belousov (1987) Map of topography of the mantle surface of Eurasia, *Physics of the USSR Academy of Sciences*, Ministry of Geology of the Russian Federation.

Lilwall R.C., and A. Douglas (1970) Estimation of P wave travel times using joint epicenter method, *Geophys. J. R. Astron. Soc.* **19**, 165-181.

Mooney, W.D., G. Laske, and T.G. Masters (1996) Crust 5.1: a revised global crustal model at 5 X 5 degrees, *EOS Trans. AGU* **77**, F483.

Mooney W. D., G. Laske, and T. G. Masters (1998) CRUST 5.1: A global crustal model at 5 by 5 degree, *J. Geophys. Res* **103**, 727-747.

Observatoire de Grenoble (1996) Rapport Preliminaire de la Mission d'Intervention suite au seisme d'Annecy (15 Juillet 1996).

Ryaboy V., D.R. Baumgardt, P. Firbas, and A.M. Dainty (1999) Application of 3-D crustal and upper mantle velocity model of North America for location of regional seismic events, submitted to *Pure Appl. Geophys.*

Sengupta M.K., and B.R. Julian (1976) P wave travel times from deep earthquakes, *Bull. Seismol. Soc. Am.* **66**, 1555-1579.

Su W., and A.M. Dziewonski (1993) Joint 3-D inversion for P- and S-velocities in the mantle, *EOS, Trans. Am. Geophys. Union.*, **72**, 557.

Vinnik L.P., and V.Z. Ryaboy (1981) Deep structure of the East European platform according to seismic data *Phys. Earth Planet. Inter.* **25**, 27–37.

Wessel P., and W.H.F. Smith (1995) The Generic Mapping Tools (GMT), Version 3. Technical Reference and Cookbook, NOAA Geodynamics Branch **N/OES12**.

Chapter 7

JOINT EVENT LOCATION - THE JHD TECHNIQUE AND APPLICATIONS TO DATA FROM LOCAL SEISMIC NETWORKS

Jose Pujol
Center for Earthquake Research and Information, The University of Memphis, Memphis, TN 38152, USA

Key words: Joint hypocenter determination, station corrections, hypocentroid, clustering, generalized inverse, normal equations, QR decomposition, Singular Value Decomposition, least squares, Loma Prieta, Nazca plate, New Madrid seismic zone, Campi Flegrei, Northridge.

Abstract: The joint hypocenter determination (JHD) technique is a relatively simple but efficient way to account for lateral velocity variations not included in the one-dimensional velocity models used to locate seismic events. The basic idea behind the technique is the simultaneous location of a group of events and the determination of a common set of station corrections. Under appropriate conditions the station corrections absorb the unmodeled velocity variations, thus improving the locations of the events. From the experience gained so far it appears that JHD always improves relative locations, while absolute locations may or may not be affected by some amount of error, depending on the nature of the lateral velocity variations. When they are very large, as when sedimentary basins and crystalline rocks are juxtaposed, the JHD locations may be affected by a systematic shift from the true locations. Here we discuss in detail four data sets which show different aspects of the JHD analysis. One is a synthetic data set based on actual data from the flat portion of the Nazca plate recorded in the Andean foreland in Argentina. These data are used to show the effect on location of exact and approximate implementations of the JHD technique. The other data sets come from the Loma Prieta, California, mainshock-aftershock sequence, from the New Madrid, central U.S.A., seismic zone, and from the Campi

C.H. Thurber and N. Rabinowitz (eds.), Advances in Seismic Event Location, 163–204.

Flegrei, a volcanic area in Italy. The Loma Prieta and New Madrid data illustrate the improvements in absolute and relative locations that can be achieved when using JHD, while the Campi Flegrei data show the significant effect of complicated velocity variations on earthquake location. Finally, analyses of actual and synthetic data show that the JHD station corrections can be used for a semi-quantitative assessment of the 3-D lateral velocity variations under a local network.

1. INTRODUCTION

In spite of much effort devoted to the subject, the determination of reliable earthquake locations is still a matter of substantial research. The routine computation of locations may be affected by at least two sources of error. One is an inadequate distribution of recording stations, with typical examples given by events recorded at regional and/or teleseismic distances, off-shore events, and events recorded by sparse local networks. The other possible source of error is the presence of significant lateral variations in the earth that are not included in the one-dimensional (1-D) velocity models routinely used. The problems introduced by inadequate networks can be solved, in principle, by increasing the number and type of stations, including ocean-bottom seismometers. The model errors, however, are more difficult to approach, and in some cases they are more serious than those introduced by the geometry of the network recording the events. The question is, then, how can these errors be minimized. Before addressing this question however, we should ask (a) what is the desired level of accuracy in earthquake location, and (b) how can the accuracy be assessed. The first question is directly related to the kind of data being analyzed. Here we will concentrate on data collected with networks having apertures of a few hundreds of kilometers at most. It is usually only in this case that earthquake epicenters and depths can be estimated with errors less than a few kilometers. If one is interested in studying subducting slabs, systematic location errors of a few kilometers may not be a serious problem, as there is much more uncertainty in the thermal structure and mineralogy of the slabs. However, if the goal of seismic monitoring is to image hidden active faults in an urban environment, such as the Los Angeles, U.S.A., area, then the location errors must be less than one kilometer, an accuracy that can presently be achieved under appropriate conditions.

It must be recognized, however, that the actual accuracy with which earthquakes are located cannot be determined using information derived from the location process itself when standard location programs are used.

The formal errors generated by the processing software only give a relative measure of the quality of the computed location, and can be used to compare the quality of results obtained under approximately the same conditions regarding the number of stations and type of arrivals used. Estimating true location errors is obviously a more difficult problem, because when the quality of the data is good, one of the most important sources of error is the model error, which is generally unknown. The introduction of techniques to simultaneously solve for event locations and 3-D velocity model addressed this problem, but reliable locations cannot be assured unless a number of conditions regarding the distribution of stations and events are satisfied.

An alternative, approximate approach is to determine event locations and station corrections simultaneously. This technique, known as joint hypocenter determination (JHD), is successful because under appropriate circumstances the station corrections account for the lateral velocity variations neglected in the 1-D model used to locate the events. This capability of the station corrections is well known, but a study of the actual relation between corrections and velocity variations was lacking until recently. It is possible to show that for clustered events there is a close relation between the equations for the joint earthquake location and 3-D velocity inversion and the JHD equations (Pujol, 1995). Unless *a priori* information is available or *a priori* assumptions are made, clustering must be avoided when using the JHD technique because in such a case the solution becomes nonunique. Therefore, the events to be located should be selected with this condition in mind.

Although the principles behind the JHD technique appear straightforward, there are a number of practical questions that need to be considered. One of them is the assessment of the quality of the joint locations. The analysis by the author of several data sets shows that the JHD locations may be quite different from the corresponding single-event locations and that the magnitudes of the station corrections are much larger than the averaged residuals referred to below. Therefore, it is reasonable to ask whether the JHD results are correct, or whether they are numerical artifacts. The evidence accumulated so far indicates that the large station corrections are not artifacts and that they represent actual velocity variations. Unfortunately, it has also been found that in the presence of large systematic velocity variations the JHD locations, although improved in a relative sense with respect to the single-event locations, may be affected in their absolute locations. As JHD does not provide an independent assessment of the quality of the locations, the user of the technique must watch for possible signs of systematic mislocations, such as a combination of large station corrections and unusually large systematic differences between JHD and single-event locations (the latter are unlikely without the former). In cases like this, establishing the quality of the JHD locations requires substantial additional effort, which includes a simultaneous event location and 3-D velocity inversion and analysis of synthetic data.

A second question that needs to be addressed relates to the use of approximate joint locations techniques. In their original formulations, the joint epicentral and joint hypocenter determination techniques (JED and JHD) (Douglas, 1967; Dewey, 1971, 1972) required large memory allocation and computing time, so that with the computers available then only small data sets could be processed. For this reason, an approximate technique was introduced (Frohlich, 1979). The basis of the technique is the individual location of the events, and the computation of each station correction as the average of the arrival time residuals for that station and for all the events. This approximation is still widely used, but as shown here, its effectiveness is questionable for local network data.

Finally, the importance of the velocity model used to relocate the events should be discussed. With few exceptions, the velocity models used to locate events individually and jointly are one-dimensional. The question here is whether the velocity model should be optimized when doing joint location, as some authors have suggested. As examples given here show, it appears that this optimization may not be required.

2. THE JOINT HYPOCENTER DETERMINATION TECHNIQUE

2.1 Statement of the problem

Let us consider the problem of determining N station corrections and the hypocenters of M earthquakes recorded by all or some of N stations. Let us assume for a moment that only P-wave first arrivals have been recorded. The only difference between single-event location and JHD is the occurrence in the latter of a correction term for each of the N stations. Therefore, the JHD equation for the j-th earthquake and the i-th station is written as

$$w_{ij}\, r_{ij} = w_{ij}\left(dT_j + \frac{\partial t}{\partial x_j} dx_j + \frac{\partial t}{\partial y_j} dy_j + \frac{\partial t}{\partial z_j} dz_j + ds_i \right); \quad i=1,N;\ j=1,M \tag{1}$$

where

$$r_{ij} = t_{ij}^{o} - (T_j + \tau_{ij} + s_i) \tag{2}$$

$t_{ij}{}^{o}$ is the observed arrival time, and the term in parentheses is the computed arrival time. T_j is the initial estimate of origin time, τ_{ij} is the travel time computed for the initial estimates (x_j,y_j,z_j) of the hypocentral coordinates, and s_i is the station correction. The w_{ij} are quality weights ($\leq$ 1). If the station did not record the earthquake it will be assumed that there is an Equation 1 with weight equal to zero. The dT_j, dx_j, dy_j, dz_j, and ds_i are adjustment terms to be determined.

Note that Equation 2 shows that there is a trade-off between origin time and station corrections. If a constant is added to all the T_j's and the same constant is subtracted from all the s_i's, the term in parentheses in (2), and r_{ij} as well, will remain unchanged. Thus, there is an intrinsic non-uniqueness in the station corrections. To handle this situation Douglas (1967) suggested fixing the value of one of the corrections, or assuming that the sum of all the corrections is zero.

Equation 1 can be written in matrix form as:

$$\mathbf{W}_j \, \mathbf{A}_j \, \mathbf{dx}_j + \mathbf{W}_j \, \mathbf{ds} = \mathbf{W}_j \, \mathbf{r}_j \; ; \; j=1,M \tag{3}$$

where $\mathbf{r}_j$ is the vector of residuals r_{ij}, $\mathbf{A}_j$ is an N x 4 matrix of partial derivatives, $\mathbf{dx}_j$ is the vector of origin time and hypocenter adjustments, $\mathbf{W}_j$ is an N x N matrix with at most one nonzero entry per row and column, which is equal to the weight for the corresponding station, and $\mathbf{ds}$ is the vector of station correction adjustments.

Incorporating other types of arrivals can be done as follows. For example, assume that S-wave arrivals have been recorded by N' stations (this number is not always equal to the number N of stations that recorded the P-wave arrivals). One efficient way to proceed is to append the corresponding residuals to those in Equation 1. Then the matrix $\mathbf{W}_j \, \mathbf{A}_j$ in Equation 3 will be (N + N') x 4, $\mathbf{W}_j$ will be (N + N') x (N + N'), and $\mathbf{ds}$ will have N + N' elements, the first N corresponding to the P-wave corrections and the last N' corresponding to the S-wave corrections. Any other arrivals that might have been recorded can be added in the same way. With this arrangement Equations 1 and 3 represent the most general case, making it unnecessary to specify the nature of the arrivals involved.

Equation 3 represents M coupled matrix equations with M + 1 unknown vectors: the $\mathbf{dx}_j$'s and $\mathbf{ds}$. To solve the JHD problem all the unknowns must be determined simultaneously. To appreciate the magnitude of the numerical aspects of this task, Equation 3 will be rewritten as a single partitioned-matrix equation (Noble and Daniel, 1977)

$$\begin{bmatrix} \mathbf{W}_1\mathbf{A}_1 & \mathbf{O} & \cdots & \mathbf{O} & \mathbf{W}_1 \\ \mathbf{O} & \mathbf{W}_2\mathbf{A}_2 & \cdots & \mathbf{O} & \mathbf{W}_2 \\ \vdots & \vdots & \ddots & \vdots & \vdots \\ \mathbf{O} & \mathbf{O} & \cdots & \mathbf{W}_M\mathbf{A}_M & \mathbf{W}_M \end{bmatrix} \begin{bmatrix} \mathbf{dx}_1 \\ \mathbf{dx}_2 \\ \vdots \\ \mathbf{dx}_M \\ \mathbf{ds} \end{bmatrix} = \begin{bmatrix} \mathbf{W}_1\mathbf{r}_1 \\ \mathbf{W}_2\mathbf{r}_2 \\ \vdots \\ \vdots \\ \mathbf{W}_M\mathbf{r}_M \end{bmatrix} \tag{4}$$

where $\mathbf{O}$ represents the zero matrix.

Equation 4 can be rewritten as

$$\mathbf{G}\,\mathbf{y} = \mathbf{b} \tag{5}$$

where $\mathbf{G}$ is the matrix of weighted partial derivatives, and $\mathbf{y}$ and $\mathbf{b}$ are the vectors of unknowns and weighted residuals, respectively. Matrix $\mathbf{G}$ is (M N) x (4 M + N), which means that it may become very large when the number of events and stations is large. However, solving the JHD problem does not require storage of the full matrix $\mathbf{G}$, as several authors have introduced solution techniques that achieve significant reductions in the size of the largest matrix that must be stored. These techniques are discussed next.

2.2 Solving the JHD problem

The JHD problem has been solved efficiently and without approximations by several authors (Herrmann et al. 1981; Smith 1982; Pavlis and Booker, 1983; Pujol, 1988). An approximate solution (Frohlich, 1979) has also been proposed. The various approaches are described below.

2.2.1 Least squares solution (normal equations)

Provided that the system 5 is overdetermined (i.e., the number of equations is larger than the number of unknowns and $\mathbf{G}$ has no zero singular values), a simple way to solve Equation 5 is to solve the least-squares normal equations

$$\mathbf{G}^T\,\mathbf{G}\,\mathbf{y} = \mathbf{G}^T\,\mathbf{b} \tag{6}$$

where the superscript T indicates matrix transposition. After performing the matrix operations indicated in Equation 6 we obtain

$$\begin{bmatrix} \mathbf{A'}_1^T\mathbf{A'}_1 & \mathbf{O} & \cdots & \mathbf{O} & \mathbf{A'}_1^T\mathbf{W}_1 \\ \mathbf{O} & \mathbf{A'}_2^T\mathbf{A'}_2 & \cdots & \mathbf{O} & \mathbf{A'}_2^T\mathbf{W}_2 \\ \vdots & \vdots & \ddots & \vdots & \vdots \\ \mathbf{O} & \mathbf{O} & \cdots & \mathbf{A'}_M^T\mathbf{A'}_M & \mathbf{A'}_M^T\mathbf{W}_M \\ \mathbf{W}_1^T\mathbf{A'}_1 & \mathbf{W}_2^T\mathbf{A'}_2 & \cdots & \mathbf{W}_M^T\mathbf{A'}_M & \sum_{j=1}^{M} \mathbf{W}_j^T\mathbf{W}_j \end{bmatrix} \begin{bmatrix} \mathbf{dx}_1 \\ \mathbf{dx}_2 \\ \vdots \\ \mathbf{dx}_M \\ \mathbf{ds} \end{bmatrix} = \begin{bmatrix} \mathbf{A'}_1^T\mathbf{W}_1\mathbf{r}_1 \\ \mathbf{A'}_2^T\mathbf{W}_2\mathbf{r}_2 \\ \vdots \\ \mathbf{A'}_M^T\mathbf{W}_M\mathbf{r}_M \\ \sum_{j=1}^{M} \mathbf{W}_j^T\mathbf{W}_j\mathbf{r}_j \end{bmatrix} \tag{7}$$

To simplify the notation the matrices $\mathbf{W}_j\mathbf{A}_j$ have been written as $\mathbf{A'}_j$. Equation 7 shows that the JHD problem has been reduced to the solution of a (4 M + N) x (4 M + N) linear system. This formulation, however, has the disadvantage that the number of equations, and unknowns, is proportional to the number of earthquakes. To reduce the size of the problem even further, Herrmann et al. (1981) proceeded as follows. Performing the operations indicated in Equation 7 and using $\mathbf{A}_j' = \mathbf{W}_j\ \mathbf{A}_j$ gives

$$\mathbf{A}_j^T\mathbf{W}_j^T\mathbf{W}_j\mathbf{A}_j\mathbf{dx}_j + \mathbf{A}_j^T\mathbf{W}_j^T\mathbf{W}_j\mathbf{ds} = \mathbf{A}_j^T\mathbf{W}_j^T\mathbf{W}_j\ \mathbf{r}_j;\ \ j=1,M \tag{8}$$

$$\sum_{j=1}^{M}\mathbf{W}_j^T\mathbf{W}_j\mathbf{A}_j\mathbf{dx}_j + \left(\sum_{j=1}^{M}\mathbf{W}_j^T\mathbf{W}_j\right)\mathbf{ds} = \sum_{j=1}^{M}\mathbf{W}_j^T\mathbf{W}_j\mathbf{r}_j \tag{9}$$

Equations 8 and 9 constitute a system of coupled equations. Herrmann et al. (1981) solved it by first solving (8) for $\mathbf{dx}_j$, which gives

$$\mathbf{dx}_j = (\mathbf{A}_j^T\mathbf{W}_j^T\mathbf{W}_j\mathbf{A}_j)^{-1}\ \mathbf{A}_j^T\mathbf{W}_j^T\mathbf{W}_j\ (\mathbf{r}_j - \mathbf{ds});\ \ j=1,M \tag{10}$$

and then introducing this expression in Equation 9. The result is an equation in which only **ds** is unknown. Once **ds** is determined, Equation 10 is used to update the $\mathbf{dx}_j$'s. In this approach the largest matrix is N x N.

2.2.2 Householder's QR decomposition solution

This solution was introduced by Smith (1982). Given an arbitrary n x m matrix **A**, it can be written as the product

$$\mathbf{A} = \mathbf{Q}\ \mathbf{R} \tag{11}$$

Q is an n x n orthogonal matrix, which means that

$$\mathbf{Q}^T \mathbf{Q} = \mathbf{Q} \mathbf{Q}^T = \mathbf{I}_n \tag{12}$$

where $\mathbf{I}_n$ indicates the n x n identity matrix. $\mathbf{R}$ is an n x m upper triangular matrix. If n ≥ m, all the elements below the diagonal elements R_{11}, R_{22}, ..., R_{mm} of $\mathbf{R}$ are equal to zero (Noble and Daniel, 1977). Note that Smith (1982) has a slightly different definition, with $\mathbf{Q}^T$ instead of the $\mathbf{Q}$ used here. Because of the special structure of $\mathbf{R}$, the last n - m columns of $\mathbf{Q}$ are multiplied by zeros, so that $\mathbf{Q}$ and $\mathbf{R}$ can be replaced by n x m and m x m matrices $\mathbf{Q}_0$ and $\mathbf{R}_0$.

The QR decomposition is very useful to solve least squares problems. Given the overdetermined system

$$\mathbf{A}\,\mathbf{x} = \mathbf{c} \tag{13}$$

the normal equations solution is obtained by solving

$$\mathbf{A}^T \mathbf{A}\,\mathbf{x} = \mathbf{A}^T \mathbf{c} \tag{14}$$

This method of solution has the disadvantage that it squares the condition number of $\mathbf{A}$, which increases the likelihood of numerical instability or errors (see below). One way to avoid this difficulty is to use the QR decomposition, in which case the solution is obtained from

$$\mathbf{R}_0\,\mathbf{x} = \mathbf{Q}_0\,\mathbf{c} \tag{15}$$

provided that $\mathbf{A}$ is not singular (Noble and Daniel, 1977). Because $\mathbf{R}_0$ is upper triangular, it is simple to solve for $\mathbf{x}$.

Smith (1982) started with equations similar to (3) and (4). Let

$$\mathbf{W}_j\,\mathbf{A}_j = \mathbf{Q}_j\,\mathbf{R}_j \tag{16}$$

Now multiply Equation 3 by $\mathbf{Q}_j^T$, and use Equations 16 and 12. This gives

$$\mathbf{R}_j\,\mathbf{dx}_j + \mathbf{Q}_j^T\,\mathbf{W}_j\,\mathbf{ds} = \mathbf{Q}_j^T\,\mathbf{W}_j\,\mathbf{r}_j;\ \ j=1,M \tag{17}$$

Note that the last N - 4 rows of $\mathbf{R}_j$ contribute a zero term to Equation 17, which means that it can be decomposed into two sets of equations, one with four rows and involving $\mathbf{dx}_j$, $\mathbf{ds}$ and $\mathbf{r}_j$, and one with N - 4 rows involving $\mathbf{ds}$ and $\mathbf{r}_j$ only. This means that Equation 4 can be replaced by a new partitioned system in which all the matrices with 4 rows are placed first, followed by all the matrices with N - 4 rows. The resulting system is

$$\begin{bmatrix} \mathbf{Q}_1{}^T\mathbf{A'}_1 & \mathbf{O} & \cdots & \mathbf{O} & \left(\mathbf{Q}_1{}^T\mathbf{W}_1\right)_4 \\ \mathbf{O} & \mathbf{Q}_2{}^T\mathbf{A'}_2 & \cdots & \mathbf{O} & \left(\mathbf{Q}_2{}^T\mathbf{W}_2\right)_4 \\ \vdots & \vdots & \ddots & \vdots & \vdots \\ \mathbf{O} & \mathbf{O} & \cdots & \mathbf{Q}_M{}^T\mathbf{A'}_M & \left(\mathbf{Q}_M{}^T\mathbf{W}_M\right)_4 \\ \mathbf{O} & \mathbf{O} & \cdots & \mathbf{O} & \left(\mathbf{Q}_1{}^T\mathbf{W}_1\right)_{N\pm4} \\ \mathbf{O} & \mathbf{O} & \cdots & \mathbf{O} & \left(\mathbf{Q}_2{}^T\mathbf{W}_2\right)_{N\pm4} \\ \vdots & \vdots & \ddots & \vdots & \vdots \\ \mathbf{O} & \mathbf{O} & \cdots & \mathbf{O} & \left(\mathbf{Q}_M{}^T\mathbf{W}_M\right)_{N\pm4} \end{bmatrix} \begin{bmatrix} \mathbf{dx}_1 \\ \mathbf{dx}_2 \\ \vdots \\ \mathbf{dx}_M \\ \mathbf{ds} \end{bmatrix} = \begin{bmatrix} \left(\mathbf{Q}_1{}^T\mathbf{r'}_1\right)_4 \\ \left(\mathbf{Q}_2{}^T\mathbf{r'}_2\right)_4 \\ \vdots \\ \left(\mathbf{Q}_M{}^T\mathbf{r'}_M\right)_4 \\ \left(\mathbf{Q}_1{}^T\mathbf{r'}_1\right)_{N\pm4} \\ \left(\mathbf{Q}_2{}^T\mathbf{r'}_2\right)_{N\pm4} \\ \vdots \\ \left(\mathbf{Q}_M{}^T\mathbf{r'}_M\right)_{N\pm4} \end{bmatrix} \tag{18}$$

where $\mathbf{A}_j' = \mathbf{W}_j \mathbf{A}_j$, $\mathbf{r}_j' = \mathbf{W}_j \mathbf{r}_j$, and the subscripts 4 and N - 4 indicate the number of rows in the matrix.

The final step in Smith's (1982) approach is to apply the QR decomposition to the submatrix in Equation 18 made of all the matrices with N - 4 rows. After that is done the whole matrix is in upper triangular form and solving for **ds** is straightforward. As only N - 1 station corrections can be determined independently, Smith (1982) set the adjustment to the N-th correction equal to minus the sum of the first N - 1 adjustments. The largest stored matrix in this approach is $(\Sigma\, n_j - 4\, M) \times N$, where n_j is the number of stations that actually recorded the j-th event and the sum is over the M events. This matrix is considerably larger than the largest matrices required by the other exact approaches described here.

2.2.3 Generalized inverse solution, I

Pavlis and Booker (1983) solved the JHD problem with the help of the singular value decomposition (SVD) of a matrix. Given an arbitrary n x m matrix **G**, it can be written as the product of three matrices

$$\mathbf{G} = \mathbf{U}\, \mathbf{\Lambda}\, \mathbf{V}^T \tag{19}$$

where **U** is an n x n matrix having as columns the eigenvectors of $\mathbf{G}\,\mathbf{G}^T$; **V** is an m x m matrix having as columns the eigenvectors of $\mathbf{G}^T\,\mathbf{G}$, and $\mathbf{\Lambda}$ is an n x m diagonal matrix whose elements $(\Lambda)_{ii}=\Lambda_i$, known as the singular values of **G**, are equal to the positive square roots of the nonzero eigenvalues of $\mathbf{G}\mathbf{G}^T$ and $\mathbf{G}^T\mathbf{G}$ (Forsythe et al., 1977; Aki and Richards, 1980). The number of nonzero singular values cannot be larger than the minimum of m and n. Because **U** and **V** are made of eigenvectors, they are orthogonal matrices, so that

$$\mathbf{U}\mathbf{U}^T = \mathbf{I}_n \quad \text{and} \quad \mathbf{V}\mathbf{V}^T = \mathbf{I}_m \tag{20}$$

Furthermore, if the number of nonzero singular values is p, then matrices **U** and **V** can be written as

$$\mathbf{U} = (\mathbf{U}_p \mid \mathbf{U}_0) \quad \text{and} \quad \mathbf{V} = (\mathbf{V}_p \mid \mathbf{V}_0) \tag{21}$$

where the vertical bars indicate matrix partitioning, and the subscripts p and 0 indicate that the vectors used to generate the corresponding matrices come from nonzero and zero singular values. Using the definition of Λ and the fact that the elements of **U** are eigenvectors, it is easy to show that

$$\mathbf{G} = \mathbf{U}_p \, \Lambda_p \, \mathbf{V}_p^T \quad \text{and} \quad \mathbf{U}_0^T \, \mathbf{U}_p = \mathbf{O} \tag{22a,b}$$

where Λ_p is the p x p diagonal matrix generated with the p nonzero singular values. Using these two equations we find that

$$\mathbf{U}_0^T \, \mathbf{G} = \mathbf{U}_0^T \, \mathbf{U}_p \, \Lambda_p \, \mathbf{V}_p^T = \mathbf{O} \tag{23}$$

This equation is of fundamental importance because it allows the separation of the station corrections from the earthquake parameters. Let **G** be $\mathbf{W}_j\mathbf{A}_j$. Then $p \leq 4$. When $p < 4$, the corresponding event location is nonunique. Events affected by this problem should not be used in the joint location. Therefore, in the following p is equal to 4. When Equation 3 is multiplied on the left by the $\mathbf{U}_0$ matrix for the j-th event, $\mathbf{U}_{j0}$, the first term becomes zero and the equation takes a simple form:

$$\mathbf{U}_{j0}^T \, \mathbf{W}_j \, \mathbf{ds} = \mathbf{U}_{j0}^T \, \mathbf{W}_j \mathbf{v}_j; \; j=1,M \tag{24}$$

It must be noted that the fact that $\mathbf{A}_j$ does not appear in Equation 24 does not mean that the station corrections are unbiased by hypocenter location errors. $\mathbf{A}_j$ enters Equation 24 implicitly via $\mathbf{U}_{j0}$ and any location error affecting the former will also affect the latter.

The M equations implied by Equation 24 can be written as a single matrix equation:

$$\mathbf{S} \, \mathbf{ds} = \bar{\mathbf{r}} \tag{25}$$

where

$$\mathbf{S}=\begin{bmatrix} \mathbf{U}_{10}^{\mathrm{T}}\mathbf{W}_1 \\ \mathbf{U}_{20}^{\mathrm{T}}\mathbf{W}_2 \\ \vdots \\ \mathbf{U}_{\mathrm{M}0}^{\mathrm{T}}\mathbf{W}_{\mathrm{M}} \end{bmatrix} \tag{26}$$

and

$$\bar{\mathbf{r}}=\begin{bmatrix} \mathbf{U}_{10}^{\mathrm{T}}\mathbf{W}_1\mathbf{r}_1 \\ \mathbf{U}_{20}^{\mathrm{T}}\mathbf{W}_2\mathbf{r}_2 \\ \vdots \\ \mathbf{U}_{\mathrm{M}0}^{\mathrm{T}}\mathbf{W}_{\mathrm{M}}\mathbf{r}_{\mathrm{M}} \end{bmatrix} \tag{27}$$

Equation 25 is similar to Equation 19 of Pavlis and Booker (1983). Matrix **S** can be very large, as it has M x (N - 4) rows and N columns. Pavlis and Booker solved this equation by computing the generalized inverse of **S** using a modification of the QR decomposition. After this step, the events are relocated using the station corrections and a conventional single-event location procedure. After the events have been relocated, a new iteration for the station corrections is started.

2.2.4 Generalized inverse solution, II

Pujol (1988) modified the method of Pavlis and Booker (1983) by solving Equation 25 by least squares:

$$\mathbf{S}^{\mathrm{T}}\,\mathbf{S}\,\mathbf{ds} = \mathbf{S}^{\mathrm{T}}\,\bar{\mathbf{r}} \tag{28}$$

In addition, the matrices $\mathbf{W}_j$ were assumed to be diagonal. There are two reasons for this. One is computational, and is based on the fact that if $\mathbf{W}_j$ is diagonal it can be treated as a vector, thus reducing computational time. The other reason is that some theoretical results have been derived under this assumption. In practice, $\mathbf{W}_j$ is not necessarily diagonal because for each earthquake the stations may not be listed in the same order, but $\mathbf{W}_j$ is made diagonal by a reshuffle of indices in the JHD program.

Equation 28 is solved by computing the generalized inverse $\mathbf{Z}^{\dagger}$ of $\mathbf{S}^{\mathrm{T}}\mathbf{S}$:

$$\mathbf{ds} = \mathbf{Z}^{\dagger}\,\mathbf{S}^{\mathrm{T}}\,\bar{\mathbf{r}} \tag{29}$$

Making use again of the properties of partitioned matrices, we can write Equation 28 as

$$\left(\sum_{j=1}^{M} \mathbf{W}_j{}^T \mathbf{U}_{j0} \mathbf{U}_{j0}{}^T \mathbf{W}_j\right) \mathbf{ds} = \sum_{j=1}^{M} \mathbf{W}_j{}^T \mathbf{U}_{j0} \mathbf{U}_{j0}{}^T \mathbf{W}_j \mathbf{r}_j = \begin{bmatrix} \mathbf{U}_{10}{}^T \mathbf{W}_1 \mathbf{r}_1 \\ \mathbf{U}_{20}{}^T \mathbf{W}_2 \mathbf{r}_2 \\ \vdots \\ \mathbf{U}_{M0}{}^T \mathbf{W}_M \mathbf{r}_M \end{bmatrix} \quad (30)$$

Because $\mathbf{W}_j$ is diagonal, it is equal to its transpose. In practice, $\mathbf{Z}^\dagger$ is the generalized inverse of the matrix on the left-hand side of Equation 30, and is applied to the vector on the right-hand side of that equation.

For an arbitrary matrix **G** with singular value decomposition given by (22a), its generalized inverse $\mathbf{G}^\dagger$ is given by

$$\mathbf{G}^\dagger = \mathbf{V}_p \, \Lambda_p{}^{-1} \, \mathbf{U}_p{}^T \quad (31)$$

where $\Lambda_p{}^{-1}$ is a diagonal matrix with

$$(\Lambda_p{}^{-1})_{ii} = \Lambda_i{}^{-1} \quad (32)$$

(Forsythe et al., 1977; Noble and Daniel, 1977; Aki and Richards, 1980).

Once **ds** has been computed, the $\mathbf{dx}_j$'s are determined from (3)

$$\mathbf{W}_j \, \mathbf{A}_j \, \mathbf{dx}_j = \mathbf{W}_j \, \mathbf{r}_j - \mathbf{W}_j \, \text{ds}; \; j=1,M \quad (33)$$

Equation 33 is solved by damped least squares, with the value of the damping parameter reduced after each iteration. After the origin time, hypocenter coordinates and station corrections have been updated a new iteration is started. To distinguish this version of the technique from others, it is indicated with JHD†.

In general, one of the concerns when solving a system of equations by least squares is that this technique squares the condition number of the original matrix, which is given by the ratio of the largest to the smallest singular value. A large condition number increases the likelihood of numerical errors in the computed solution (Forsythe et al., 1977). For this reason the numerical aspects of the JHD† solution were extensively studied, with the conclusion that because of the special properties of the matrices $\mathbf{U}_{j0}$, Equation 30 is numerically well behaved. Because these matters are discussed in detail in Pujol (1988), here only a summary of the main results will be presented.

a) One of the singular values of $\mathbf{S}^T\mathbf{S}$ is identically equal to zero. This zero value arises from the presence of a column of ones (and possibly zeros) in matrices $\mathbf{A}_j$. If the earthquakes form a point cluster, three additional singular values will also be zero. The fact that the generalized inverse solution automatically takes care of the presence of zero singular values makes it unnecessary to impose external constraints (e.g., fixing the value of one station correction, using a master event).

b) The generalized inverse solution is a minimum-length solution (Forsythe et al., 1977). As a consequence, the solution **ds** computed using Equation 29 has zero mean value. Therefore, if the initial estimates of the station corrections have a given mean value, then the final values of the corrections will have the same mean value. If the initial estimates are all equal to zero (or sum to zero), then the sum of the final values of the corrections will also be zero. Recall that this is one of the conditions proposed by Douglas (1967) to constrain the JED solution.

c) The largest singular value of $\mathbf{S}^T\mathbf{S}$ cannot exceed M, the total number of earthquakes. In addition, the smallest singular value of the sum of the first k terms in the left-hand side of Equation 30 cannot be smaller than the smallest singular value of the sum of the first k - 1 terms. This means that the smallest singular value cannot decrease as the left sum in Equation 30 is computed.

These results show that as long as the events are not too tightly clustered, the JHD† solution is not affected by intrinsic numerical errors. Furthermore, the condition that the average value of the station corrections should remain constant through all the iterations constitutes a strong test to assess the numerical stability of the solution. In addition, the technique is very efficient, allowing the simultaneous processing of thousands of events in a time comparable to that required to locate the events individually.

2.3 Approximate solution

The computation of approximate JHD station corrections is based on the following procedure:

1) Locate each earthquake individually.

2) For the i-th station average the residuals r_{ij} for all the events.

3) Relocate the events using the average residuals as station corrections.

4) Optionally, iterate steps 1-3.

This is the basis of Frohlich's (1979) method, which also includes the constraint that the sum of the average residuals be equal to zero. Unlike the exact least-squares solution, which takes into account the coupling of Equations 8 and 9, the approximate solution ignores it. Step (1) is equivalent to solving Equation 8 with $\mathbf{ds} = \mathbf{0}$, or solving Equation 10 with $\mathbf{ds}$ estimated in the previous iteration. Step (2) is equivalent to solving Equation 9 with the first term on the left-hand side ignored. From that equation and using the fact that $\mathbf{W}_j^T\mathbf{W}_j$ is a diagonal matrix (whether $\mathbf{W}_j$ is diagonal or not) with the i-th diagonal entry equal to w_{ij}^2, the approximation a_i to the i-th component of $\mathbf{ds}$ is given by

$$ds_i \approx a_i = \frac{1}{\left(\sum_{j=1}^{M} w_{ij}^2\right)} \sum_{j=1}^{M} w_{ij}^2 \, r_{ij} \qquad (34)$$

Equation 34 shows that ds_i is approximated by a weighted average residual.

Two questions arise in connection with this approximate procedure. One, when, if ever, is it a good approximation? Two, if it is not, what is the effect of the approximation on the computed earthquake locations?

These questions can be addressed from two perspectives. One is purely mathematical, and is based on the comparison of Equations 30 and 34. The two equations would be equivalent if $\mathbf{U}_{j0}\mathbf{U}_{j0}^T = \mathbf{I}$ for all the earthquakes, but it has been shown (Pujol, 1988) that this equality cannot be satisfied exactly, and although it may be approximately valid for large values of N and M, there is no way to predict how large they should be for a reasonable approximation.

The second perspective is more intuitive. The basic idea behind single-event location techniques is to minimize arrival time residuals, not to minimize location errors. If the velocity model used to locate the events is incorrect, two situations may arise. If the velocity errors have an approximately 3-D random distribution (on a scale smaller than the interstation distances), then it is likely that the computed locations will be close to the true locations, with the magnitude of the arrival time residuals reflecting the magnitude of the velocity variations. Then, depending on the distributions of earthquakes and stations, using a_i to estimate ds_i may be appropriate. Of course, it is not possible to know *a priori* when the errors in a velocity model will be random. In addition, the existing evidence seems to indicate that velocity errors are seldom random (although this appears to be the case in the Loma Prieta area, discussed below). The most common occurrence is the juxtaposition of low and high velocities, corresponding, for example, to areas with sedimentary and metamorphic or igneous rocks. In these cases, the single event locations will be biased, and the computed time residuals will not be representative of the actual velocity errors.

The problem of bias in the single-event locations was recognized by Douglas (1967), who realized that at least some of the bias could be removed with the joint location technique that he introduced. However, the use of average residuals in lieu of JHD station corrections is widespread. Therefore, it is important to establish whether the average residuals and the JHD station corrections have similar values for a given data set, and, if not, what is the effect of each set of values on earthquake location. For the first task it is sufficient to compare the average residuals and station corrections computed for the same data set. For the second task synthetic data are required, as this is generally the only way to assure that the "true" locations are known. This analysis has been carried out in few instances. One of them is a study of earthquake location in subduction zones, with examples taken from the Alaskan Wadati-Benioff zone (McLaren and Frohlich, 1985). This study used synthetic arrival times computed for velocity models that included the lateral variations introduced by the subducting slab to demonstrate that ignoring these variations causes significant distortion in the shape of the Wadati-Benioff zone. Interestingly, when the average residuals were used to relocate a particular set of synthetic events, the resulting locations were "nearly indistinguishable from those determined by single event methods" (McLaren and Frohlich, 1985), although the RMS residuals were significantly reduced for most of the relocated events. On the other hand, Ratchkovsky et al. (1997) carried out an analysis of synthetic data for the same area, and found that the JHD† locations represent a significant improvement over the single-event locations.

Another synthetic data set that demonstrates a major difference between JHD corrections and average residuals is based on results of the relocation of events from the near-horizontal portion of the Nazca plate, which subducts under South America. The data were recorded in the Andean foreland in Argentina and are discussed in Pujol (1992) and references therein. The major conclusion of the JHD† analysis of these events, as well as crustal events from the same area, is that there are major lateral velocity variations in the upper crust that have a large distorting effect on the locations of Nazca plate events. The synthetic arrival times were generated for a velocity model with lateral velocity anomalies of -20%, 0% and 10% in the upper 10 km of the crust in a roughly west-east direction. The horizontal velocity boundaries

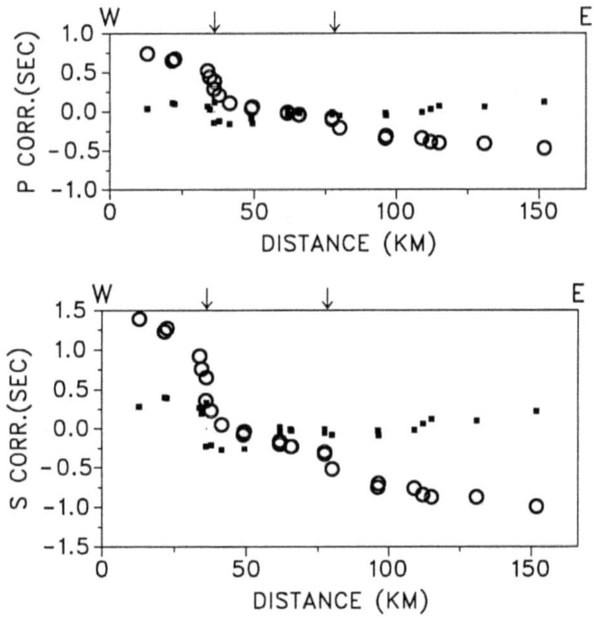

Figure 1. P- and S-wave station corrections (circles) computed by application of JHD† to synthetic arrival times generated for the locations shown in Figure 2. The two arrows indicate velocity variations from west to east: -20%, 0% (between arrows) and 10% in the upper 10 km of the crust. Below 10 km a 1-D model was used. Also shown are the average station residuals (squares), computed using Equation 34. Note the large range spanned by the corrections and their pattern, which indicates an eastern increase in velocity. In contrast, the average residuals are much smaller and from their pattern it is not possible to infer the nature of the velocity variations (modified from Pujol, 1992).

are shown in Figure 1, which also shows the JHD† station corrections obtained using the synthetic data. These corrections match the pattern of the actual corrections very well.

The locations of the events used to generate the synthetic data are shown in Figure 2. To simplify a visual comparison all the depths were set at 100 km. When the actual events were located using a single-event location technique and a 1-D velocity model, the locations showed a spurious dip to the west (Pujol, 1992), similar to that observed in Figure 2a. Application of JHD† (with the same 1-D velocity model) to the synthetic data produces locations with much less error, although a slight dip to the east has been introduced (Figure 2b). These results should be compared to those obtained when average residuals are used (Figure 2c). The following differences should be noted. (a) The average residuals are much smaller than the JHD† corrections (Figure 1). (b) JHD† correctly detects the eastward increase in velocity, as evidenced by increasingly negative corrections towards the east.

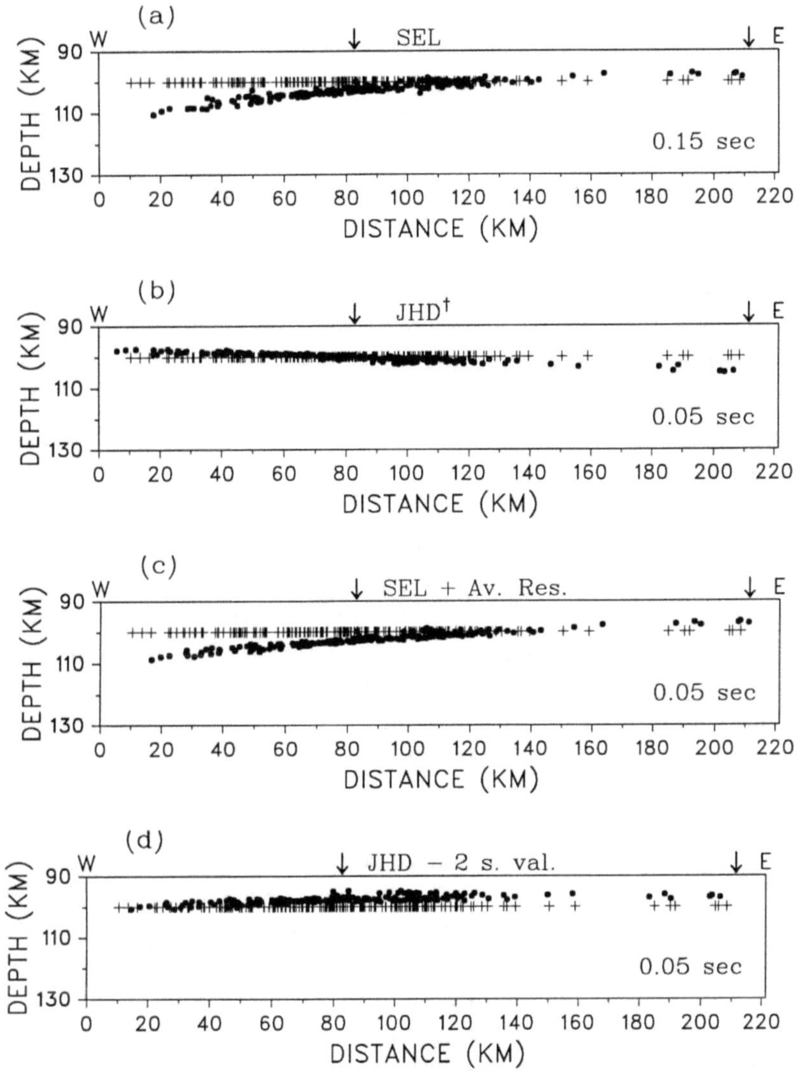

Figure 2. Initial locations of 199 events used to generate synthetic arrival times (crosses) and locations obtained using the synthetic times and different location techniques (dots), as follows. Single event location (SEL), JHD†, SEL with the average residuals shown in Figure 1, and JHD† with the contribution of the two smallest nonzero singular values cancelled in the computation of the generalized inverse $Z^\dagger$ (modified from Pujol, 1992). The number on the lower right corner of each plot is the average of the RMS residuals for all the events.

On the other hand, some of the average residuals increase to the east, which would be interpreted as a decrease in velocity. The behavior of the average residuals computed using the actual data is similar. (c) The locations obtained using the average residuals as station corrections are only slightly different from the single-event locations (Figure 2a and c). (d) Locating the events with the average residuals as station corrections reduces the individual RMS residuals to values similar to those obtained when using JHD†, which confirms that small RMS residuals do not necessarily mean well-located events (see, e.g., McLaren and Frohlich, 1985).

Several other comparisons similar to that described above have been carried out by the author. Some of the results are discussed below and some others can be found in Pujol (1992, 1996b). The main conclusion of these studies is that the average residuals do not appear to be an adequate substitute for the exact JHD station corrections for local network studies, and that the reduction in RMS residuals that accompanies the use of average residuals gives a misleading sense of improvement. The only case where the two sets of values are very close to each other is for the Loma Prieta area, discussed below.

2.4 Practical considerations

The JHD technique attempts to account for unmodeled velocity variations by assigning a common correction term to each station regardless of the distribution of events. The JHD equations can be applied, in principle, to events widely distributed under the network, but in this case the JHD locations may represent an improvement over single-event locations only if the velocity variations are restricted to the portions of the raypaths close to the stations (Viret et al., 1984). We know, however, that in many cases of interest the major sources of lateral velocity variation are large-scale, which means that large portions of the raypaths may be affected. In such cases the JHD technique will produce improved locations only if the distribution of events is such that each station correction represents approximately the same traveltime anomaly (generated by the difference between actual and model velocities) for all the events. This condition can be achieved if the events are relatively clustered, but for this kind of event distribution other problems arise, as discussed below. Early papers on joint location (e.g., Billington and Isacks, 1975; Cardwell and Isacks, 1976) make explicit reference to closely grouped hypocenters when discussing the assumptions of the JHD and JED techniques.

The problem of clustered events was recognized by Douglas (1967) in the context of teleseismic event location. This author noted that in this case the station corrections and traveltimes are virtually linearly dependent and indicated that the method could be adapted to obtain relative locations by fixing the location and origin time of a reference event. The problem was analyzed in detail by Jordan and Sverdrup (1981), who applied a number of ideas similar to those used by Pavlis and Booker (1983). Jordan and

Sverdrup (1981) introduced the concepts of hypocentroid of a cluster, which is the average location of the events in a cluster, and of cluster vectors, which are the deviations of the individual hypocenters with respect to the hypocentroid. An algorithm for the computation of these quantities was derived, but the station corrections, although explicitly considered in the formulation, were not computed. Rather, they were estimated using single-event travel-time residuals.

The problem of clustered events in the context of local data was discussed by Pavlis and Hokanson (1985) and Pujol (1988). The effect of reducing the size of a hypocentral volume is to reduce the size of the smallest nonzero singular values of $\mathbf{S}^T\mathbf{S}$ (Equation 28). If the singular values become so small that they must be taken as zero in the computation of $\mathbf{Z}^\dagger$ (see Equations 31 and 32), then it is well known that the solution $\mathbf{ds}$ becomes nonunique, which in turn may affect the quality of the locations. The synthetic data used earlier will be used to show this effect. The JHD† locations when two nonzero relatively small singular values were neglected in the computation of $\mathbf{Z}^\dagger$ are shown in Figure 2d. The difference from the initial locations is larger than when all the nonzero singular values were used, although the average RMS residuals are equal. The station corrections are different from those in Figure 1, with a considerably smaller range and a more irregular variation.

The actual Nazca plate events were also useful because they showed that the intuitive idea of a cluster as a distribution of events much smaller than the aperture of a network may not be applicable in the context of the JHD technique. When only the events under the network were jointly relocated, some of the singular values were so small that they had to be neglected when computing $\mathbf{Z}^\dagger$ (Pujol, 1992). Including events outside of the network, as done for the synthetic events of Figure 2, allows inclusion of all the singular values, but even in this case the condition number of $\mathbf{S}^T\mathbf{S}$ is relatively high (750), indicating that in some cases a distribution of events as wide as the network may be relatively clustered in a mathematical sense.

3. APPLICATIONS OF THE JHD† TECHNIQUE

The JHD† technique has been applied to numerous data sets recorded in a variety of tectonic environments, and in each case important insights regarding earthquake location and velocity variations have been gained. The following examples have been selected to show that the JHD technique can produce improved event locations (Loma Prieta and New Madrid seismic zone examples), but that there may be cases (Campi Flegrei example) where establishing the best set of locations is not straightforward.

3.1 Loma Prieta, California, mainshock-aftershock sequence

The magnitude 7.1 mainshock occurred on October 17, 1989, on the San Andreas fault, and was followed by thousands of aftershocks over a period of several months. As one of the largest earthquakes in the United States in a populated area, it was the subject of numerous and extensive studies. In this process, a number of questions arose regarding the actual location of the sequence (e.g., Eberhart-Phillips and Stuart, 1992), thus exemplifying the subtle points that must be considered when highly accurate locations are sought.

Figure 3 shows two sets of locations. One was released by the U.S. Geological Survey (U.S.G.S.) in 1992. These locations were computed using a velocity model and station corrections (delays) based on those determined by Dietz and Ellsworth (1990) (see Pujol, 1995, for details). These authors used a joint location technique and two velocity models, one for the area to the southwest of the San Andreas fault and one for the area to the northeast of it (Figure 4). The second set of locations was determined using JHD† (Pujol, 1995). The main difference between the two sets was a quasi-systematic SW shift of 1 km on average of the JHD† locations with respect to the U.S.G.S. locations. This difference may seem small, but is important for at least two reasons. One is the relation between seismicity and faulting, particularly in the vicinity of the intersection of the San Andreas and Sargent faults (Figure 3). The other reason was a puzzling discrepancy between the U.S.G.S. locations and the fault plane determined by inversion of geodetic data. This question is addressed below.

More recently, Dietz and Ellsworth (1997) relocated the events with a new velocity model (Figure 4) and station corrections. A comparison of mainshock locations determined by these and other authors is presented in Table 1. Inspection of this table shows that the new Dietz and Ellsworth and JHD† locations agree very well, with epicentral and depth differences of less than 0.4 km. On the other hand, the new location is shifted 0.8 km in a roughly SW direction with respect to the original Dietz and Ellsworth (1990) location. A similar shift applies to the other events in the sequence, in agreement with the shift observed by Pujol (1995).

To investigate the effect of the velocity model on the joint locations, the JHD† computations were repeated with a velocity model obtained by averaging the two models of Dietz and Ellsworth (1997) (Figure 4). Comparison of the old and new results show only minor differences in epicentral locations (0.13 km on average for 3542 events), negligible differences in station corrections (none exceeded 0.02 s in absolute value),

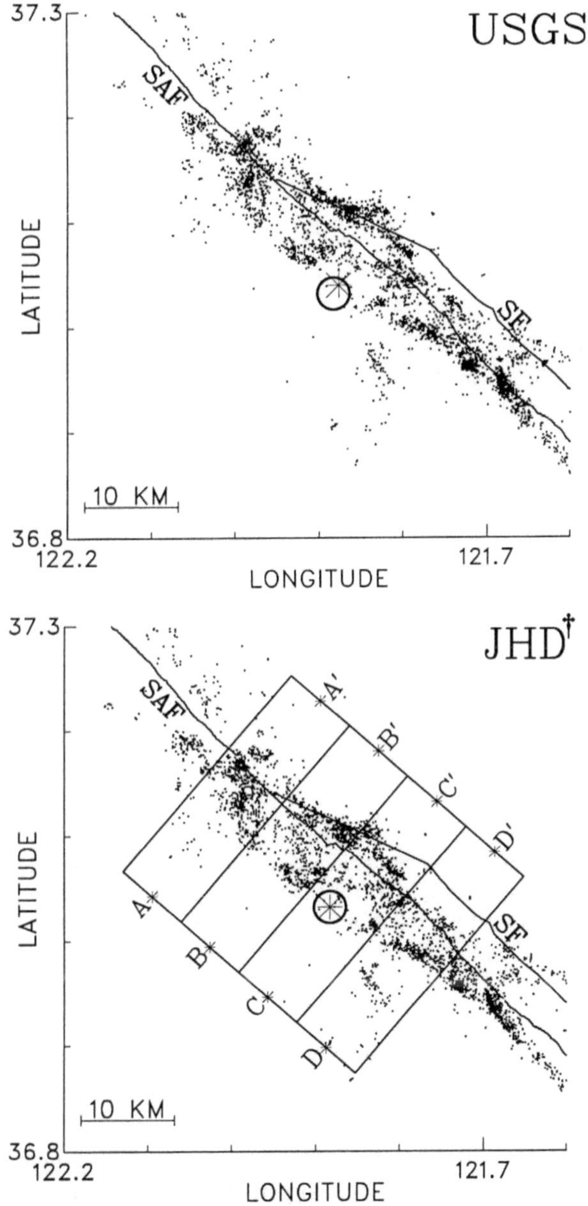

Figure 3. Epicenters of selected Loma Prieta events as determined by the U.S.G.S. (see text, 3648 events) and by using the JHD† technique (3666 events). The large asterisk represents the corresponding mainshock epicenter. The large circle is centered at the JHD† mainshock epicenter. Labels SAF and SF indicate the San Andreas and Sargent faults. The events within the boxes are plotted in cross section in Figure 6. Asterisks identify cross section end points. After Pujol (1996a).

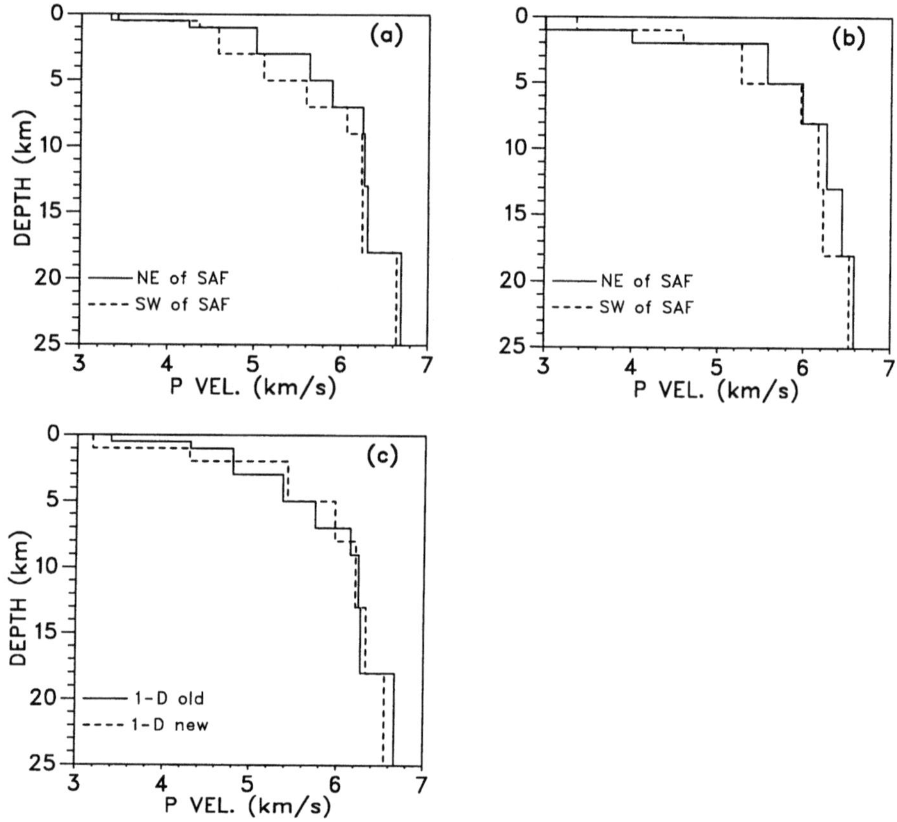

Figure 4. (a) and (b): P-wave velocity models used by Dietz and Ellsworth (1990) and Dietz and Ellsworth (1997), respectively, for stations northeast and southwest of the San Andreas fault. (c): The two 1-D velocity models obtained by averaging of the two models in (a) (old) and in (b) (new), used by Pujol (1995) and in this study, respectively.

and a systematic difference in depth of 0.5 km on average, with the new locations shallower. The new mainshock hypocentral coordinates are also listed in Table 1.

These results show that in this case fine-tuning the 1-D model is not a prerequisite for a successful application of the JHD technique. In the Loma Prieta area the lateral velocity variations are quite large, and cannot be represented by two 1-D juxtaposed velocity models. The JHD† station corrections (Figure 5) give a good indication of the variability of the velocity field. For example, there are significant low- and high-velocity anomalies (positive and negative corrections, respectively) in directions parallel and perpendicular to the San Andreas fault. Note in particular the two stations

Table 1. Summary of Loma Prieta mainshock hypocenter locations

Source	LAT	LON	Depth	O.T.	dx	dy	dxy	dz
D-E97	121.885	37.034	15.94	15.28	0.00	0.00	0.00	0.00
D-E90	121.880	37.040	17.85	15.21	0.40	-0.67	0.78	1.91
JHD†a	121.883	37.032	16.30	15.26	0.15	0.17	0.22	0.36
JHD†b	121.882	37.032	15.80	15.26	0.24	0.17	0.29	-0.14
GS90	121.883	37.036	17.60	15.25	0.15	-0.33	0.36	1.66
E-S92	121.881	37.034	17.16	15.34	0.34	-0.02	0.34	1.22

Sources: D-E90, D-E97: Dietz and Ellsworth (1990, 1997); JHD†a: Pujol (1995); JHD†b: this study; GS90: U.S. Geological Survey Staff (1990); E-S92: Eberhart-Phillips and Stuart (1992). O.T.: origin time (seconds); dx, dy, dxy, dz: differences in longitude, latitude, epicenter and depth (in km) with respect to the values of DE97 (dx and dy computed before rounding off of longitude and latitude).

near the SE corner of Figure 5 with corrections of -0.63 and 0.54 s. This large variation over a distance of just 14 km is the result of a velocity variation on the order of 40-50% down to a depth of about 3 km (Pujol, 1995).

Given the large velocity variations in the Loma Prieta area and the existence of conflicting earthquake locations, it appeared necessary to have an objective way to assess the quality of the locations and to estimate realistic error bounds. To accomplish this goal Pujol (1995) proposed the following integrated approach:

a) Locate the events using JHD†.

b) Relocate a subset of well-recorded events using a 3-D velocity inversion program (as part of a joint structure-hypocenter inversion).

c) Using the 3-D velocity model determined in step b, generate synthetic arrival times. To assure a realistic simulation, these times should be weighted exactly as the actual times.

d) Apply JHD† to the synthetic data, using the 1-D velocity model used with the actual data.

e) Compare the station corrections and locations determined in steps a and d. As the velocity inversion software requires damping, an indication of overdamping will be actual station corrections larger in magnitude than the synthetic corrections. On the other hand, comparison of the two sets of locations will give a good indication of any systematic errors that JHD† may introduce.

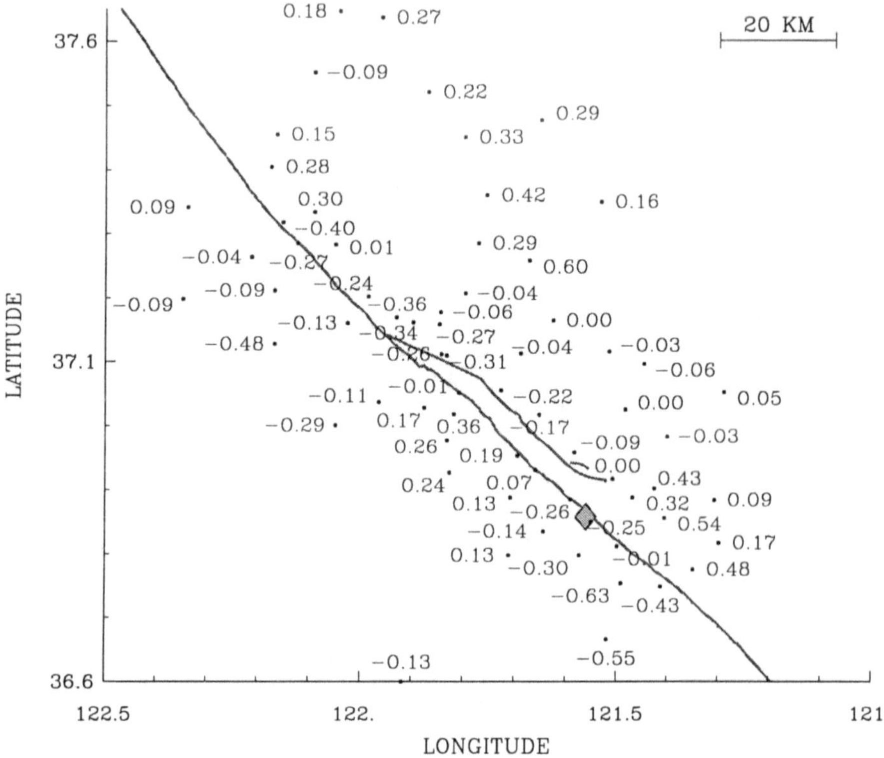

Figure 5. Stations used to relocate the Loma Prieta sequence and JHD† station corrections (in seconds). The gray lines correspond to the San Andreas and Sargent faults. The gray diamond near the SE corner represents the location of the explosion discussed in the text. Note the large contrast in station corrections in that area.

When this procedure was applied to the Loma Prieta data (Pujol, 1995) it was found that the JHD† technique was capable of locating the events with average epicentral errors of 0.5 km, and systematic depth errors of about 1 km. In addition, the actual and synthetic station corrections agreed very well except for a few stations. The differences in station corrections are indicative of low spatial resolution in the velocity model introduced by the relatively large size of the inversion blocks.

Dietz and Ellsworth (1997) used explosions with known locations to select their new velocity model. However, because the raypaths for earthquakes and explosions are extremely different, if a given velocity model leads to an explosion location with small error, it may not mean that an earthquake will also be located with a small error. For the Loma Prieta area there is also the fact that one of the explosions has epicentral and depth errors of 2.6 and 3.5 km, respectively. The location of the explosion (Figure

5) is within the area with the large velocity variation. Therefore, it is important to establish whether the JHD† locations are affected by as much error as the explosion. To answer this question the locations of the events used to generate the synthetic 3-D arrival times (as described above) were compared to the corresponding JHD† locations. The comparison was restricted to 61 events with epicenters below 36.9° N, i.e., close to the explosion. Four of the events had epicentral errors between 1.1 and 3.2 km, while forty-four had errors of 0.5 km or less. The four events with large errors have latitudes of 36.78° N or less. Regarding depth, the average error is 1 km, with the largest error equal to 2.3 km (corresponding to the event with epicentral error of 3.2 km). Therefore, a great majority of the events are located with errors much smaller than those of the explosion.

Another aspect of the Loma Prieta sequence that received considerable attention was the discrepancy between the fault planes inferred from the seismicity and from the inversion of geodetic data, with the latter offset to the SW. The offset could be as high as about 3 km, depending on the data and method used to determine the geodetic fault plane (see Àrnadòttir and Segall (1994) for a summary) and the position along the rupture zone. Most of this discrepancy, however, was a result of the mislocation of the sequence. When the JHD† locations are used (Figure 6), the discrepancy is removed, or is minor, along most of the rupture zone and reduces to about 1.2 km in its southern part (Pujol, 1996a). The Dietz and Ellsworth (1997) locations also remove most of the discrepancy (Àrnadòttir and Segall, 1996).

Finally, brief reference will be made to the average station residuals obtained using equation 34. These residuals and the JHD† corrections are very close to each other, and, as a consequence, the locations obtained using the station corrections and the average residuals agree extremely well (Pujol, 1995). This agreement has not been observed in any other data set, and may be the result of quasi-random velocity variations in the area.

3.2 New Madrid, central U.S.A., events.

The New Madrid seismic zone (NMSZ) is the most seismically active area in the central and eastern United States, and was the site of three earthquakes with magnitudes between about 7 and 8 (Nuttli, 1973; Johnston and Schweig, 1996) that occurred in the winter of 1811-1812. In addition, paleoseismological studies (e.g., Russ, 1979; Schweig and Ellis, 1994) indicate that large pre-1811 earthquakes also occurred in the area with a recurrence interval of 1000 years of less. This means that there is a potential for future large earthquakes, although the time of occurrence and magnitude of a damaging earthquake are almost impossible to predict. Currently the

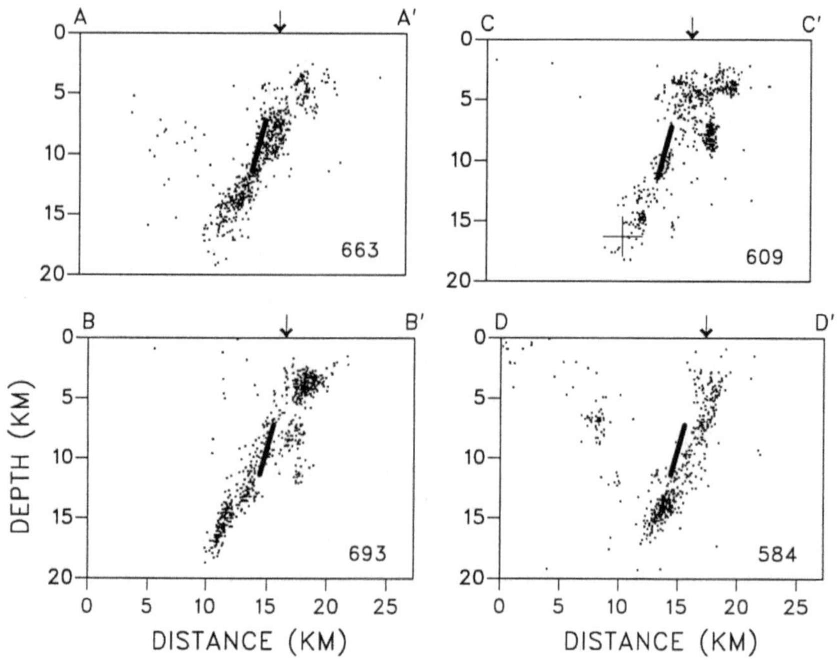

Figure 6. Depth cross sections for the events in the boxes in Figure 3 (JHD† locations). The mainshock is indicated by a cross. The arrows indicate the intersections of the San Andreas fault with the center of the boxes. The number in the lower right corner is the number of events in the cross section. The bold line represents the geodetic fault plane of Àrnadòttir and Segall (1994). In DD' the difference between the plane and the center of the seismicity is about 1.2 km. After Pujol (1996a).

seismic activity level is low, with one event of magnitude 5 of higher every 10-12 years (Johnston and Nava, 1990).

An important source of uncertainty in the early seismic hazard studies of the NMSZ was the lack of evidence for a relation between seismicity and causative faults. The installation in 1974 of the Central Mississippi Valley seismic network, operated by Saint Louis University, was critical to establish the spatial pattern of seismicity (Stauder et al., 1976), but the sparseness of the network and the fact that most of the seismometers recorded the vertical component only prevented the determination of well-constrained depths. This situation changed dramatically with the deployment during 1989-1992 of the University of Memphis three-component portable network (PANDA). The availability of reliable S-wave arrivals and the dense station distribution gave this network an unprecedented location capability. As a result, it was possible, for the first time, to image seismogenic faults in the NMSZ (Chiu

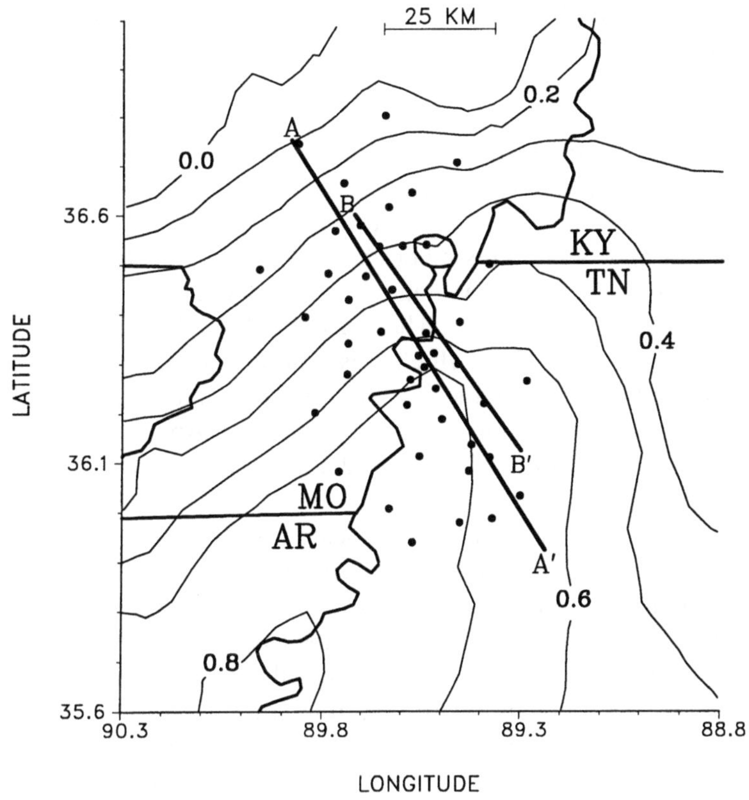

Figure 7. New Madrid seismic zone. Labeled lines: depth (km) to the Paleozoic rocks underlying the Mississippi embayment. Adding 0.09 km gives an approximate thickness for the sedimentary cover. Two-letter symbols: state name abbreviations. Dots: locations of PANDA stations. Lines AA' and BB': locations of station correction and earthquake cross sections, respectively, used in subsequent figures. Modified from Pujol et al. (1997).

et al., 1992). However, it was not until the relocation of the events using JHD† that the finer structure of the fault zone was determined.

The NMSZ is covered by a southwest-plunging trough (known as the Mississippi embayment) filled with low-velocity unconsolidated to poorly-consolidated Late Cretaceous to Cenozoic sediments. The underlying rocks are high-velocity clastic and carbonate Paleozoic rocks (Andrews et al., 1985). Although the thickness of the sediments increases to the SW, in the area where the PANDA stations were deployed (Figure 7) the variation is predominantly NW-SE, with a difference of about 0.6 km over a distance of about 70 km. This difference is quite small, but because of the low velocities involved, ignoring it has a significant effect on the earthquake depths determined using standard single-event location techniques.

As discussed in Pujol et al. (1997) there is some uncertainty in the velocity model for the NMSZ. To study its effect on the JHD† locations, four different 1-D models were considered. One of them, labeled EMB, is based on a model used in regional studies (Nuttli et al., 1969) and does not include the low-velocity sediment layer. This model is clearly unsuitable to locate the PANDA events, but was used to investigate the effect of gross model errors. Two other models (AM2 and AM3) are slight variations of the model of Andrews et al. (1985). This model has a Vp/Vs of 3.2 in the upper layer (0.65 km thick), while the AM2 and AM3 models have Vp/Vs ratios of 2 and 3, respectively, in that layer. The fourth model (AM4) was based on the model of Chiu et al. (1992), which was determined by velocity inversion with the P- and S-wave velocities for the first layer fixed (1.8 and 0.6 km/s). The inversion model is similar to the model of Andrews et al. (1985) although it has a Vp/Vs of 1.5 between 2.5 and 5 km, instead of a Vp/Vs of 1.75.

These four models were used to relocate the PANDA events (Figure 8) using a single-event location (SEL) program and JHD†. The major difference among the four sets of SEL results is an overall depth shift. Figure 9 shows the locations obtained using the EMB and AM3 models. For these two sets of locations the average of the absolute values of the depth differences is 2.7 km. Interestingly, the corresponding average RMS residuals are 0.14 and 0.13 s, so that there is no objective way to establish which set of locations is affected by more error. For the other two models the depth differences with respect to the AM3 locations are smaller. When the events are located with JHD†, the depth differences became very small (Figure 9). The average depth differences (in absolute value) for the EMB, AM2, and AM4 locations with respect to the AM3 locations are 0.4, 0.3 and 0.2 km. The average epicentral differences are even smaller, 0.1 km for all models, and the average RMS residuals are between 0.05 and 0.06 s.

The most important difference between the SEL and JHD† locations is the depth of the events that are between 0 and about 30 km along the cross section of Figure 9. The events are shallower when located individually, with the effect becoming more important as the events get closer to the origin of the cross section. This "pull up" effect is due to the NW decrease in the thickness of the unconsolidated sediments, which is not accounted for in the 1-D velocity model. This effect was confirmed with synthetic data generated for velocity models that included the 3-D variation of the thickness of the embayment sediments.

The relocation of the synthetic events also shows that the JHD† locations are little affected by the variation in sediment thickness. Regardless of the velocity model used, the average epicentral errors are 0.2 km, and the

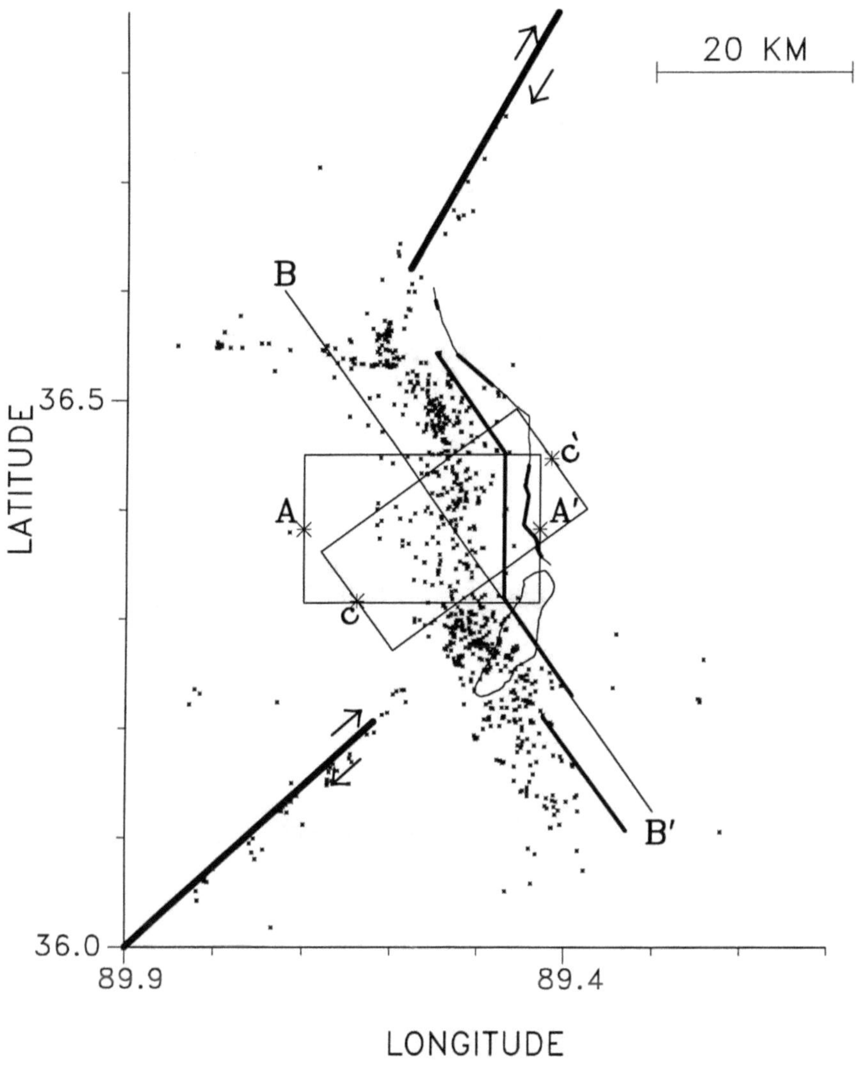

Figure 8. Epicenters of 839 events recorded by the PANDA stations in the New Madrid seismic zone. Locations determined using station corrections derived from the application of JHD† to a subset of 580 better located events (Pujol et al., 1997). Line BB' as in Figure 7. The two large heavy lines represents the roughly SW-NE-trending arms of the seismicity; the double arrows indicate inferred right-lateral strike-slip faults. Events within the boxes are plotted in cross-section in Figure 12. Asterisks identify cross-section ends. The heavy lines parallel to line BB' and the N-S line in box AA' represent the intersection of inferred faults with a horizontal plane at a depth of about 5 km. The irregular line with heavy and thin portions represent the Reelfoot and Kentucky bend scarps (thin where inferred). The closed elongated contour represents the Ridgely ridge (Russ, 1982). The southernmost NW-SE fault segment was proposed by Liu (1997). Modified from Pujol et al. (1997).

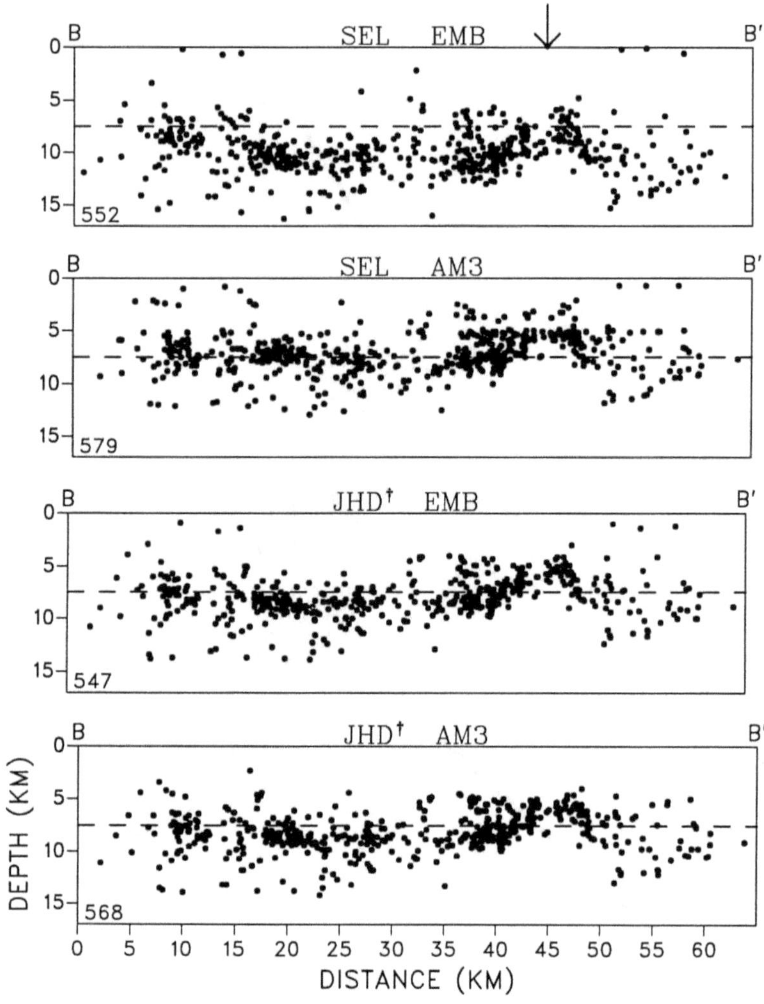

Figure 9. Depth cross sections for the events in Figure 8 within 10 km of line BB', determined by SEL and JHD† and two velocity models, EMB and AM3 (see text for details). The number in the lower left corner is the number of events in the cross section. The dashed line at 7.5 km was drawn for reference. The SEL-EMB locations are generally below this line and the SEL-AM3 locations are roughly centered on this line. The two sets of JHD† locations are almost indistinguishable. The peculiar wedge devoid of seismicity indicated by the arrow is spatially correlated with the Ridgely ridge. Modified from Pujol et al. (1997).

average depth errors do not exceed 0.5 km. A representative example is shown in Figure 10. A recent 3-D velocity inversion using the same PANDA data (Gao, 1999; Gao et al., 1999) confirms the pull-up effect described above and produces event locations very similar to the JHD† locations.

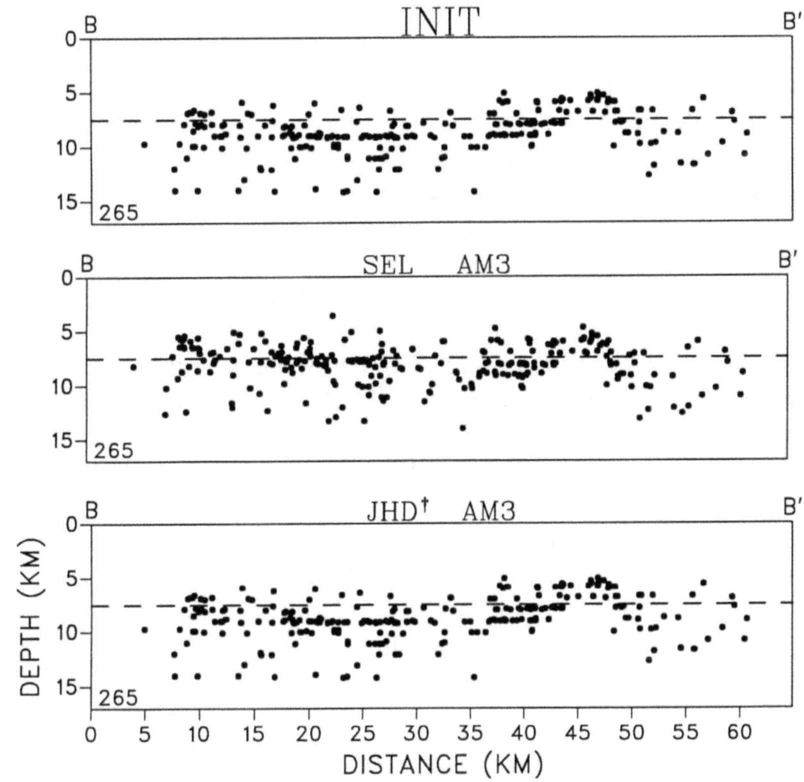

Figure 10. Depth cross sections for synthetic events generated for a velocity model that includes the 3-D variation of the Mississippi embayment sedimentary cover (Test 4 of Pujol et al. (1997)). The initial locations (INIT) were rounded to the nearest integer. Note the mislocation affecting the SEL locations and the excellent agreement between initial and JHD† locations.

How is it possible to have extremely close sets of JHD† locations when the velocity models used to generate them are so different? The answer is that the differences in velocities are compensated by systematic shifts in origin times and station corrections. As shown in Figure 11, the P-and S-wave corrections obtained with the EMB and AM3 are almost coincident after a constant shift has been added to the EMB corrections. The differences between the shifted corrections for the AM2 and AM4 models and the AM3 corrections are even smaller.

The station corrections are worth discussing for two other reasons. First note that they follow a NW-SE pattern and span wide ranges, equal to 0.35 and 0.63 s, respectively, for the AM3 model. For comparison, the average station residuals (Equation 34) computed for model AM3 are much smaller and do not show an obvious pattern (Figure 11). The average station residuals are even smaller for the AM4 model, which should not be surprising as the model is close to that obtained by velocity inversion.

The analysis of the synthetic data referred to above is important not only because it confirmed that the actual JHD† locations are well constrained, but also because they show that a Vp/Vs of 3 for the unconsolidated sediments generates synthetic S-wave station corrections that are much larger that the observed. Only when a Vp/Vs of about 2.3 is used is it possible to obtain a good match. This result is consistent with the observed station corrections. As noted before, the thickness of the sediments under the PANDA stations increases in a NW-SE direction (Figure 7). Because of the large velocity contrast, the raypaths in the low-velocity layers are, to a first approximation, vertical. For a thickness difference of 0.6 km and for P- and S-wave velocities of 1.8 and 0.6 km/s, the vertical traveltime differences are 0.33 s and 1.0 s, respectively. The 0.33 s difference is basically equal to the range of P-wave corrections noted above (0.35 s), but the 1.0 s difference is significantly larger than the range of observed S-wave corrections (0.63 s). This discrepancy shows that a Vp/Vs of 3 for the whole upper layer is too high. On the other hand, recent results obtained using shallow borehole data and the difference between arrival times of direct S-waves and S-to-P

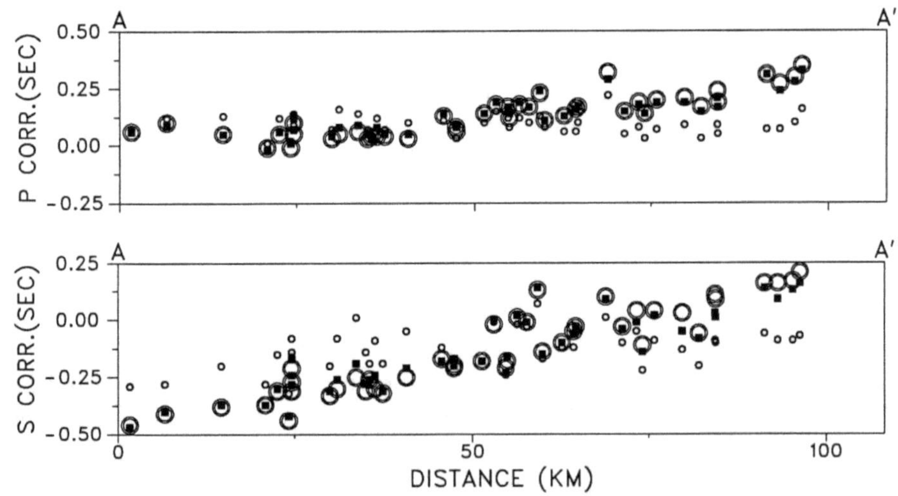

Figure 11. JHD† P- and S-wave station corrections obtained using the AM3 and EMB velocity models (squares and large circles, respectively). There is a 0.28 s shift between the corrections obtained using the two models. For cross section locations see Figure 7. The small circles represent average residuals computed using Equation 34. Note the large range spanned by the corrections and their clear pattern, which coincides with the increase of the thickness of the sediments in the direction of line AA'. The average residuals span a much smaller range and do not show the same pattern. Modified from Pujol et al. (1997).

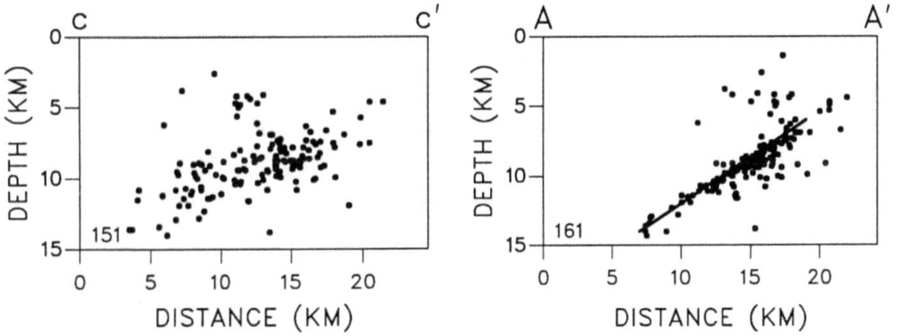

Figure 12. Depth cross sections for the events in the boxes of Figure 8. The bold line in the right plot represents an inferred fault. Modified from Pujol et al. (1997).

converted waves (Gao, 1999) show that the average Vp/Vs for the layer is between about 3 and 3.2. Thus, here we have a paradox: one set of data indicates a high Vp/Vs, while another suggests a much lower value. So far we have not been able to solve it.

The final question to consider is whether the JHD† locations represent a substantial improvement over those obtained using SEL. The answer is yes, as they help refine the fault model established by Chiu et al. (1992). In that model, faulting takes place along a direction roughly NW-SE. The refined model (Pujol et al., 1997), on the other hand, has two en-echelon NW-SE thrust faults dipping to the SW, connected by a N-S west-dipping fault (Figs. 8, 12). This feature is fuzzy when projected along a roughly NE line (Figure 12). The model was slightly modified by Liu (1997), who subdivided the original southern NW-SE segment into two others offset in a NE-SW direction (Figure 8). The en-echelon model was based on the JHD† locations only, but happens to agree with changes in the trends of the Kentucky bend and Reelfoot scarps (Figure 8), which have been interpreted as the surface expression of the fault that ruptured during one of the 1810-1811 earthquakes (Van Arsdale et al., 1995). Focal solutions determined by waveform modeling of events recorded by PANDA stations (Liu et al., 1996; Liu, 1997) also support a segmented fault model. Finally, the importance of the west-dipping segment has recently been recognized (Mueller et al., 1999).

3.3 Campi Flegrei events, Italy

The Campi Flegrei, near the city of Naples, is a volcanic caldera characterized by numerous episodes of ground uplift and subsidence (Bianchi et al., 1990; Corrado et al., 1976-1977; Lirer et al., 1987). The most recent unrest took place in 1970-1972 and 1982-1984, with highly localized and pronounced inflation episodes. The second one generated a 1.7 m uplift and was accompanied by several earthquakes with magnitudes of about 4 and by thousands of microearthquakes that forced a partial evacuation of the town of Pozzuoli (Figure 13), which was near the point of maximum uplift. The seismic activity was recorded in 1983-1984 by a portable three-component seismic network deployed by the University of

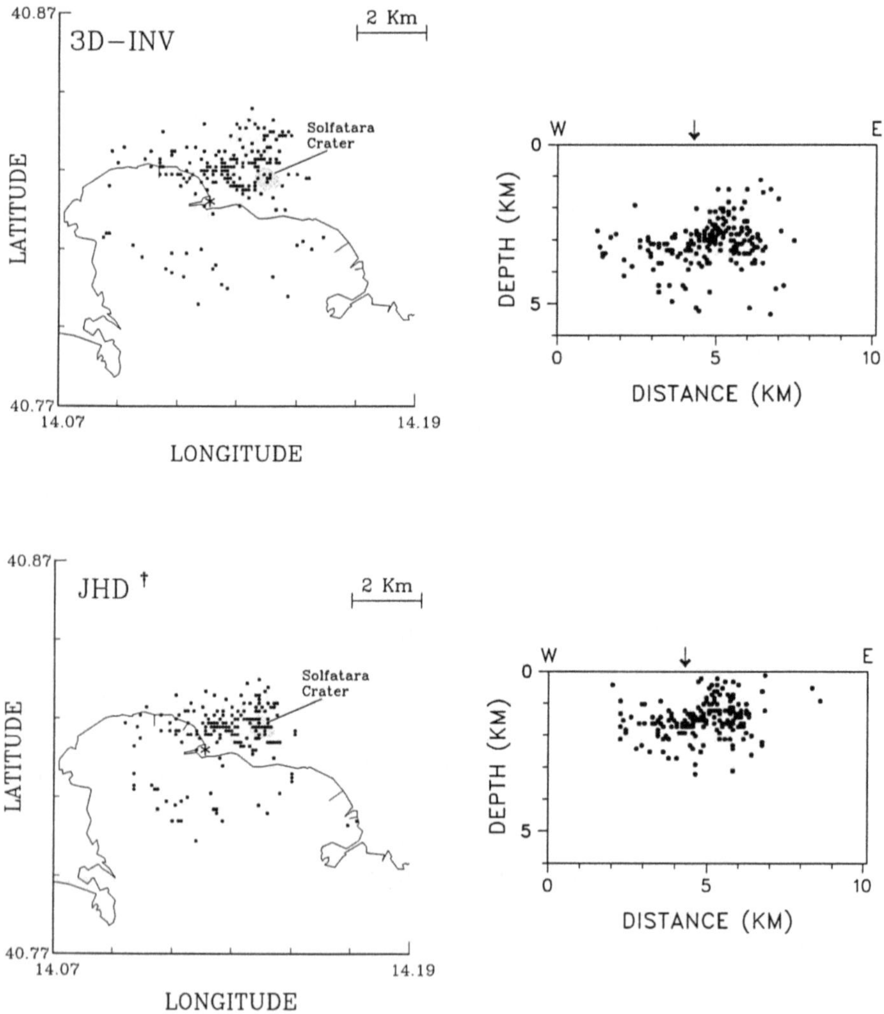

Figure 13. Epicenters of Campi Flegrei events (left) and W-E cross sections (right). Locations from Pujol and Aster (1990). The asterisk near the center of the epicentral map represents the town of Pozzuoli (shown by an arrow in the cross sections). The Solfatara crater was the only volcanic feature active during the uplift episode. Modified from Pujol (2000).

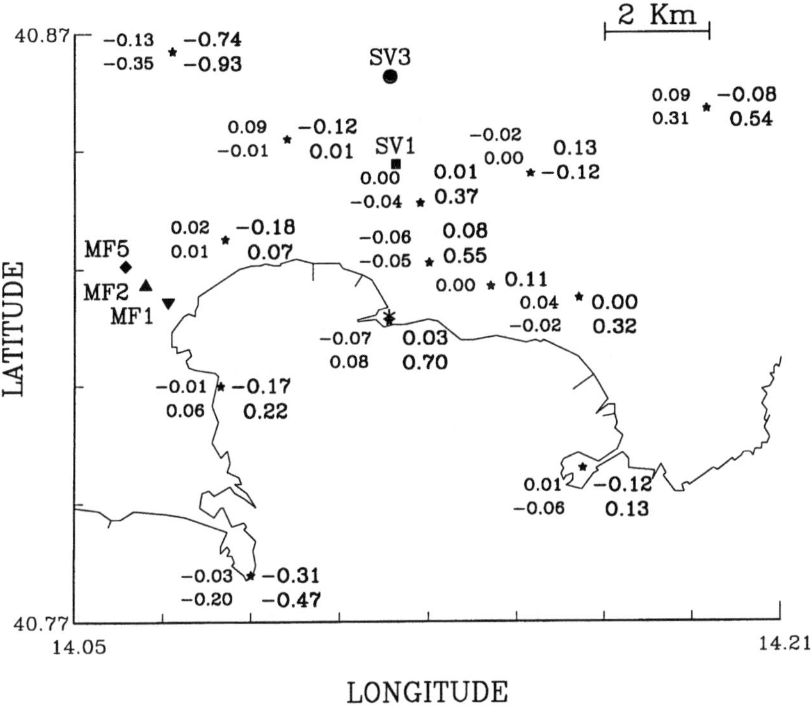

Figure 14. Stations used to locate the Campi Flegrei events (stars). The upper and lower numbers on the right of the stations are P- and S-wave JHD† station corrections (in seconds), respectively. The numbers on the left are the corresponding average residuals (Equation 34). One station does not have S-wave information. Note the large magnitude of some of the corrections and the much smaller magnitude of the corresponding residuals. The dot, square, diamond and triangles represent boreholes with P-wave sonic velocities available in the San Vito (SV) and Mofete (MF) areas. Modified from Pujol (2000).

Wisconsin-Madison. Aster and Meyer (1988) used a subset of the events recorded by this network to determine a 3-D velocity model for the area, which was used to relocate the events. The same events were subsequently relocated by Pujol and Aster (1990) using JHD†.

The two sets of locations (Figure 13) are substantially different, with those determined by JHD† less scattered and substantially shallower. In addition, the range of the station corrections is surprisingly large considering the small area involved and the overall small magnitude of the average station residuals (Figure 14). For example, the S-wave corrections for the stations in Pozzuoli and in the NW corner of the network are equal to 0.7 and -0.9 s, a difference of 1.6 s over a distance of 6.6 km. The large difference in location was interpreted as the result of the underestimation of the actual velocity variations in the 3-D velocity model (Pujol, 1992). Simulations with synthetic data indicate that lateral P- and S-wave velocity variations of about 25% and 33%, respectively, in the upper 2 km are required in order to match most of the observed station corrections, although even larger variations would be required to match the corrections for some of the outer stations. These velocity variations are much larger than those determined by 3-D velocity inversion.

Determining the actual depths of the events has more than academic importance. The 1982-1984 episode presented a significant potential hazard to the population of Pozzuoli. As future unrest is possible, a realistic assessment of the associated risk requires an accurate modeling of the source of the inflation, whose nature is still a matter of debate (De Natale and Pingue, 1992). In this context, seismology plays an important role because the maximum depth of the seismicity is one of the factors used to constrain the nature of the process(es) leading to inflation. Thus, accurate earthquake locations, particularly depths, are needed. However, as the latter are strongly dependent on the velocity model used (i.e., 1-D, 3-D) it is necessary to have an independent assessment of the velocities obtained by 3-D inversion. This task has been basically carried out by Zamora et al. (1994), who relied on velocities measured in boreholes and in the laboratory. Their conclusion is that the 3-D velocity model is not applicable in the Mofete area and may be accurate only at shallow depths (~ 1 km) in the San Vito area (Figure 14).

The borehole and laboratory measurements show that the velocities in the Campi Flegrei area have strong variations both in depth and laterally. As Figure 15 shows, the sonic P-wave velocities range between 2.3 and 5.3 km/s. In contrast, the range for the 3-D inversion velocities is 2.5-3.2 km/s. This difference in the two sets of velocities is not the only significant feature. Equally important is the decrease in velocities towards the center of the caldera. For example, at depths of 0.8 and 2.1 km the velocity at SV1 is

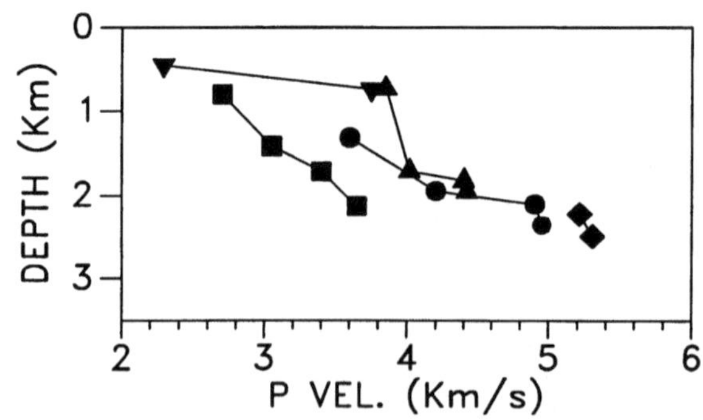

Figure 15. P-wave sonic velocities for the boreholes shown in Figure 14 (same symbols used). Information from Zamora et al. (1994) and from G. Sartoris (University of Naples, personal comm.). Note the significant velocity variations both in depth and laterally. Modified from Pujol (2000).

43% and 34% smaller than at MF2 and SV3, respectively (see Figure 14 for locations). It must be noted, however, that the central part of the caldera has not been sampled, so that even larger lateral variations cannot be ruled out. In any case, the information available confirms the presence of the velocity anomalies higher than 25% inferred from the analysis of synthetic JHD† corrections.

The availability of borehole and laboratory velocity measurements has a number of implications for earthquake location. The fact that the 3-D velocity model does not represent the actual velocity variations in the Campi Flegrei cast doubts on the quality of the locations determined using that model. In addition, if the velocities at the center of the caldera are even lower, as suggested by the synthetic data, then the 3-D locations may be deeper and more scattered than the actual locations. Regarding the JHD† locations, the analysis of the synthetic data showed that the technique introduces a small (~ 0.5 km) artificial epicentral clustering and that the computed depths are shallower than the initial depths by up to about 1 km (Pujol and Aster, 1990). These results indicate that the difference between the JHD† and 3-D inversion locations are not an artifact of the JHD† technique. What might be questionable, however, is the velocity model used to relocate the events.

The 1-D velocity model used for single-event and JHD† location was the best-fit half-space model determined by Aster and Meyer (1988), which has P-wave velocity equal to 2.86 km/s and a Vp/Vs of 1.76. In view of the velocities in Figure 15, it appears that the 1-D model has serious shortcomings. The effect on earthquake location of a layered model created using the information in Zamora et al. (1994) was investigated by Pujol (2000), but his results are not particularly encouraging. The single-event locations are more scattered than those obtained using the half-space model, while the average RMS residual was the same. This confirms earlier results (Aster and Meyer, 1988) regarding half-space versus layered models. The new JHD† results show locations in between the two sets of locations shown in Figure 13, and station corrections very different from those in Figure 14. In particular, the pattern showing a central low-velocity anomaly is not present.

In summary, the actual location of the Campi Flegrei seismicity and the velocity variations near the center of the caldera are affected by considerable uncertainty. Removing it is necessary before an important scientific problem, and its societal implications, can be better understood.

4. CONCLUDING REMARKS

The publication of the Douglas (1967) paper was a landmark in the literature on joint earthquake location, and was followed by important contributions by several other authors who proposed different exact approaches to reduce the size of the largest JHD matrix involved. The contribution of Pavlis and Booker (1983) is particularly relevant because of the introduction of the generalized inverse solution, which has a number of advantages. One of them is the control the user has over the resolution of the solution, which is a function of the magnitude of the smallest nonzero singular value included in the computation of the generalized inverse of the matrix $\mathbf{S}^T\mathbf{S}$.

In the examples given here, the generalized inverse includes all the N - 1 singular values, thus assuring maximum resolution. In general, this practice may introduce numerical problems when some of the singular values are too small, but in the author's experience with various data sets, this condition results in a sum of the station corrections not equal to the expected value of zero. This problem did not affect any of the data sets described here. It might be argued, however, that a high-resolution solution may introduce spurious results. This possibility was thoroughly studied using synthetic data. The conclusion from these studies is that the large range of station corrections observed in some cases is not an artifact of the technique. On the contrary, it is the result of two circumstances: the presence of large lateral velocity variations and the relative clustering of the events. What is interesting is the capability of the JHD technique to extract high-amplitude station corrections from small-magnitude residuals. To appreciate it, compare the residuals and corrections in Figures 1, 11 and 14. Paraphrasing the title of a well known lecture (Wigner, 1960) it can be said that the JHD technique has proven to be unreasonably effective in its capability to detect lateral variations. This makes the joint-hypocenter determination an inexpensive, semi-quantitative method for 3-D velocity assessment.

The analysis of synthetic data also showed that large lateral velocity variations may have an adverse effect on the JHD locations, particularly in an absolute sense. For the Loma Prieta and New Madrid events the location errors are not important, but there are examples of more significant errors. One is provided by the events in the 1994 Northridge (California) mainshock-aftershock sequence (Pujol, 1996b). In this case, the synthetic tests show that the JHD† locations have small relative errors but are affected by an overall epicentral shift of about 2.7 km to the NW. The single-event locations, on the other hand, are affected by a shift to the SE of as much as 2.0 km and by considerable relative location errors. Another example is provided by the Nazca plate events referred to earlier. The single-event locations (synthetic events) are affected by a systematic shift of about 4.7 km to the SE while the JHD† locations have a systematic NW shift of about 4.3 km on average (see Figures 2a and b). From these two examples, it appears that the main source of location bias is the juxtaposition of large-scale high- and low-velocity areas. Note, however, that this bias is not what is expected when small singular values are neglected in the computation of the generalized inverse. Of course, the velocity variations also affect the single-event locations, which generally have much larger relative errors than the JHD locations. When a bias is suspected, a full 3-D velocity inversion is required to determine whether it exists and ultimately to quantify it (Pujol, 1996b).

Another significant result from the analysis of actual and synthetic data is the critical difference in the performance of the exact JHD technique and its approximate counterpart. Except for one documented case (Loma Prieta), the average residuals are much smaller than the JHD station corrections and the locations determined with the former often times are not significantly different from the single-event locations. Furthermore, the RMS residuals obtained when the events are relocated with the average residuals generally are as small as those obtained when the JHD technique is used. This means that small RMS residuals should not be viewed as evidence of well constrained locations.

The JHD technique has another application that has not been discussed here for lack of space, namely its use as a tool to detect inconsistent arrivals in a phase list. This feature was noted by Viret et al. (1984), and can be very useful in the relocation of events using catalog data, which do not always have the quality required for JHD analysis. An important application can be the identification of converted waves erroneously reported as S waves. When the misidentification is consistent, the corresponding corrections will have clearly anomalous values (Pujol et al., 1998; Ratchkovsky et al., 1997).

ACKNOWLEDGEMENTS

I gratefully acknowledge the continued support provided by the State of Tennessee Centers of Excellence Program, without which the research reported here would not have been possible. Prof. G. Sartoris, University of Naples, kindly provided some of the Campi Flegrei sonic log velocities. The paper benefited from the reviews by C. Thurber and G. Pavlis. CERI contribution No. 369.

REFERENCES

Aki, K., and P. Richards (1980) *Quantitative Seismology*, Vol. 2, W.H. Freeman.
Andrews, M., W. Mooney, and R. Meyer (1985) The relocation of microearthquakes in the northern Mississippi embayment, *J. Geophys. Res.* **90,** 10,223-10,236.
Àrnadòttir, T., and P. Segall (1994) The 1989 Loma Prieta earthquake imaged from inversion of geodetic data, *J. Geophys. Res.* **99**, 21,835-21,855.
Àrnadòttir, T., and P. Segall (1996) Reply to comment, *J. Geophys. Res.* **101**, 20,137-20,140.
Aster, R., and R. Meyer (1988) Three-dimensional velocity structure and hypocenter distribution in the Campi Flegrei caldera, Italy, *Tectonophysics* **149**, 195-218.
Bianchi, R., R. Casacchia, A. Coradini, A. Duncan, J. Guest, A. Kahle, P. Lanciano, D. Pieri, and M. Poscolieri (1990) Remote sensing of Italian volcanos, *EOS, Trans. Am. Geophys. Union* **71**, 1789-1791.
Billington, S., and B. Isacks (1975) Identification of fault planes associated with deep earthquakes, *Geophys. Res. Lett.* **2**, 63-66.
Cardwell, R., and B. Isacks (1976) Investigation of the 1966 earthquake series in northern China using the method of joint epicenter determination, *Bull. Seism. Soc. Am.* **66**, 1965-1982.
Chiu, J. M., A. Johnston, and Y. T. Yang (1992) Imaging the active faults of the central New Madrid seismic zone using PANDA array data, *Seism. Res. Lett.* **63**, 375-393.
Corrado, G., I. Guerra, A. Lo Bascio, G. Luongo, and R. Rampoldi (1976-1977) Inflation and microearthquake activity of Phlegraean Fields, Italy, *Bull. Volcanol.* **40-3**, 169-188.
De Natale, G., and F. Pingue (1992) Seismological and geodetic data at Campi Flegrei (southern Italy): constraints on volcanological models, in: P. Gasparini, R. Scarpa, and K. Aki (Eds.), *Volcanic seismology*, Springer-Verlag, 484-502.
Dewey, J. (1971) Seismicity studies with the method of joint hypocenter determination, Ph.D. Dissertation, University of California, Berkeley.
Dewey, J. (1972) Seismicity and tectonics of western Venezuela, *Bull. Seism. Soc. Am.* **62,** 1711-1751.
Dietz, L. and W. Ellsworth (1990) The October 17, 1989, Loma Prieta, California, earthquake and its aftershocks: Geometry of the sequence from high-resolution locations, *Geophys. Res. Lett.* **17**, 1417-1420.
Dietz, L. and W. Ellsworth (1997) Aftershocks of the Loma Prieta earthquake and their tectonic implications, in P. Reasenberg (Ed.), *The Loma Prieta, California, earthquake of October 17, 1989 – Aftershocks and postseismic effects, U. S. Geol. Surv. Prof. Pap.* **1550-D**, D5-D47.
Douglas, A. (1967) Joint epicentre determination, *Nature* **215**, 47-48.
Eberhart-Phillips, D. and W. Stuart (1992) Material heterogeneity simplifies the picture: Loma Prieta, *Bull. Seism. Soc. Am.* **82**, 1964-1968.
Forsythe, G., M. Malcom, and C. Moler (1977) *Computer methods for mathematical computations,* Prentice-Hall.
Frohlich, C. (1979) An efficient method for joint hypocenter determination for large groups of earthquakes, *Comput. and Geosci.* **5**, 387-389.
Gao, F. (1999) High resolution upper crustal P and S velocity structures in the upper Mississippi embayment, M.S. thesis, The University of Memphis.
Gao, F., J. M. Chiu, E. Schweig, J. Pujol, and R. Street (1999) Three-dimensional upper crustal Vp and Vs structures in the central New Madrid seismic zone, *Eastern Section Seism. Soc. Am.*, 75-th annual meeting, Memphis.

Herrmann, R., S. Park, and C. Wang (1981) The Denver earthquakes of 1967-1968, *Bull. Seism. Soc. Am.* **71**, 731-745.
Johnston, A. and S. Nava (1990) Seismic-hazard assessment in the central United States, in E. Krinitzsky and D. Slemmons (Eds.), *Neotectonics in earthquake evaluation, Geol. Soc. Am. Rev. Eng. Geol.* **8**, 47-58.
Johnston, A. and E. Schweig (1996) The enigma of the New Madrid earthquakes of 1811-1812, *Ann. Rev. Earth Planet. Sci.* **24**, 339-384.
Jordan, T. and K. Sverdrup (1981) Teleseismic location techniques and their application to earthquake clusters in the south-central Pacific, *Bull. Seism. Soc. Am.* **71**, 1105-1130.
Lilwall, R. and A. Douglas (1970) Estimation of P-wave travel times using the joint epicentre method, *Geophys. J. R. Astr. Soc.* **19**, 165-181.
Lirer, L., G. Luongo and R. Scandone (1987) On the volcanological evolution of Campi Flegrei, *EOS, Trans. Am. Geophys. Union* **68**, 226-234.
Liu, Z. (1997) Earthquake modeling and active faulting in the New Madrid seismic zone, Ph.D. Dissertation, Saint Louis University.
Liu, Z., R. Herrmann, and J. K. Xie (1996) Microearthquake source parameters in the central New Madrid seismic zone from waveform modeling, *Seism. Res. Lett.* **67**, No. 2, 45.
Mc Laren J. and C. Frohlich (1985) Model calculations of regional network locations for earthquakes in subduction zones, *Bull. Seis. Soc. Am.* **75**, 397-413.
Mueller, K., J. Champion, M. Guccione, and K. Kelson (1999) Fault slip rates in the modern New Madrid seismic zone, *Science* **286**, 1135-1138.
Noble, B., and J. Daniel (1977) *Applied Linear Algebra,* Prentice-Hall.
Nuttli, O. (1973) The Mississippi valley earthquakes of 1811 and 1812: intensities, ground motion and magnitude, *Bull. Seis. Soc. Am.* **63**, 227-248.
Nuttli, O., W. Stauder, and C. Kisslinger (1969) Travel time tables for earthquakes in the Central United States, *Earthquake Notes* **40**, No. 4, 19-28.
Pavlis, G., and J. Booker (1983) Progressive multiple event location (PMEL), *Bull. Seism. Soc. Am.* **73**, 1753-1777.
Pavlis, G. and N. Hokanson (1985) Separated earthquake location, *J. Geophys. Res.* **90,** 12,777-12,789.
Pujol, J. (1988) Comments on the joint determination of hypocenters and station corrections, *Bull. Seism. Soc. Am.* **78**, 1179-1189.
Pujol, J. (1992) Joint hypocentral location in media with lateral velocity variations and interpretation of the station corrections, *Phys. Earth Plan. Int.* **75**, 7-24.
Pujol, J. (1995) Application of the JHD technique to the Loma Prieta, California, mainshock-aftershock sequence and implications for earthquake location, *Bull. Seism. Soc. Am.* **85,** 129-150.
Pujol, J. (1996a) Comment on: "The 1989 Loma Prieta earthquake imaged from inversion of geodetic data" by Àrnadòttir, T. and Segall, P., *J. Geophys. Res.* **101**, 20,133-20,136.
Pujol, J. (1996b) An integrated 3D velocity inversion-joint hypocentral determination relocation analysis of events in the Northridge area, *Bull. Seism. Soc. Am.* **86**, S138-S155.
Pujol, J. (2000) Comment on "Laboratory measurements of ultrasonic wave velocities in rocks from the Campi Flegrei volcanic system and their relation to other field data" by Zamora, M., Sartoris, G. and Chelini, W., *J. Geophys. Res.*, in press.
Pujol, J. and R. Aster (1990) Joint hypocentral determination and the detection of low-velocity anomalies. An example from the Phlegraean Fields earthquakes, *Bull. Seism. Soc. Am.* **80**, 129-139.
Pujol, J., A Johnston., J. M. Chiu, and Y. T. Yang (1997) Refinement of thrust faulting models for the central New Madrid seismic zone, *Engineering Geology* **46**, 281-298.

Pujol, J., R. Herrmann, S. C. Chiu, and J. M. Chiu (1998) Constrained relocation of New Madrid seismic zone earthquakes, *Seism. Res. Lett.* **69**, 56-68.

Ratchkovsky, N., J. Pujol and N. Biswas (1997) Relocation of earthquakes in the Cook Inlet area, south-central Alaska, using the joint hypocenter determination method, *Bull. Seism. Soc. Am.* **87**, 620-636.

Russ, D. (1979) Late Holocene faulting and earthquake recurrence in the Reelfoot Lake area, northwestern Tennessee, *Geol. Soc. Am. Bull., Part 1* **90**, 1013-1018.

Russ, D. (1982) Style and significance of surface deformation in the vicinity of New Madrid, Missouri, in F. McKeown and L. Pakiser (Eds.), *Investigations of the New Madrid, Missouri, Earthquake Region, U. S. Geol. Surv. Prof. Paper* **1236**, 95-114.

Schweig, E., and M. Ellis (1994) Reconciling short recurrence intervals with minor deformation in the New Madrid seismic zone, *Science* **264**, 1308-1311.

Smith, E. (1982) An efficient algorithm for routine joint hypocentre determination, *Phys. Earth Plan. Int.* **30**, 135-144.

Stauder, W., M. Kramer, G. Fischer, S. Schaefer and S. Morrissey (1976) Seismic characteristics of southeast Missouri as indicated by a regional telemetered microearthquake array, *Bull. Seism. Soc. Am.* **66**, 1953-1964

U. S. Geological Survey Staff (1990) The Loma Prieta, California, earthquake: an anticipated event, *Science* **247**, 286-293.

Van Arsdale, R., R. Kelson, and C. Lumsden (1995) Northern extension of the Tennessee Reelfoot scarp into Kentucky and Missouri, *Seism. Res. Lett.* **66**, 57-62.

Viret, M., G. Bollinger, J. Snoke, and J. Dewey (1984) Joint hypocenter relocation studies with sparse data sets – A case history: Virginia earthquakes, *Bull. Seism. Soc. Am.* **74**, 2297-2311.

Wigner, E. (1960) The unreasonable effectiveness of mathematics in the natural sciences, *Comm. Pure Appl. Math.* **13**, 1-14.

Zamora, M., G. Sartoris, and W. Chelini (1994) Laboratory measurements of ultrasonic wave velocities in rocks from the Campi Flegrei volcanic system and their relation to other field data, *J. Geophys. Res.* **99**, 13,553-13,561.

Chapter 8

AUTOMATED EVENT LOCATION BY SEISMIC ARRAYS AND RECENT METHODS FOR ENHANCEMENT

Manfred Joswig
Tel Aviv University, Israel

Key words: Seismic array, CTBT, pattern recognition, ARCESS, automatic location, pluralistic reasoning, f-k analysis, beamforming, rule-based system

Abstract: This paper describes procedures to locate seismic events at regional distances by single seismic arrays. First it reviews the methods used for interactive analysis and compares their performance using array data to that of single three-component stations. Then the paper focuses on automated location. Ten days of continuous-operation ARCESS data are analyzed. They are evaluated by standard NORSAR routines of beamforming and *f-k* analysis, and by a complementary, independent approach. This approach utilizes pattern recognition (PR) and rule-based system (RBS) techniques to derive an automated bulletin from a subset of four array stations. The content of both lists, i.e., the ARCESS log of NORSAR and the new, PR-based list, are combined to form a joint bulletin. Its enhanced quality is achieved by rules for consistency check or "pluralistic reasoning."

1. INTRODUCTION

Seismic arrays are sensitive tools to detect and locate seismic events from local to teleseismic distances. They operate like phased arrays in radar surveillance (Whiteway, 1965) and were introduced to seismology in the 1960's (see Lay and Wallace, 1995). "Beamforming," i.e. the delay-and-sum of individual traces to a "beam" trace, offers signal-to-noise improvements approximately proportional to the square root of number of stations (Green

C.H. Thurber and N. Rabinowitz (eds.), Advances in Seismic Event Location, 205–230.

et al., 1965). The enhanced sensitivity of seismic arrays is crucial for the network of International Monitoring System (IMS) stations as determined in the Comprehensive Nuclear Test Ban Treaty (CTBT). Tuned for regional monitoring, the high-frequency arrays such as NORESS, ARCESS, FINESSA or GERESS reliably detect small earthquakes and explosions over thousands of kilometers distance and allow for enhanced event identification capability.

2. EVENT LOCATION BY SEISMIC ARRAYS

The usefulness of seismic arrays is based on the concept that propagating plane waves over the array aperture cause coherent signals at all stations. Thus, an array represents just one site for any attempt of location calculation despite its many stations. The utilized method equals that for a single, three-component station: the backazimuth points from station to event and constrains the direction, while the S-P time yields a radius that determines the distance.

In an array, both parameters can be derived and confirmed by *f-k* analysis utilizing the spatial sampling of the wave field (Capon, 1969). This approach of wave number filtering allows for separation of signal and noise even when they cover the same frequency band. Figure 1 shows the result of a sliding *f-k* analysis in 1-s windows over a regional seismogram; just the slowness value of maximum energy is given. The example exhibits a clear increase above 0.20 s/km once the first shear wave energy arrives, resolving much of the ambiguity usually associated with S-onset determination on weak signals. In the "teleseismic window" from 30 – 90°, the slowness of P constantly decreases from 0.08 – 0.04 s/km with approximately linear slope, which gives a second, independent estimate for the epicentral distance (Birtill and Whiteway, 1965).

The estimation of backazimuth also relies on the *f-k* analysis. Figure 2 displays the results for another regional event where maxima are given over **k** by contour lines; the chosen frequency bands depend on phase type. Backazimuth and slowness are determined per single phase allowing for consistency checks on common azimuths and correct order of phase arrivals. Thus even complex seismograms composed by different overlapping events may be resolved with ease.

The error limits for the estimated parameters can be derived from the width of the main lobe in the array response, which resembles the distribution of stations over the aperture (Harjes and Henger, 1973). Two constraints for optimization contradict each other: achieving significant time

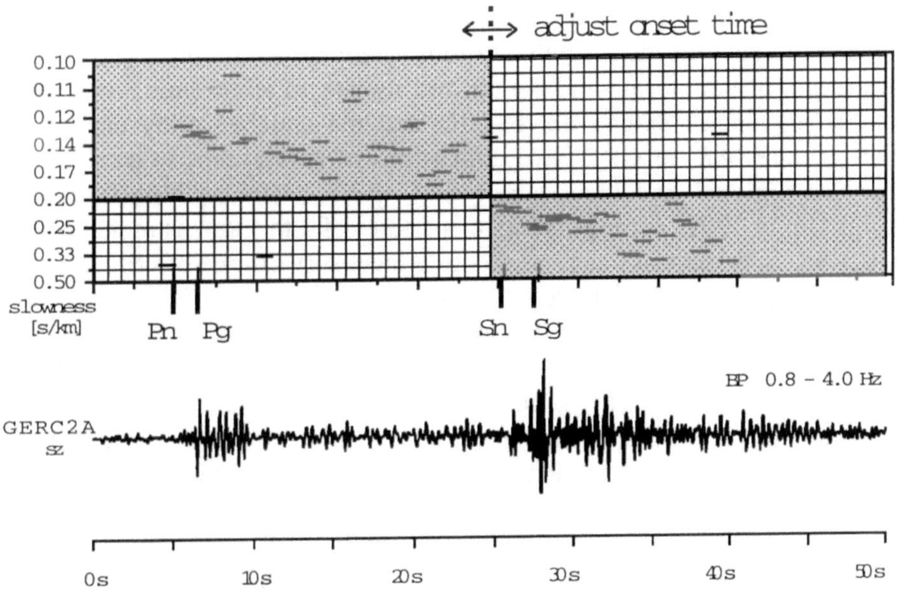

Figure 1. Constraining the S-onset by *f-k* analysis (modified from Gestermann.,1993). Horizontal bars indicate the slowness with maximum arriving energy.

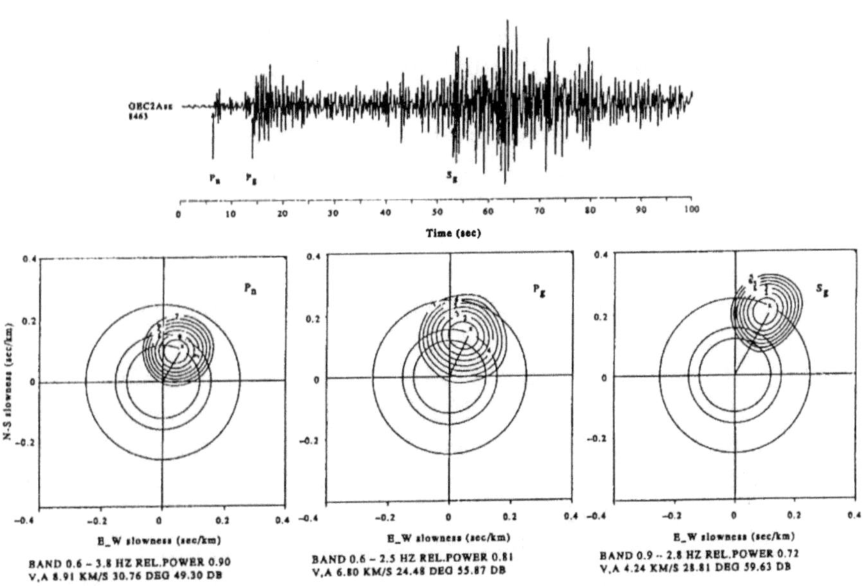

Figure 2. Testing on common azimuths for Pn, Pg, and Sg (from Harjes et al., 1993).

differences within the array demands a large aperture, while maximizing signal coherency needs a small aperture. The chosen compromise depends on the aim of monitoring and the utilized signals. In teleseismic monitoring, signal frequencies extend up to 1 Hz and the matched array designs of LASA, NORSAR, or GRF have dimensions of several tens to hundreds of km. For regional and local monitoring, the focus is on high-frequency designs up to 30 Hz corresponding to apertures of 1 to 5 km, as in the case of ARCESS. Specific tasks such as the near-local surveillance of critical target areas may even require just 100 m aperture (Bartal et al., 2000).

The precision of phase picking is generally limited by the sampling interval, yielding errors of a few degrees for the backazimuth in high-performance arrays. The precision could be increased to subsample resolution by interpolation techniques, e.g., the cross spectral analysis (Scherbaum and Wendler, 1986); however, these techniques are not in routine use to date.

For any single 3-C station, the backazimuth is determined by the trajectory of particle motion during the first onset, while S is the first appearance of significant signal amplitudes on the horizontal traces. Both estimates are error-prone due to noise; the backazimuth as the ratio of horizontals suffers from small angles of incidence in P, while S does not exhibit the clear, linear polarization above 1 Hz that would distinguish it from noise or P coda (Menke et al., 1990; Roberts and Christoffersson, 1990; Menke and Lerner-Lam, 1991; Klumpen and Joswig, 1993; Der et al., 1993). Additionally, Buchbinder and Haddon (1990) report on up to 40° deviation of P backazimuth from the direction to the epicenter. In broader investigations, Jarpe and Dowla (1991) and Walck and Chael (1991) determine a typical error of ±6°, given that the optimum frequency band is found and the signal-to-noise ratio (SNR) is sufficiently large.

Due to the differences in approach and in the utilized amount of information, large discrepancies exist in the performance of array versus 3-C location results. Suteau-Henson (1991), Kvaerna and Ringdal (1992), Christoffersson and Roberts (1996) compare both methods and consistently find the array location superior to 3-C location. Reports on array locations are given by Ringdal and Husebye (1982), Mykkeltveit and Bungum (1984), Der et al. (1990), Cansi et al. (1993), and Jost et al. (1996), while the performance of 3-C approaches is described by Magotra et al. (1987), Ruud et al. (1988), Roberts et al. (1989), Cassidy et al. (1990), Ruud and Husebye (1992), and Tarvainen (1992).

Thus, deriving event locations from single 3-C stations is largely an academic exercise, while the results of array processing truly enhance the seismic monitoring of surrounding regions. Rabinowitz and Joswig (1994) find the detection and locations results of one high performance, digital, tripartite array comparable to six analog network stations distributed over 5000 km^2 in Northern Israel. The evaluation of array traces was reliably possible to -12 dB below the signal-to-noise ratio at comparable network traces due to signal correlation and the backazimuth and slowness constraints (see Figs. 1 and 2).

Some principal restrictions remain from the single array or station approach that makes even the most advanced array inferior to network locations: (1) Depth can not routinely be determined which restricts array locations to the evaluation of epicenters; only additional depth phases like pP could constraint the hypocenter (Bowers, 1999; Pearce, 1999). (2) The location uncertainty arises by constant azimuthal error limits, thus it increases linearly with distance. (3) The array measurement is affected by local inhomogeneities of the crust beneath its aperture (Glover and Alexander, 1969; Flatte and Wu, 1988), while a network most probably averages out some different effects. The last shortcoming results in the concept of time delay corrections to the plane wave propagation which also compensate for elevation differences (Berteussen, 1974). Another approach introduces slowness corrections depending on the approximately determined source region (Harjes et al., 1994; Schweitzer, 1995). Besides this "array calibration", further improvements in location accuracy are possible by the concepts of array relocation (Kvaerna and Ringdal, 1993) and non-linear master event correlation (Joswig and Schulte-Theis, 1994). Both approaches apply to restricted regions for which additional knowledge must be defined in terms of tuned parameter sets or representative seismograms.

Finally, the combination of location results by different arrays closes the loop in our comparison of location methods. The arrays perform as single sites in a larger, possibly mixed network with single 3-C stations. The arrays provide the network location program with a full set of onset times, and attached slowness and backazimuth values that must be weighted in sophisticated ways (Bratt and Bache, 1988).

3. AUTOMATED ARRAY PROCESSING

The high-frequency arrays of NORSAR, such as ARCESS, are routinely evaluated by the RONAPP processing system (Mykkeltveit and Bungum, 1984). To date it consists of the three modules, "Detection Processing" (DP) for short-term-average/long-term average (STA/LTA) triggering, "Signal Processing" (SP) to perform signal feature extraction, and "Event Processing" (EP) that associates phases, calculates location and plots the events (Fyen, 1996). The left part of Figure 3 shows the flow of data and parameters. The "Generalized Beam Former" (GBF) at the third stage is an alternate location module that evaluates the joint phase report of all six NORSAR arrays (Kvaerna and Ringdal, 1996). Although different in detail, this scheme of data processing also applies to the prototype International Data Center (pIDC) with modules like DFX or SigPro (IDC, 1999).

In our study, the NORSAR listings for 10 days of continuous recording (April 12-April 21, 1996) at ARCESS were analyzed. The initial beam-forming weights the 25 ARCESS station traces to produce 76 beams. From these beams, some 20 detections per hour are found by the STA/LTA. This averages to 480 phases per day. Taking any phase detection as an event, however, would result in a high number of false alarms. Instead, the declaration of an event demands one Pn/Pg *and* one S/Lg phase constrained by a time window *dt* of 240 s and by an azimuthal difference $d\phi$ of 30°. Three typical errors may occur despite the timing and azimuth constraints:

(1) If Sn and Lg are emergent, they will be missed regardless of signal-to-noise ratio (SNR).
(2) False alarms ("ghost events") are declared, sometimes with large magnitude. This happens by the casual combination of single phases; if they are at the limits of *dt*, the erroneous results for distance and magnitude become large.
(3) For mixed events, e.g. a sequence of quarry blasts with overlapping seismograms, phases may be grouped incorrectly, causing wrong distances.

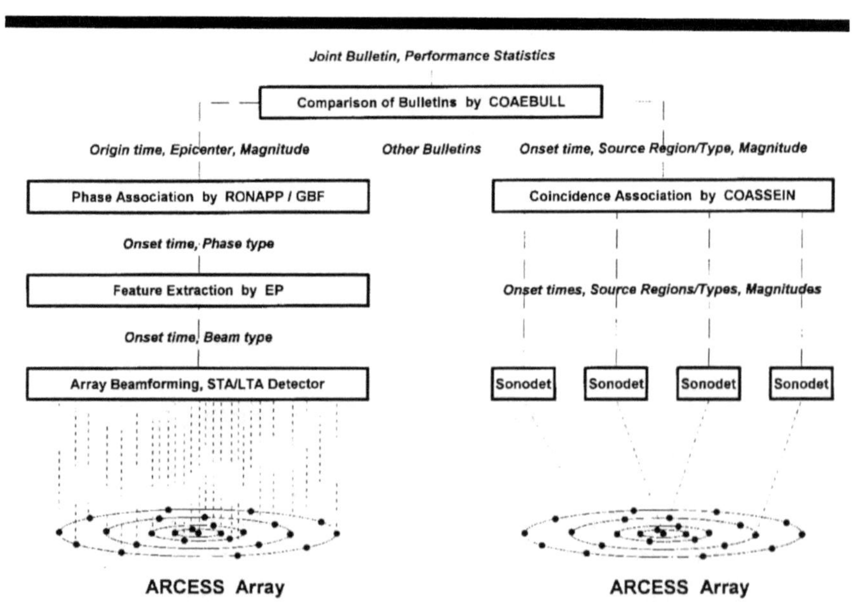

Figure 3. Signal flow in NORSAR processing (left) and for SparseNet processing (right).

Figure 4 displays examples of these kinds of processing errors. For comparison, all six seismograms are shown at the same scale even though the fifth trace is clipped. All associated phases are given; RONAPP utilized all unprimed phases. The first trace was missed because Sn/Lg is emergent, whereas the second case was excluded due to a low-SNR-criterion. The false

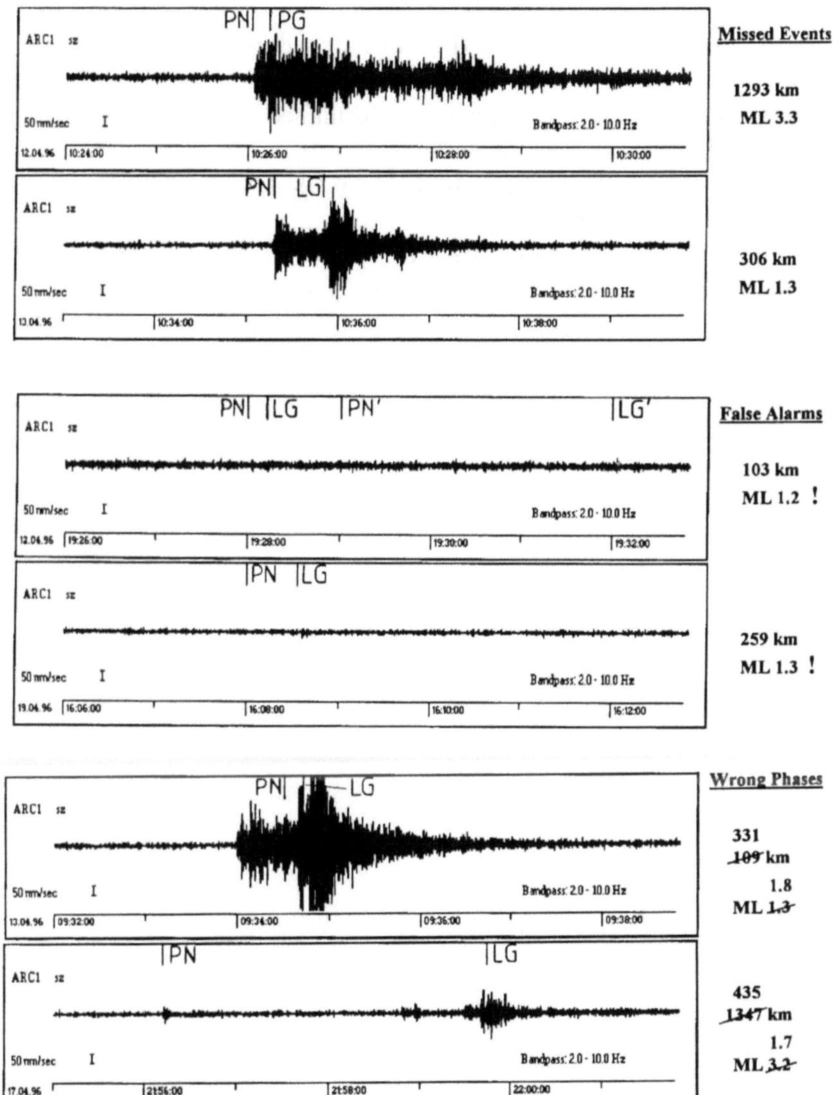

Figure 4. Examples of types of errors that can be made in routine automated processing (top - missed events, middle - false alarms, bottom - phase misidentification). See discussion in the text.

alarms in traces 3 and 4 have associated magnitudes that would demand seismogram amplitudes as large as those above. In the fifth trace, the detected first onset was ignored during phase association while in the sixth trace the 'early' Pn is in fact a teleseismic P with 14.8 km/s apparent velocity from Hindu Kush with the same backazimuth as the regional event.

In Figure 5, the automated bulletin is resolved by magnitude versus distance and divided into four classes. Each class corresponds to some minimum, local SNR (bandpass 2-10 Hz) as listed in Table 1; *Mc*(100) is the completeness threshold specified at 100 km. Per class, the automated processing should meet the expected performance to effectively relieve the human analyst from some routine tasks.

However, even in the highest class (*Mc*(100)=0.0), 2 events were missed, each day 1 ghost event was created, and 2 events were significantly mislocated. All examples in Figure 4 are from this class; the missed events had magnitude M_L 1.2 at 306 km and M_L 3.3 at 1293 km. Results would degrade when either the local noise level is above the favorable conditions at ARCESS or when the array consists of less than 25 elements.

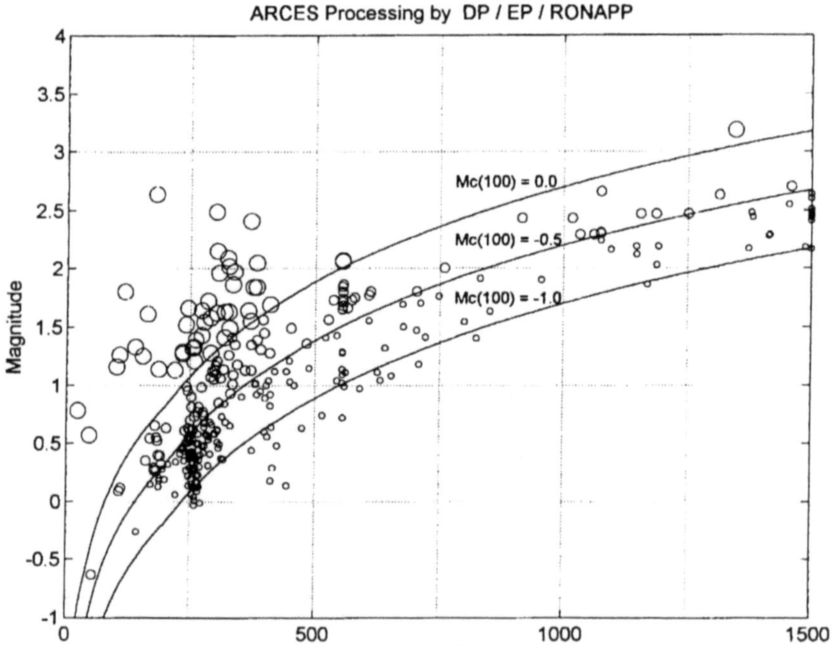

Figure 5. Detection capability by distance achieved by the standard automated processing algorithm for the 10-day testbed (circles) and separation into performance classes by different completeness thresholds.

Table 1. SNR and completeness thresholds Mc at ARCESS.

$M_C(100)$	SNR_{ST} on single trace	SNR_{AB} on array beam	expected performance
0.0	5.0 : 1 (14 dB)	25 : 1 (28 dB)	complete without errors
-0.5	1.5 : 1 (3 dB)	7.5 : 1 (17 dB)	complete, with errors
-1.0	0.4 : 1 (-8 dB)	2.0 : 1 (6 dB)	not complete, with errors

4. *SPARSENET* PROCESSING

The monitoring by pattern recognition in the processing software package *SparseNet* (Joswig, 1999) originally stems from limitations when shrinking the number of array stations below, for example, six sites. A common scenario is that funds do not support installation and operation of a conventional, *dense* array like ARCESS with 25 sites. The alternative, a so-called *sparse* array, consists of a minimum of 3 to 4 stations arranged as an equidistant triangle with possible center site. For example, the BUG sparse array was set up to monitor the mining induced seismicity in the Ruhr district of NW Germany (Joswig, 1993). The potential benefit of the sparse array concept for CTBT monitoring has been investigated by Kvaerna (1992) and Kvaerna and Ringdal (1992).

To automate the seismogram processing in sparse arrays, the authors found most advantages of dense arrays either not existent or severely degraded:

- SNR is usually better for the best single station than for any reasonably weighted array beam. Optimal detection is achieved by voting on single traces instead of beamforming.
- The performance of *f-k* analysis for phase association is severely degraded due to the ambiguity by many side lobes caused by the insufficient sampling of the wave field. Also the maxima are very flat due to the marginal gain in SNR.

Additional approaches for automated processing were necessary to benefit from the sparse array concept. Many of these approaches utilize the power of artificial intelligence (AI) techniques to stabilize signal processing given the decreased SNR (see Joswig, 1995a, 1996). These concepts result in a layered approach (Joswig, 1993) consisting of:

- Pattern Recognition (PR) detection based on recognizing full seismogram signatures from P onset to surface wave coda which are coded into sonogram patterns and shifted over the data sonogram (Joswig, 1990 and 1995b). This approach is realized in SONODET and described subsequently.
- Rule-based system (RBS) voting to exclude the casual coincidence of noise bursts at different stations. This approach of COASSEIN is based on rules that check the SONODET messages for common event type (Joswig, 1995b) and will also be described below.
- Non-linear cross-correlation, calculated between array traces to determine relative delay times for backazimuth and slowness, and performed with available master events of known site to refine the locations (Joswig and Schulte-Theis, 1994).
- 3-component phase picking of P- and S-onsets, based on spectral images such as sonograms and contour patterns (Klumpen and Joswig, 1993).

The first two tasks are performed by modules SONODET and COASSEIN of the IASPEI shareware library contribution *SparseNet* (Joswig, 1999).

Although both AI principles - subsymbolic PR and symbolic RBS - are utilized to enhance sparse array processing, the clear focus is on PR. This is different from the AI applications for dense arrays which concentrate on RBS (Searfus, 1989; Bache et al., 1990; Bratt et al., 1990; Bache et al., 1993). The same distinction holds for processing the incoherent seismograms of seismic networks. Sparse networks with up to six stations are successfully evaluated by PR (Leonard et al., 1999), while AI solutions for dense networks favor the RBS approach (Chiaruttini et al., 1989; Roberto and Chiaruttini, 1992).

It is worth emphasizing that the simple change in the *quantity* of sites converts to a distinct *quality* for the preferred AI approaches in automated processing. This separation, shown in Figure 6, is between dense arrays and dense networks versus sparse arrays and sparse networks (Joswig, 1992). The reason is the different availability of data: dense station distributions ease the parameter extraction, either by good SNR of neighboring stations or by beamforming and *f-k* analysis. However, the many parameters will contain mutual contradictions that must be resolved by sophisticated reasoning, which is the essence of RBS. For the case of few stations and low SNR, the derived parameters such as P-onset will directly determine the location; the lack of redundancy prevents consistency checks. Hence some effort must be made to ensure reliable parameter extraction, favoring PR applications.

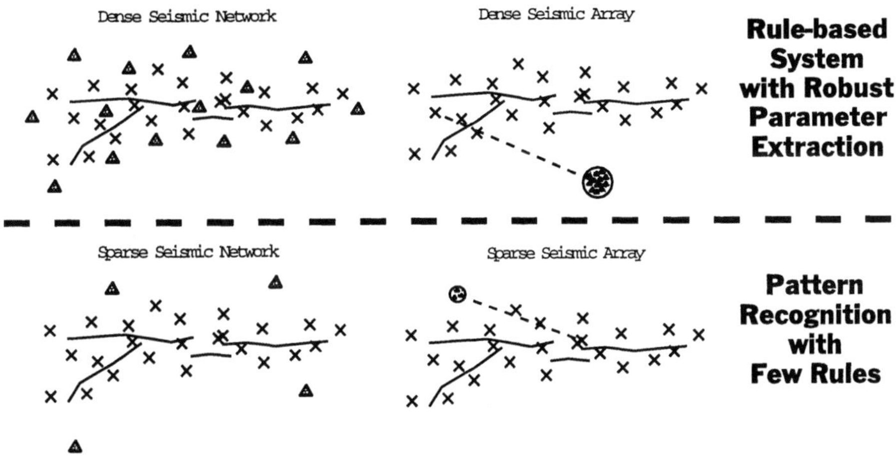

Figure 6. Monitoring of a fault zone with earthquakes (X) by dense and sparse array/network layouts of stations (Δ) and preferred AI schemes for automated processing. Dense station distributions ease the parameter extraction, either by good SNR of neighboring stations or by beamforming and *f-k* analysis (top). For few stations and low SNR, the derived parameters will directly determine the location, so some effort must be made to ensure reliable parameter extraction, favoring PR applications.

The application of *SparseNet* modules to the ARCESS data follows the processing scheme in the right part of Figure 3. The stations ARC1, ARC3, and ARC6 are selected to form a sparse, tripartite array with ARA1 as the fourth, center station; the array aperture is an equidistant triangle of 1.5 km. The PR detection is performed by SONODET on single traces, the results are checked for array-wide coincidence by COASSEIN exploiting RBS principles.

4.1 SONODET

The sonogram detector works on denoised, prewhitened spectrograms or "sonograms" to recognize the similarity or "fit" with predefined, adapted patterns by a sliding 2-D cross-correlation (Joswig, 1995b). It utilizes the signature of full event seismograms from P-onset to surface wave coda decay; the STA/LTA instead would trigger on each distinct amplitude increase. Figure 7 shows from top to bottom the non-linearly scaled sonogram, the seismogram, the STA/LTA ratio, and the fit with the two selected patterns Kiruna_I, Kiruna_II. This source region association is correct as the signal originated from the Kiruna quarry region at 250 km distance (see Figure 8).

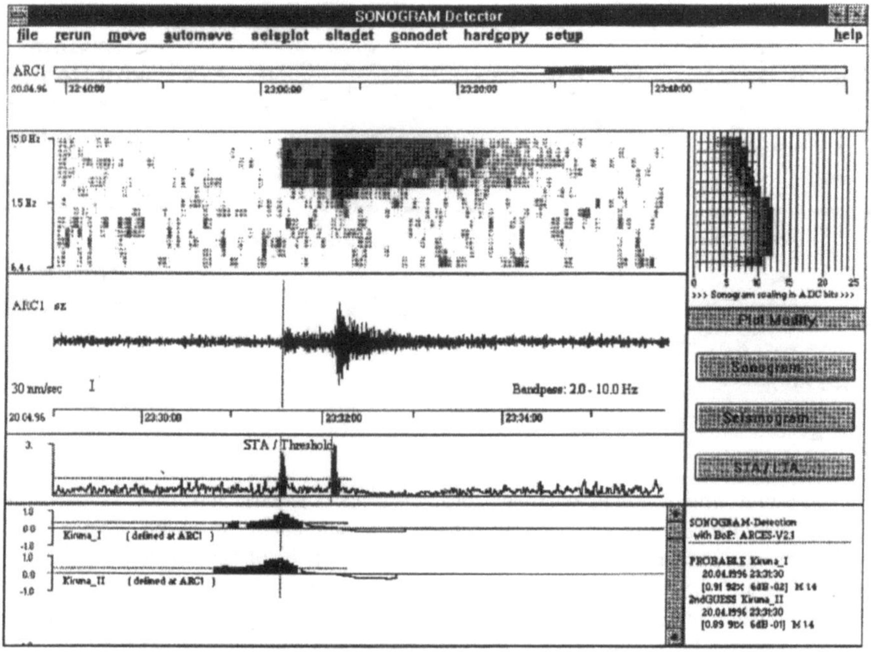

Figure 7. SONODET processing for a Kiruna quarry blast at 250 km from ARCESS (Top left – sonogram, top right – noise statistics with frequency, middle – STA/LTA detector, bottom – pattern fit by SONODET).

SONODET was successfully applied to local data where the spectral shift of energy over time is a strong discriminant between earthquakes and noise bursts. For the regional seismograms at ARCESS, however, the situation was more difficult. In Figure 7, the recorded wave train has nearly constant frequency content with virtually no surface wave energy in the lower frequency bands: the shape of the sonogram is close to a box. This feature is valid for all types of signals except teleseisms; comparing it to sonograms of regional events in Western Europe and California, we suggest this lack of low frequencies might be a distinctive feature of records in the Scandinavian Shield. Obviously, we have lost one of the assets for sonogram recognition.

Another difficulty for PR comes from the fact that many source regions fall into the same distance range. Figure 8 displays the most frequent clusters of Fennoscandian seismicity. Quarries from Kiruna, Kovdor, and the events east of Nikel all have the same epicentral distance to ARCESS; the same is true for Paakkola and Khibiny. The blasts south of Kostomuksha are usually masked by noise at ARCESS; the recognized earthquakes are located at the

North Atlantic ridge (outside this map) and at the Fjord area in northern Norway.

SONODET will recognize this seismicity according to its pattern base. This pattern base must be tuned per region but stays constant for all array stations. Figure 9 compiles the set of patterns used for this study. Note that we had to characterize the quarry regions Nikel, Kiruna and Khibiny by two patterns each to cover some fraction of variations caused by the delayed firing of multiple explosions. This heuristic scheme does, of course, not cover all cases of mixed quarry events where the number of shots, the time delays and the magnitude differences are free parameters. Nonetheless, this approach did enhance the PR results to the subsequently reported performance level. The source region Norwegian_Sea was so large that we had to select two patterns from distinct epicenters for sufficient characterization.

Figure 8. 18 months of seismicity in Fennoscandia (modified from Norwegian National Data Center).

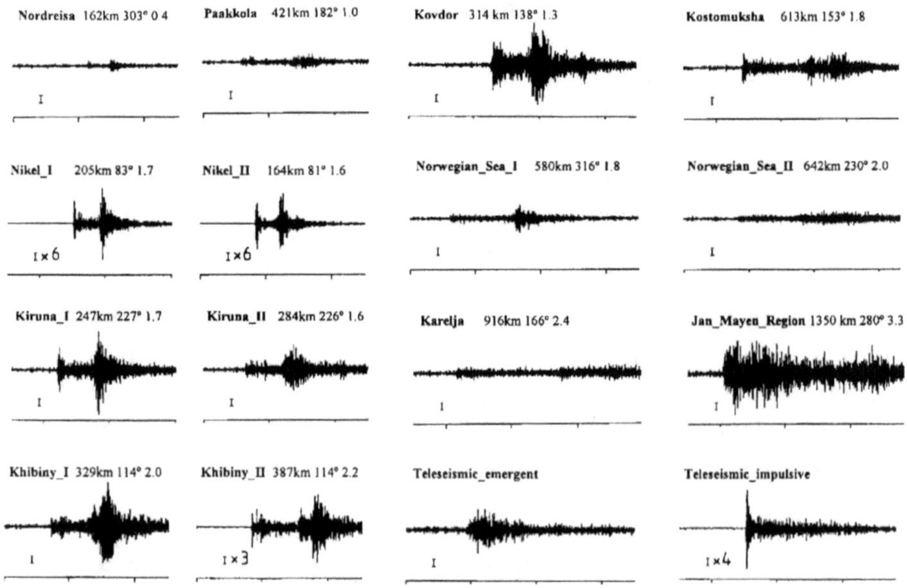

Figure 9. Pattern set of SONODET at ARCESS for Fennoscandian seismicity.

4.2 COASSEIN

Once the detection lists of ARA1, ARC1, ARC3, and ARC6 are derived, COASSEIN as the second module in *SparseNet* performs the rule-based coincidence analysis (Joswig, 1995b, 1999). The rules will check on common source region in the SONODET results; they also evaluate any differences in reported onset time or magnitude. Achieving sufficient performance demands the resolution of many potential contradictions in the detection lists; the necessary strategies occupy about half of the rule set. The key contribution is by the "cluster exchange" rule, a penalty scheme which alters initial identifications of SONODET according to Table 2 if no joint identification can be found.

Table 2. Cluster exchange pairs of COASSEIN at ARCESS (see text for details).

Kiruna_I	<->	Kiruna_II_
Kovdor, Paakkola, Khibiny_I	->	Kiruna_II
Khibiny_I	<->	Khibiny_II
Norwegian_Sea_I	->	Khibiny_II
Norwegian_Sea_I	<->	Norwegian_Sea_II
Nikel_I	<->	Nikel_II
Nordreisa	->	Nikel_II
Teleseism_impulsive	<->	Teleseism_emergent

The information of Table 2 must match the specific regions defined by the pattern set of Figure 9. The exchange is permitted between source regions of similar distance but different backazimuth, and for neighboring regions since both conditions cause similar seismograms. Another interpretation is based on the principles of Bayesian or "statistical" PR (see Bishop, 1995). The cluster exchange resembles the multiplication of SONODET's pattern fit values with "prior probabilities," i.e., the fraction of different source regions in the bulletin. In case COASSEIN encounters uncertainty by any contradiction in reported event types, the most probable choice is selected. This guideline is necessary, since SONODET has performed a template matching that did not prefer frequent clusters over exotic events.

The *SparseNet* approach suffers from some typical weaknesses that can be explained by Figure 10. Once again we have selected the most important class with Mc(100)=0.0 equivalent to 14 dB SNR at single traces. In this class, using the *SparseNet* approach, there were no missed events and no false alarms for the whole 10-day testbed, so only wrong locations are shown.

The first event is from the cluster east of Nikel. There is no appropriate pattern defined since all other events – with one exception – are very weak signals of different shape. The other challenge is the small Pg or Pn arrival 15 s before the true onset. It has the same backazimuth and may stem from a weak foreshock whose remainder is masked in the coda of the main event. In this case, *SparseNet* adjusted the most similar pattern to the overall shape of the whole sonogram and obtained the wrong region and even the wrong distance range.

The other examples are from clusters well represented in the pattern set of Figure 9. The difficulties once again come from overlapping events. In the second case, two blasts from Nikel are erroneously combined to Kostomuksha, four Kiruna blasts are jointly taken as Khibiny, and finally a Khibiny event is misclassified as Norwegian_Sea. The single trace results of SONODET at ARC1 are given to the right of the sonogram; in the cases 2 to 4 they also contain the correct solution. COASSEIN has outvoted the correct solution in favor of more agreement.

The examples in Figure 10 are the same 4 cases which Pluralistic Processing identifies as problematic to suggest human re-evaluation (see below). The errors are "reasonable" as changes in event signature have indeed moved these examples close to other types.

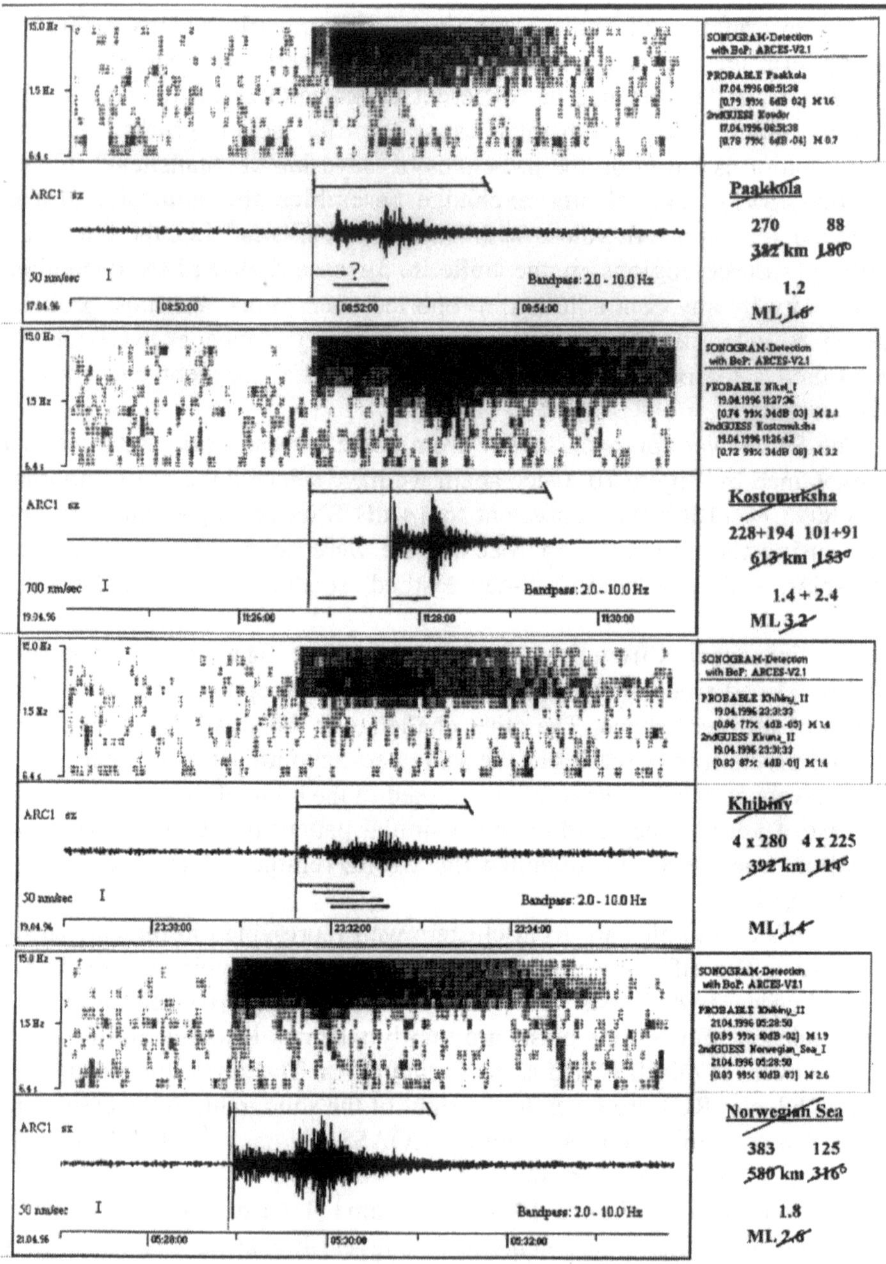

Figure 10. Processing weaknesses by SparseNet (see text for details).

4.3 COAEBULL

SparseNet contains a third software module COAEBULL intended to compare different bulletins (Joswig, 1999). It is not the envisioned, rule-based system needed for pluralistic reasoning that will weight the strengths and weaknesses of independent processing approaches. Still we benefit from COAEBULL as it collects the performance statistics presented in the next section. COAEBULL takes one to four bulletins as reference to evaluate an additional one.

The comparison is done on all events within a short time window; classification results are based on the definition of source regions. They may be *match*, not matched but *close* (within predefined "gray zones" around the source regions), wrong but *equidistant*, *wrong* region, *false alarm* and *missed*. COAEBULL implements a concept of completeness threshold that considers as *missed* only those events that are above the magnitude thresholds *Mc* of Figure 5.

5. RESULTS OF TESTBED PROCESSING

To establish ground-truth information, the available, human-reviewed bulletins from pIDC and Helsinki/Bergen University were complemented by our own, manual re-evaluation of the entire testbed April 12 - April 21, 1996. This re-evaluation was necessary because the automated detection results included many weak events, some of them composed by phase onsets not even visible at single traces. In this case, the reported event was assumed to be correct if any of the three conditions holds:

- the *SparseNet* identification by source region matches one phase from DP by time, type, and backazimuth,
- the declaring phases of RONAPP fit the seismogram signature,
- there is at least one additional phase in the onset list from other NORSAR arrays.

A typical situation was to confirm *SparseNet* reporting an event from the Kiruna quarry region when DP has missed the Pn/Pg onset but reported a Sn/Lg arrival with 225° backazimuth that occurred some 35 s behind the reported onset time of SONODET. The event is confirmed because it matches condition (1) – the timing of SONODET onset to Sn/Lg fits the Kiruna distance of 250 km, the backazimuth is correct also.

A reviewed bulletin was established that was complete for all events above *Mc*(100) = -0.5 within 1500 km radius around ARCESS. According to

Table 1, this corresponds to +3 dB SNR_{ST} on single traces. Below, we could only test if any reported detection conforms to the event confirmation criteria described above, so the bulletin will definitely be incomplete.

5.1 Classes for Performance Statistics

The available ground truth led to four different conditions for the performance statistics compiled by COAEBULL:

- events above Mc(100) = 0.0. This will summarize success and failure for the large, clearly visible events. Only ghost events of sufficiently large magnitude will be considered as false alarms. One would like to arrive at automated processing that masters this dataset without any errors.
- events above Mc(100) = -0.5. Reports on the maximum available data set with complete ground truth. Again the number of false alarms is limited by Mc. Now we reach the limits of human routine processing. The automated system will be worse than an analyst but it is important to quantify the decrease compared to case 0.0.
- as before with Mc(100) = -0.5 +trg. Includes all triggers, i.e., all declared events, regardless of magnitude. This compiles the performance under some more realistic assumption of observatory routine works since all false alarms bothering the analyst will be considered.
- extend to Mc(100) = -1.0 +trg. There can't be a change in the absolute number of successfully or erroneously declared events, as all of them were already included in the previous case. However, we consider even weaker events for detection now and the increase of missed events will quantify the ultimate limits of sensitivity.

5.2 Standard automated processing results

Table 3 lists the results of standard automated processing for the 10-day testbed. In the highest class with 28 dB SNR_{AB} on the array beam, 2 events were missed, and every day 1 ghost event was reported and 2 events were severely mislocated. The total error rate (*close+equidistant+wrong+missed* but excluding *false alarms*) is 39%.

In the next class with 17 dB SNRAB, the daily error rates are 3 missed events, 3 false alarms, and 4 mislocations. This gives a total error rate of 36%. Taking all events in column [–0.5 +trg], one gets 5 false alarms and 5 mislocations each day; the total error decreases to 29%. Lowering the detection threshold for half a magnitude will add another 50 missed events.

Investigating the cause of mislocations, we found a typical scenario quite independent of magnitude. Often they were mixed events, i.e., a sequence of blasts from the same quarry region with seismogram overlap. Then phases are grouped incorrectly which causes wrong distance. In the epicenter map of Figure 11, this error will introduce some artificial source regions like for the arc (210° - 240°, 550 km) around ARCESS. The size of circles conforms to Figure 5 to indicate magnitude. Actually these events are quarry blasts from Kiruna lying at half the distance; in an extreme case one weak Kiruna blast M_L 0.9 has been mislocated to Northern France with M_L 2.2. Other obvious clusters exist for Kiruna (300 km SW), Nikel (200 km E) and Khibiny (400 km SE). The remaining seismicity seems very scattered with many ghost events in Barents Sea, i.e., the NE quadrant. Note that the estimated backazimuths were correct for most of the erroneous cases. The average success rate for all events was 71% but decreased for the highest class to 61%. In low SNR, only P and Lg are visible and there are no ambiguities in phase association. For good SNR, more phases per event were detected, which complicates their association.

Table 3. Processing performance for 10 days of ARCESS data, comparing SparseNet (left columns) to standard processing (right columns).

	SparseNet on 4 stations				Standard processing, 25 stations			
Compl. Thresh. m100	0.0	-0.5	-.5+trg	-1.+trg	0.0	-0.5	-.5+trg	-1.+trg
number of events	67	177	228	307	67	180	262	312
regions matched	55	113	144	144	41	116	187	187
not matched but close	1	3	3	3	15	23	30	30
wrong but equidistant	6	28	42	42	0	0	0	0
wrong regions	5	16	21	21	9	14	18	18
FAs by region	0	17	28	28	8	32	48	48
detected by voting	0	0	1	1				
FAs by voting	0	0	5	5				
missed events	0	17	17	96	2	27	27	77

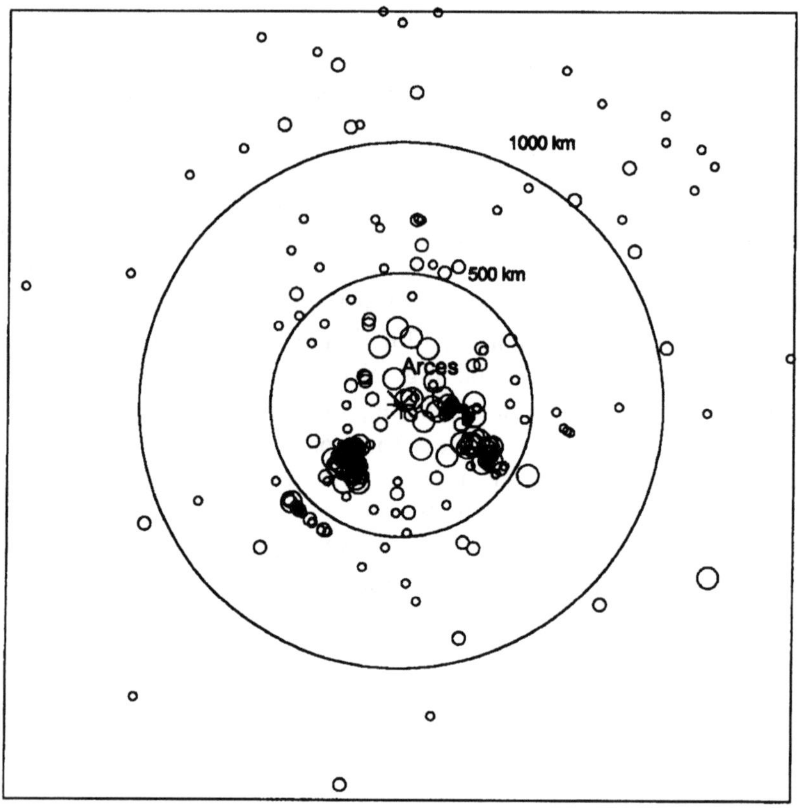

Figure 11. Automatically derived epicenter map obtained by standard processing for the 10-day testbed.

5.3 SparseNet Results

The *SparseNet* results are also compiled in Table 3. In order to obtain an equal number of processed events, all teleseisms have been excluded, although the success rate for these arrivals is generally above 90%. The remaining differences in number of events come from slight magnitude biases. The *SparseNet* statistics have two new rows that summarize event detections without location, e.g. by STA/LTA or the SONODET warning: *Significant energy outside pattern base* issued for larger events. *SparseNet* has systematically fewer false alarms than the standard processing; surprisingly it also missed less events indicating the high sensitivity of our specific sonogram scalings and the pattern adaptation process. Only for $Mc(100)$ = -1.0, the standard processing has fewer missed events, but this class corresponds to -8 dB SNR_{ST} for SONODET.

In the highest class, *SparseNet* had just 1 mislocation per day, no false alarm, and no missed event. The most frequent error is the assignment to a source region with same distance but different backazimuth; this remains true for all other classes. The overall error rate is 18% excluding teleseisms. This success demands an appropriate knowledge base and sufficiently clustered seismicity. The next class exhibits 2 missed events, 2 false alarms, and 5 mislocations per day, for an error rate of 36%. Adding all events declared by *SparseNet*, we see in column [-0.5 +trg] only a slight change to 37% error rate and 51 more events; the standard processing has added 82. Thus 3 dB SNR_{ST} brings *SparseNet* down to the limits of its processing capabilities.

6. PLURALISTIC PROCESSING

The examples and performance statistics given so far have demonstrated that the ideal solution for array processing is neither the exclusive application of beamforming/*f-k* analysis nor the sole use of AI techniques like pattern recognition. Obviously, most appealing is the intelligent combination of both approaches to utilize their respective strengths: the sensitivity of beamforming and the plausibility check on overall event signatures by PR. The integration of PR into the processing scheme of arrays is done by a pluralistic approach shown in Figure 3: both methods work in parallel on the same stream of raw data. Thus the picked phases with their reliable backazimuths from *f-k* analysis get augmented by a robust estimate of event distance and relative phase timing derived by PR. This joint information forms the base for better locations, suppression of false alarms and detection of events with emergent seismograms. PR can also support the regionalized discrimination by defining master events from active faults, known quarries and mining areas in the same epicenter area (Wüster, 1993). However, this aspect is not covered in our paper.

Pluralistic processing was realized here by the preliminary set of rules in Table 4 restricted to the evaluation of existing information in the standard processing list and SONODET/COASSEIN. There is also the possibility of a guided re-evaluation of the original data, but this processing path was not considered for the purposes of this study.

The first rule limits sensitivity to the capabilities of the weaker partner. As *SparseNet* performed surprisingly well for low SNR, this is not a real restriction. The bonus is a quite complete suppression of all false alarms. Rule 2 defines the ideal state when both approaches independently find the same solution. The result will have high reliability. The third rule will exclude distance errors by wrong phase timing in the standard processing; it will not help for cases where seismicity occurs outside the defined source regions. Rule 4 performs a plausibility check for the selected phases against the overall seismogram signature; it suppresses SparseNet solutions with wrong backazimuth. The next rule mirrors our human event confirmation when an emergent phase was missed by the standard detector. As seen by Figure 4, strong events can also be affected. The situation of rule 6 is counted as *detected as Problem* in the statistics of Table 5.

Inherently we followed a very conservative approach. Rules 3 to 5 consider only the SparseNet solutions although you may find better results by SONODET at single stations as in Figure 10. Also we did not try other solutions from the standard processing by selecting a different combination of phases for distance determination. These two decisions favor robustness and reliability at a cost of a higher number of problematic cases as can be seen in Table 5. Still, for the highest class we reach the quite ideal result of correct or marked for human inspection, no miss and no false alarm. The four problematic cases are just the ones of Figure 10; in cases two and three the standard processing also failed by distance.

One could decrease the number of problematic cases, especially in the lower classes, by more sophisticated rules designed in a less conservative manner. This could also handle a fraction of actual errors causes by the large variation of multiple events from Kiruna and Khibiny and the clusters East of Kiruna, NNW and East of Nikel which are not represented by patterns. However, this goes beyond the focus of our study.

Table 4. Ruleset for Pluralistic Processing.

Rule 1	An event is declared if it was considered by both detectors: at least one single site SONODET detection and one phase from the standard detector.
Rule 2	If both approaches derive full solutions that agree, the solution is found.
Rule 3	If both approaches derive full solutions that contradict, attempt to confirm the *SparseNet* source region with the standard processing solution's backazimuths.
Rule 4	Accept the standard processing solution if its distance is within the "gray zone" of the presumed *SparseNet* region.
Rule 5	Accept the *SparseNet* solution if standard processing does not declare a event but one phase from the standard detector supports by type, time and backazimuth.
Rule 6	If none of the conditions succeeds, the automated processing is considered *problematic* and flag the event for human analysis.

Table 5. Joint performance by Pluralistic Processing of 10 days ARCESS data.

	Pluralistic Processing			
Compl. Thresh. m100	0.0	-0.5	-.5+trg	-1.+trg
number of events	67	179	228	307
regions matched	63	134	171	171
not matched but close	0	3	5	5
wrong but equidistant	0	0	0	0
wrong regions	0	5	6	6
FAs by region	0	0	1	1
detected as Problem	4	20	29	29
FAs by Problem	0	0	0	1
missed events	0	17	17	96

7. CONCLUSIONS

Conventional array processing proceeds as a one-way street: one detector, one phase associator, one event declarator/locator. If any member of the chain fails, there is little chance for recovery. Evaluating 10 days of continuous ARCESS seismograms for a 4-station subarray, the *SparseNet* modules SONODET and COASSEIN achieved comparable performance to standard processing down to 3 dB SNR_{ST} at single stations; only below this threshold was the conventional beamforming/*f-k* processing of the full 25 station array clearly superior. More important, the SparseNet modules failed because of other reasons: events are missed when obscured by noise bursts, and mislocation typically affects the azimuth while distance estimates are mostly correct.

Combining both independent approaches performs like multi-computer control in an airplane; any mission-critical task must be secured by cross-checking. This adds redundancy by complementary processing, which is more than just another parameter evaluation based on the same preprocessing. In our case, a simple version of pluralistic evaluation successfully reduced errors to an acceptable minimum. It did not automatically solve all problems, but marked questionable cases for human inspection.

ACKNOWLEDGMENTS

This work was supported by Dr. Weiler and Dr. Leonard from the Israel Atomic Energy Commission. Jan Fyen at NORSAR compiled the data set and supplied detailed information about RONAPP and the Fennoscandian seismicity. The editor helped improve the manuscript by valuable comments.

REFERENCES

Bache, T.C., S.R. Bratt, J. Wang, R.M. Fung, C. Kobryn, and J.W. Given (1990) The Intelligent Monitoring System, *Bull. Seism. Soc. Am.* **80,** 1833-1851.

Bache, T.C., S.R. Bratt, H.J. Swanger, G.W. Beall, and F.K. Dashiell (1993) Knowledge-based interpretation of seismic data in the Intelligent Monitoring System, *Bull. Seism. Soc. Am.* **83,** 1507-1526.

Bartal, Y., M. Villagran, Y. Ben-Horin, G. Leonard and M. Joswig (2000) Definition of exclusion zones using seismic data, *Pure Appl. Geophys.*, in press.

Berteussen, K. A. (1974) NORSAR location calibrations and time delay corrections, *NORSAR Sci. Rep.* 2-73/74.

Birtill, J. W. and F. E. Whiteway (1965) The application of phased arrays to the analysis of seismic body waves, *Phil. Trans. Roy. Soc. Lond.* **A 258,** 421-493.

Bishop, C. M. (1995) *Neural networks for pattern recognition,* Oxford University Press, Oxford.

Bowers, D. (1999) Depth phase identification for event screening: Future improvements. Proc. IDC Technical Experts' Meeting on Event Screening, CTBTO, Vienna.

Bratt, S. R. and T. C. Bache (1988) Locating events with a sparse network of regional arrays, *Bull. Seism. Soc. Am.* **78,** 780-798.

Buchbinder, G. G. R. and R. A. W. Haddon (1990) Azimuthal anomalies of short-period P-wave arrivals from Nahanni aftershocks, Northwest Territories, Canada, and effects of surface topography, *Bull. Seism. Soc. Am.* **80,** 1272-1283.

Cansi, Y., J.-L. Plantet and B. Massinon (1993) Earthquake location applied to a mini-array: k-spectrum versus correlation method, *Geophys. Res. Lett.* **20,** 1819-1822.

Capon, J. (1969) High-resolution frequency-wavenumber spectrum analysis, *Proc. IEEE* **57,** 1408-1418.

Cassidy, F., A. Christoffersson, E. S,. Husebye and B. O. Ruud (1990) Robust and reliable techniques for epicenter location using time and slowness observations, *Bull. Seism. Soc. Am.* **80,** 140-149.

Chiaruttini, C., V. Roberto and F. Saitta (1989) Artificial intelligence techniques in seismic signal interpretation, *Geophys. J. Int.* **98,** 223-232.

Christoffersson, A. and R. Roberts (1996) Trinity or verity? in *Monitoring a Comprehensive Test Ban Treaty,* eds. E. S. Husebye and A. M. Dainty, NATO ASI, Series E, 303, 611-628, Kluwer, Dordrecht.

Der, Z. A., D. R. Baumgardt and R. H. Shumway (1993) The nature of particle motion in regional seismograms and its utilization for phase identification, *Geophys. J. Int.* **115,** 1012-1024.

Der, Z. A., M. R. Hirano and R. H. Shumway (1990) Coherent processing of regional signals at small seismic arrays, *Bull. Seism. Soc. Am.* **80,** 2161-2176.

Flatte, S. M. and R.-S. Wu (1988) Small-scale structure in the Lithosphere and Asthenosphere deduced from arrival time and amplitude fluctuations at NORSAR, *J. Geophys. Res.* **93B,** 6601-6614.

Fyen, J. (1996) Personal communications.

Gestermann, N. (1993) Unpublished report.

Green, P. E., R. A. Frosch and C. F. Romney (1965) Principles of an experimental large aperture seismic array (LASA), *Proc. IEEE* **53,** 1821-1833.

Harjes, H.-P. and M. Henger (1973) Array-Seismologie, *Z. Geophysik* **39,** 865-905 (in German)

Harjes, H.-P., M. L. Jost, J. Schweitzer and N. Gestermann (1993) Automatic seismogram analysis at GERESS, *Computers and Geosciences* **19**, 157-166.

Harjes, H.-P., M. Jost *and* J. Schweitzer (1994) Preliminary calibration of candidate alpha stations in the GSETT-3 network, *Ann. Geophys.* **37**, 383-396.

Haubrich, R. A. (1968) Array design, *Bull. Seism. Soc. Am.* **58**, 977-991.

IDC (1999) IDC processing of seismic, hydroacoustic, and infrasonic data, IDC-5.2.1, CTBTO, Vienna.

Jarpe, S. and F. Dowla (1991) Performance of high-frequency three-component stations for azimuth estimation from regional seismic phases, *Bull. Seism. Soc. Am.* **81**, 987-999.

Jost, M. L., J. Schweitzer and H.-P. Harjes (1996) Monitoring nuclear test sites with GERESS, *Bull. Seism. Soc. Am.* **86**, 172-190.

Joswig, M. (1990) Pattern recognition for earthquake detection, *Bull. Seism. Soc. Am.*, **80**, 170-186.

Joswig, M. (1992) System architecture of seismic networks and its implications to network automatization, *Cahiers Centre Europ. Geodyn. Seism.* **5**, 75-84.

Joswig, M. (1993) Automated seismogram analysis for the tripartite BUG array: an introduction, *Computers* and *Geosciences*, **19**, 203-206.

Joswig, M. (1995a) Artificial Intelligence techniques applied to seismic signal analysis, in Dynamical systems and artificial intelligence applied to data banks in Geophysics, ed. A. Gvishiani and J. Dubois, *Cahiers Centre Europ. Geodyn. Seism.*, **9**, 5-15.

Joswig, M. (1995b) Automated classification of local earthquake data in the BUG small array, *Geophys. J. Int.*, **120**, 262-286.

Joswig, M. (1996) Pattern recognition techniques in seismic signal processing, in Proc. 2. workshop AI in seismology and engineering seismology, *Cahiers Centre Europ. Geodyn. Seism.*, **12**, 37-56.

Joswig, M. (1999) Automated Processing of seismograms by SparseNet, *Seism. Res. Letters* **70**, 705-711.

Joswig, M. and Schulte-Theis, H. (1994) Master-event correlations of weak local earthquakes by dynamic waveform matching, *Geophys. J. Int.*, **113**, 562-574.

Klumpen, E. and Joswig, M. (1993) Automated reevaluation of local earthquake data by application of generic polarization patterns for P- and S-onsets, *Computers & Geosciences*, **19**, 223-231.

Kvaerna, T. (1992) Automatic phase association and event location using data from a network of seismic microarrays, *NORSAR Sci. Rep.* 2-91/92, 96-121.

Kvaerna, T. and F. Ringdal (1992) Integrated array ad three-component processing using a seismic microarray, *Bull. Seism. Soc. Am.* **82**, 870-882.

Kvaerna, T. and F. Ringdal (1993) Intelligent post-processing of seismic events – part 3: Precise relocation of events in a known target region, *NORSAR Sci. Rep.* 2-92/93, 93-104.

Kvaerna, T. and F. Ringdal (1996) Generalized beamforming, phase association and threshold monitoring using a global seismic network, in *Monitoring a Comprehensive Test Ban Treaty*, eds. E. S. Husebye and A. M. Dainty, NATO ASI, Series E, 303, 447-466, Kluwer, Dordrecht.

Lay, T. and Wallace, T. C. (1995) *Modern global seismology*, Academic Press, San Diego CA.

Leonard, G., M. Villagran, M. Joswig, Y. Bartal, N. Rabinowitz and A. Saya, (1999) Seismic source classification in Israel by signal imaging and rule-based coincidence evaluation, *Bull. Seism. Soc. Am.* **89**, 960-969.

Magotra, N., N. Ahmed and E. Chael (1987) Seismic event detection and source location using single-station (three-component) data, *Bull. Seism. Soc. Am.* **77**, 958-971.

Menke, W. and A. Lerner-Lam (1991) Observations of the transition from linear polarization to complex polarization in short-period compressional waves, *Bull. Seism. Soc. Am.* **81**, 611-621.

Menke, W., A. L. Lerner-Lam, B. Dubendorff and J. Pacheco (1990) Polarization and coherence of 5 to 30 Hz seismic wave fields at a hard-rock site and their relevance to velocity heterogeneities in the crust, *Bull. Seism. Soc. Am.* **80**, 430-449.

Mykkeltveit, S. and Bungum, H. (1984) Processing of regional seismic events using data from small-aperture arrays, *Bull. Seism. Soc. Am.*, **74**, 2313-2333.

Pearce, R. G. (1999) Depth phase identification for event screening: Current practice and some theory, Proc. IDC Technical Experts' Meeting on Event Screening, Vienna.

Rabinowitz, N. and M. Joswig (1994) Detection, localization, and identification of local seismic events by a small array in Israel, 1. Scientific Progress Report, GIF I-0914-123.08/91.

Ringdal, F. and E. S. Husebye (1982) Application of arrays in the detection, location, and identification of seismic events, *Bull. Seism. Soc. Am.* **72**, S201-S224.

Roberto, V. and C. Chiaruttini (1992) Seismic signal understanding: A knowledge-based recognition system, *IEEE-Trans. Sig. Proc.* **40**, 1787-1806.

Roberts, R. G., A. Christoffersson and F. Cassidy (1989) Real-time detection, phase identification and source location estimation using single station three-component seismic data, *Geophys. J.* **97**, 471-480.

Roberts, R. G. and A. Christoffersson (1990) Decomposition of complex single-station three-component seismograms, *Geophys. J. Int.* **103**, 55-74.

Ruud, B. O. and E. S. Husebye (1992) A new three-component detector and automatic single-station bulletin production, *Bull. Seism. Soc. Am.* **82**, 221-237.

Ruud, B. O., E. S. Husebye, S. F. Ingate and A. Christofferssson (1988) Event location at any distance using seismic data from a single, three-component station, *Bull. Seism. Soc. Am.* **78**, 308-325.

Scherbaum, F. and J. Wendler (1986) Cross spectral analysis of Swabian Jura (SW Germany) three-component microearthquake recordings, *J. Geophys.* **60**, 157-166.

Schweitzer, J. (1995) Mislocation vectors for small aperture arrays – a step towards calibrating GSETT-3 stations, *NORSAR Sci. Rep.* 1-94/95.

Searfus, R. M. (1989) Using expert systems in treaty verification, *Energy and Techn. Review*, No 1, 26-33, Livermore CA.

Suteau-Henson, A. (1991) Three-component analysis of regional phases at NORESS and ARCESS: polarization and phase identification, *Bull. Seism. Soc. Am.* **81**, 2419-2440.

Tarvainen, M. (1992) Automatic seismogram analysis: statistical phase picking and locating methods using one-station three-component data, *Bull. Seism. Soc. Am.* **82**, 860-869.

Walck, M. C. and E. P. Chael (1991) Optimal backazimuth estimation for three-component recordings of regional seismic events, *Bull. Seism. Soc. Am.* **81**, 643-666.

Whiteway, F. E. (1965) The recording and analysis of seismic body waves using linear cross arrays, *Radio Electr. Eng.* **29**, 33-48.

Wüster, J., 1993. Discrimination of chemical explosions and earthquakes in Central Europe - a case study, *Bull. Seism. Soc. Am.*, **83**, 1184-1212.

Chapter 9

AUTOMATIC PHASE PICK REFINEMENT AND SIMILAR EVENT ASSOCIATION IN LARGE SEISMIC DATASETS

Richard Aster and Charlotte Rowe
Department of Earth and Environmental Science and Geophysical Research Center, New Mexico Institute of Mining and Technology, Socorro, New Mexico, USA

Key words: microearthquake, clustering, cross-correlation, relocation, coherency

Abstract: Large digital seismic data sets typically have approximate and inconsistent P and S wave arrival estimates. If not corrected, this situation frequently blurs the true hypocenter distribution so as to mask important fault, volcanic, or other seismogenic structure. Recent applications which address this problem for increasingly large data sets have resulted in dramatic illumination of seismogenic features in hydrocarbon and geothermal reservoirs, as well as in tectonic and volcanic earthquake regions. Existing algorithms frequently require substantial analyst interaction, which can become prohibitive once more than a few tens of events are considered. We summarize progress towards developing a fully automatic technique for removing pick inconsistencies and subsequently re-estimating hypocenters, and discuss our development of an adaptive, correlation-based algorithm which points the way towards techniques that appear suitable for the largest data sets presently existing.

1. INTRODUCTION

Earthquake location has historically depended upon the ability of human analysts to estimate body wave arrival times. Standard network operations generally involve the manual estimation of such arrivals, or computer estimation of them using automatic software pickers. The most common human or computer picking approach is done one event at a time, on a gather

C.H. Thurber and N. Rabinowitz (eds.), Advances in Seismic Event Location, 231–263.

across several or all recording stations. This method, although well suited for the near-real-time processing demands of network operation, is disadvantageous for producing highly consistent picks because source radiation pattern and propagation path differences among the gathered traces are large. Arrivals estimated from this heterogeneous suite of waveforms are subsequently used to estimate the hypocenter and the event is then catalogued and archived. Seldom are any but the most egregious picking errors noted and corrected prior to moving on to the next earthquake; hence, picking inconsistencies between even similar events remain unresolved. Such network operations have produced very large sets of hypocenter locations (e.g., in excess of four million events with over a hundred million associated picks from 1981-1999 for southern California) with 1-σ location error estimates of generally several to a few tens of kilometers for regional-scale networks with good azimuthal coverage. Standard catalogue locations have served to document general seismicity levels and the gross geometry of larger-scale seismogenic features, but, within the diffuse scatter of locations which result from picking inconsistencies, fine details remain unresolved.

We summarize a methodology for reducing this source of error prior to location and/or velocity model re-estimation processing. The fundamental observation is that, rather than examining traces across all stations on an event-by-event basis, it is better for pick consistency to simultaneously analyze traces from many nearby events on a station by station basis (e.g. Dodge et al., 1995; Shearer, 1997; Shearer, 1998). This data regrouping allows one to constrain the consistency and accuracy of pick times between nearby events by exploiting increased waveform similarity due to reduced propagation path variability (and reduced source variability, in some cases). The tightest constraints can be expected to be found between the most similar earthquake waveforms, which arise from events with the most consistent radiation patterns, moment rate functions and propagation paths. This approach is similar to the statics correction problem in reflection seismology (e.g., Sheriff and Geldart, 1995), although in such applications the spatial sampling is sufficient to limit propagation path variability between nearby receivers so that processing may still be done on an event-by-event basis. In addition, the resulting catalogue of similar events can be used in reciprocity-based constraining of near-receiver structures (Spudich and Bostwick, 1987) and analyzing temporal variability in Earth properties (e.g., Poupinet et al., 1984; Nadeau et al., 1995; Haase et al., 1995; Beroza et al., 1996; Aster et al., 1996; Antolik et al., 1996).

In many cases, significant improvement in hypocenter location precision, and resulting delineation of fine seismogenic details, has been achieved through visually-assisted cross-correlation and repicking of phases (e.g., Ito, 1985; Fremont and Malone, 1987; Phillips et al., 1997; Lees, 1998; Phillips,

2000) and/or by pattern recognition and collapsing of spatial structures (e.g., Fehler et al., 1987; Jones and Stewart, 1997; Fehler et al., 2000). This is an extremely time-intensive undertaking, however, that is limited to small, focused subsets of the larger catalogs. Dodge et al. (1995) developed a semi-automatic correlation approach which estimates individual pick lags within event ensembles, and uses weighted inter-event phase lag constraints to solve for consistent picks. This approach is a significant improvement over some earlier efforts which rely upon a master event method, in that highly similar multiplets are not required, as dissimilar event pairs will have limited influence over one another. Shearer (1997) demonstrated significant improvements in delineating the seismogenic features associated with the Whittier Narrows aftershock sequence in California, also invoking a pick lag estimation method with preliminary segregation of events into clusters meeting joint minimum cross-correlation and proximity criteria.

Quantitative, waveform correlation-based phase repicking and/or relative doublet and multiplet relocations have produced impressive resolution of seismogenic structures within families of similar events, and are providing new insight into faulting and volcanic processes. Fremont and Malone (1987) used relative relocations based on multiplet cross-correlation lags at Mount St. Helens to delineate source regions on the order of a few tens of meters. Got et al. (1994) relocated decollement seismicity at Kilauea volcano, Hawaii, using multiplets chosen based on cross-spectral coherency for individual event pairs. Gillard et al. (1996) and Rubin et al. (1998) used highly automated similar event methods to reduce quasi-linear "cigars" of Kilauea microearthquake foci into pencil-thin lines of seismicity. Nadeau et al. (1995) relocated small clusters at Parkfield and demonstrated that the seismogenic regions of the San Andreas fault in this region are primarily characterized by repeating microearthquakes on very small patches surrounded by large, nearly aseismic, probably creeping, regions. Rubin et al. (1999) identified seismogenic streaks on creeping segments of the San Andreas fault system north of Parkfield which resemble along-strike smearing of the Parkfield seismicity, and Waldhauser et al. (1999) found similar lineations on the northern Hayward fault.

The above methods, although progressively more quantitative and efficient, still rely to varying degrees on substantial user interaction and have been generally applied to limited special studies of catalogue subsets chosen either through spatial restrictions or genetic assumptions (e.g., a limited box of data, a specific aftershock sequence). We seek to obtain high-accuracy locations through retroactive analysis of large existing data sets and in near-real-time processing of data. It is thus desirable to have methods which combine the most advantageous features of the techniques already described in a highly automated package which can easily be implemented for even the largest data sets with a manageable amount of analyst interaction. Such a technique will have to be adaptive to stationary and nonstationary sources of noise, to a wide range of signal frequency content and signal-to-noise, and to the wide range of similar event rates observed in different settings (e.g., Aster and Scott, 1993; Nadeau et al., 1995).

We first outline our technique with a summary of signal processing tools and an explanation of how we apply them to the data, followed by a discussion of the iterative calculation of lags from the inter-event constraints. Throughout, we also point out aspects where the technique may be enhanced in the future. This is followed by an example application to hydrofracture microseismicity at the Soultz-sous-Forêts geothermal reservoir, France, where we can compare our automatic results with painstakingly estimated manual cross-correlation picks.

2. TECHNIQUE

2.1 Preliminary Data Organization

Events are organized into individual event directories, each of which contains individual time series files. Currently we operate on SAC data file format (Goldstein et al., 1998). Prior to processing, all waveforms must have phase picks produced by an analyst or autopicker. Event directories are scanned by a C-shell script to identify, for each phase, sets of events recorded at each station.

2.2 Relative Lag Estimation

Relative correlation lag estimation between pairs of station-common traces proceeds in two steps: a coarse (integer sample) correlation step, which provides an estimate of relative lag to the nearest sample, and a fine (subsample) correlation step which provides a sub-sample relative lag refinement. A user-specified M-sample time window, including a fractional pre-pick offset, is established about the preliminary picks for each pair of events to be compared. We currently choose M to accommodate several cycles of the phase to be compared; further work is needed to set this parameter more adaptively. Within each subset of events recorded at a given station, we compare each potentially similar event pair for each phase (e.g., P and S for local earthquakes) to measure waveform similarity and to estimate relative lag. Calculating the entire suite of $N(N-1)/2$ relative lags for a data set of N events occupying a region with a spatial extent comparable to or larger than the station-to-event distance is usually unproductive, as it results in a large percentage of low-correlation (and hence unusable) comparisons due to large waveform variations. It is thus desirable to limit the number of comparisons via a nearest neighbor approach (e.g., Aster and Scott, 1993; Shearer, 1997) based on preliminary hypocenter locations, although this may orphan events with inaccurate initial locations from the process. A full catalogue processing algorithm may thus frequently require a final step to appropriately integrate such non-correlating events into the final set of relocated events.

For stations with multiple components (usually 3), we apply polarity filtering to improve phase arrival signal-to-noise prior to waveform comparison. An average covariance matrix is calculated from the sum of the energy-normalized multicomponent signal for each event in the pair, as

$$\mathbf{C} = \frac{1}{2}\left(\frac{\mathbf{x}_1{}^T\mathbf{x}_1}{E_1} + \frac{\mathbf{x}_2{}^T\mathbf{x}_2}{E_2}\right) \tag{1}$$

where x_i is a matrix with columns which are individual component time series for event *j*, and

$$E_j = \text{trace}\left(\mathbf{x}_j{}^T\mathbf{x}_j\right) \tag{2}$$

The diagonalization of the positive definite **C** gives a unit eigenvector λ_1 characterizing the best data projection for mutually linearizing the particle motions of the two traces in the vicinity of the phase of interest. Subsequent analysis is performed on the projected (1-dimensional) data:

$$\mathbf{x}'_j = \mathbf{x}_j \cdot \hat{\lambda}_1 \tag{3}$$

Each waveform pair is next transformed into the frequency domain, where $X_j = \text{FFT}(\mathbf{x}_j)$. The cross-spectrum for the two traces is the index-wise conjugate product of their respective Fourier Transforms

$$s_k = X_{1k} X^*_{2k} \tag{4}$$

and the corresponding coherency is

$$c_k = \frac{\sum_{l=j-m}^{j+m} s_l}{\sum_{l=j-m}^{j+m} |s_l|} \tag{5}$$

where the frequency index is j, and the corresponding frequency is $f_j = j f_s/M$, where f_s is the sampling rate and M is the length of the time series. The frequency smoothing bandwidth over which the coherency is calculated is $2m+1$ positive frequency bins, and correspondingly less than this near the zero and Nyquist frequencies, which occur at frequency indices j=0 and $j=M/2$, respectively. m is chosen to be five frequency bins. We use the coherency to zero-phase prefilter the two seismograms under consideration, which emphasizes coherent and/or high energy frequency bands prior to cross-correlation lag estimation. The spectral weighting used is

$$\gamma_k = \left(|X_{1k}||X_{1k}|\right)^{\alpha/2} |c_k|^{\beta} \tag{6}$$

where the exponents α and β are chosen to control the desired amount of filtering and the balance between its coherency- and power-based components. The goal is to down-weight incoherent frequency bands while avoiding removal of useful signal, which would be a risk if an *a priori* bandpass filter choice were used (Figure 1).

The coherency-filtered signals, with spectra

$$Y_{jk} = X_{jk}\ \gamma_k \tag{7}$$

are next correlated using a fast frequency-domain calculation (with M-length zero padding to eliminate circular correlation wrap-around effects), and the maximum of the cross-correlation function is chosen as the preferred estimate for the coarse inter-event pick lag. Because it is common for the initial pick inconsistency to be an appreciable fraction of the window length, M, we symmetrically rewindow the two traces (keeping track of the total window shift) until the final coarse lag between the re-windowed segments is 1, 0, or -1 sample, and the resulting integer sample lag estimate relative to the original picks, l_1, is stored.

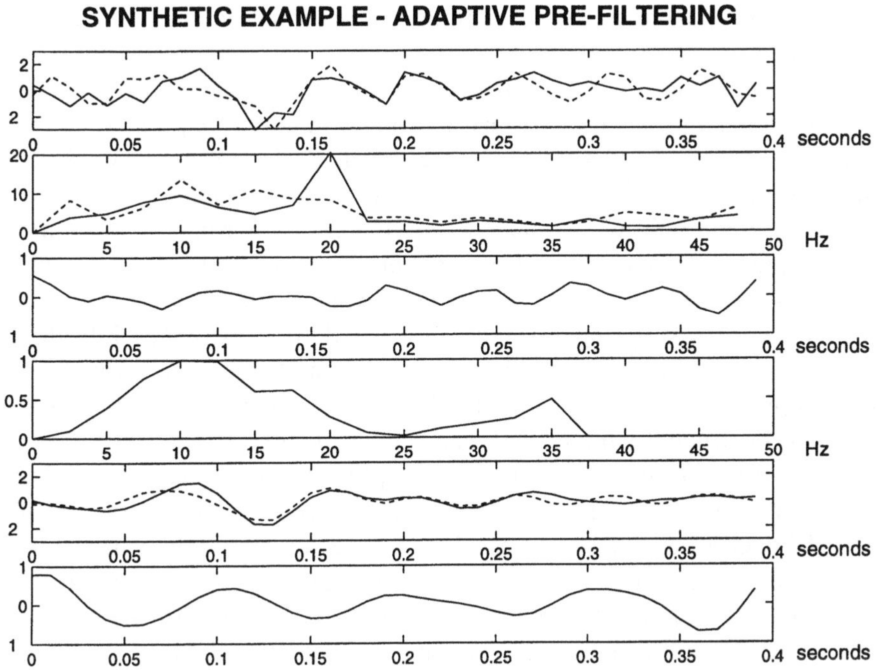

Figure 1. Example adaptive pre-filtering. a) Two synthetic seismograms lagged by one sample with independent broadband Gaussian noise and a sampling rate of 100 Hz. b) Amplitude spectra of unfiltered seismograms. c) Cross-correlation function for seismograms of (a). The maximum cross-correlation coefficient is approximately 0.5. and the maximum cross-correlation lag is at zero samples. d) Coherency calculated from spectra of 1b. Although most coherent energy resides below 20 Hz, cross-coherency has a small peak at 35 Hz. *A priori* low pass filtering may reject this energy, which could have an important effect on the cross-correlation. e) Adaptively filtered seismograms (Eq. 5). Note significant reduction in the random noise constituent and overall increased waveform similarity. f) Re-computed cross-correlation function for the waveform pair, showing a maximum cross-correlation coefficient > 0.9 and a retrieved correct coarse lag of 1 sample.

To estimate a standard deviation for these coarse (integer sample) correlation lags, we have devised a band-pass algorithm in which we perform a set of eight narrow-band correlation lags encompassing the entire Nyquist range and estimate the standard deviation about the mean, where each term is weighted by the cross-spectral power in that band. The mean of these lag estimates is identical with that obtained by an unfiltered correlation, and the standard deviation reflects the power-weighted degree of lag consistency throughout the available frequency range. The coarse correlation standard deviation is used as a discriminator to determine which waveforms are sufficiently similar to merit subsample lag estimation. Waveform pairs with a sufficiently high correlation function maximum are subsequently passed to the subsample cross-correlation step for further lag refinement.

For highly similar seismograms, the attainable level of relative lag resolution is significantly less than the sample interval (e.g., Poupinet et al., 1984). Furthermore, this level of precision is desirable to accurately image fault structures, which may have spatial scales of a meter or less. For example, in a region where the P-wave velocity is approximately 5 km/s and the sampling rate is 100 samples/s, the integer sample arrival time resolution for high-signal-to-noise data can be as poor as 5 ms, even for perfect data. This level of quantizing error in the arrival time estimates introduces a worst-case location error of up to 25 m. The ability to align waveforms to subsample precision can dramatically improve resolution of small-scale features if sufficiently high quality data are available.

One method of obtaining sub-sample precision lags is to Fourier interpolate the cross-correlation function by padding its Fourier transform with *KM* zeros (e.g., Aster et al., 1990). The inverse Fourier transform of the padded frequency-domain representation is an interpolated cross-correlation function with a new time resolution $1/K$ that of the original; it further has no new frequency content, consistent with the initial Fourier assumption of an unaliased time series. Fourier interpolation is robust, but in its most straightforward implementation it does not adapt for differences in frequency content and/or coherency in the signals under consideration and has time resolution limited to $1/K$ samples. Significant increase in the sample density also leads to a noticeable loss in computational efficiency, as the number of samples in a given time window multiply by a factor K and the most efficient spectral-based correlation methods typically scale in computational time by $K \ln K$.

An alternative method for determining subsample lags is to use the zero-intercept slope of the cross-spectral phase (Eq. 4), where the slope-based sub-sample lag estimate

$$l_2 = \frac{1}{2\pi}\frac{d\phi(f)}{df} \tag{8}$$

may be found from the discrete cross-spectral phase

$$\phi_j = \tan^{-1}\left(\frac{imag(s_j)}{real(s_j)}\right) \tag{9}$$

In many phase slope-fitting applications (e.g., Poupinet et al., 1984; Got et al., 1994), the relative weights of the phase values used in the slope estimate are calculated from the coherency (Eq. 4). This provides useful relative weights for the phase points, but requires ad-hoc scaling to give meaningful standard deviations to calculate error statistics, as the calculated weights are simply heuristic coherency mappings.

Because proper estimation of the cross-spectral phase slope is essential to obtaining the highest-accuracy estimates of relative lag and its standard deviation, we apply multitaper spectral estimation. In the multitaper formalism, statistically independent spectral estimates are obtained using orthogonal prolate spheroidal eigentaper windows (Figure 2). These tapering functions can be precomputed for a particular data length, M, and a specified time-bandwidth product, W. W is typically chosen to be four frequency bins, and $2W$ approximately specifies the resolution of the resulting spectral estimates. A suite of such eigenspectra may be combined as an average or in a weighted sum to provide a low spectral leakage (outside of the $[j-W, j+W]$ frequency band), low variance spectral estimate and its standard deviation. Multitaper-based estimates have superior properties for spectral estimation, both in terms of data variance and spectral leakage, over traditional tapering methods (Thomson, 1982; Park et al., 1987). Here, we calculate multitaper cross-spectral estimates from two seismograms through the conjugate multiplication (3) of corresponding multitaper spectra.

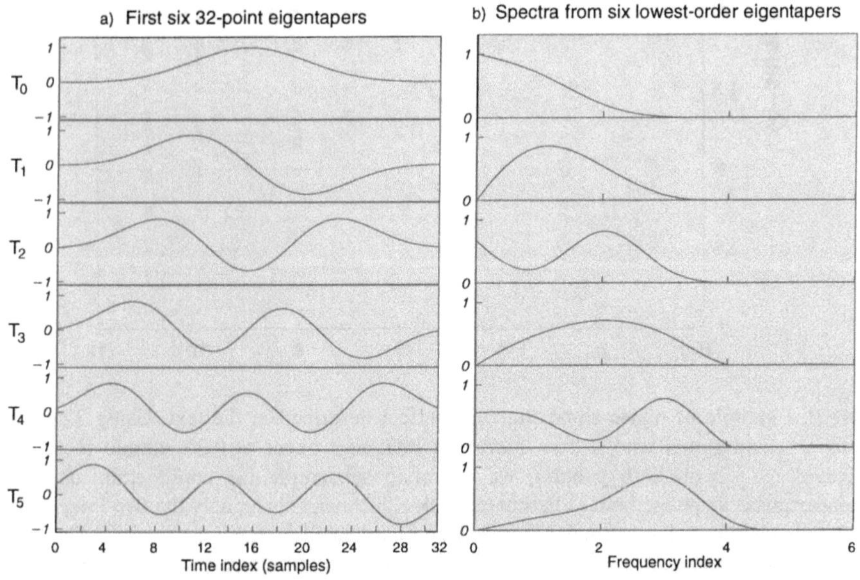

Figure 2. a) First six prolate spheroidal multitaper orthogonal windows for a time-bandwidth product of 4 and a window length of 32 samples. b) Amplitude spectra of the windowing functions of (a). Note that leakage for spectral estimates obtained using each of these windowing functions will be nearly confined to ±4 frequency bins.

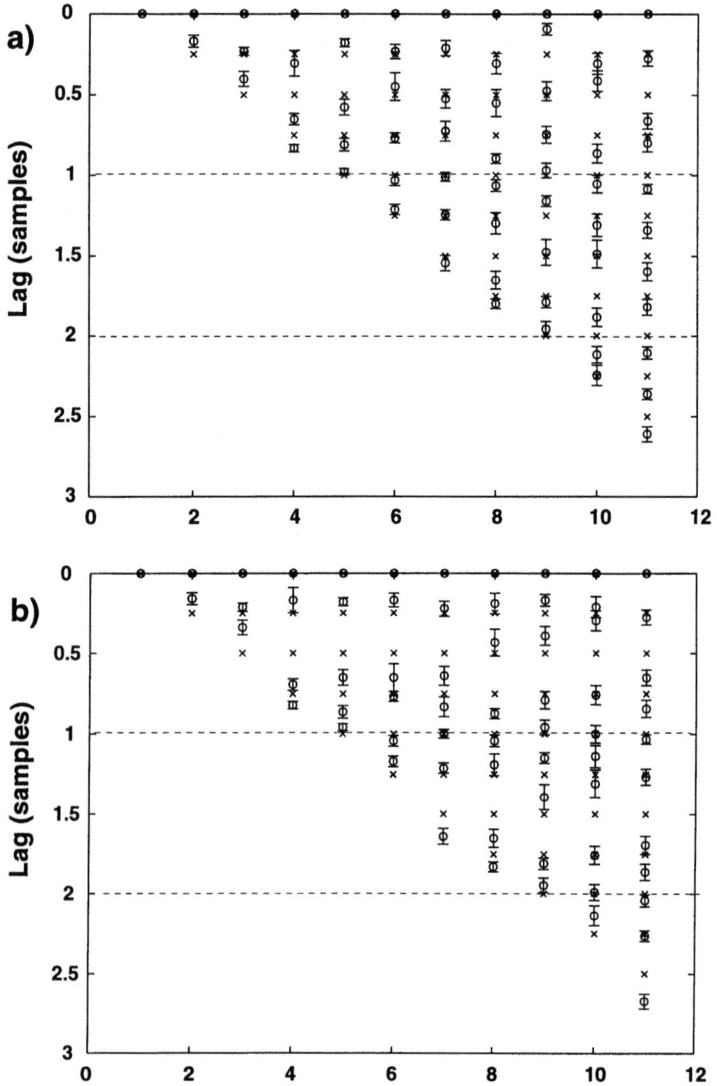

Figure 3. Example of phase slope flattening effect described in the text. Using 32-sample synthetic seismograms whose true interevent pick lags occur at 0.25 sample increments (indicated by x's on both panels), we calculate subsample lag components using the multitaper phase slope method. a) Estimated pick adjustments using only the two lowest-order eigentapers showing lag estimates (circles) and associated 1-sigma error bars. The misfit between the true and estimated lags is generally consistent with the error bars. b) Subsample lag estimates calculated using eight lowest order eigentapers. Note that although the integer sample corrections have been properly applied, subsample shifts away from integer lag values are underestimated (circles tend to cluster about the integer lag values). This bias is caused by flattening of the phase slope at higher frequencies due to spectral leakage from the higher order tapers (Figure 2).

The lowest order eigentapers (especially 0 and 1) have very low spectral leakage outside of the specified resolution (2*W*). Higher order tapers have progressively worse spectral leakage characteristics which may unacceptably flatten the estimated phase slope, and hence underestimate the subsample correlation lag (Eq. 7). This effect is especially significant for smaller window lengths (typically M<64, where the maximum number of usable phase points is less than 31). The flattening results from spectral leakage contaminating phase estimates with out-of-band phase contributions which are constrained to the first branch of the inverse tangent function (Eq. 8). Because the phase at high frequencies is generally random due to the absence of strong coherent signal, phase unwrapping does not solve the problem. Figure 3 shows a synthetic example demonstrating this systematic underestimation of the lag. We have determined that an acceptable trade-off between the advantages of the multitaper method (quantitative error bars and low spectral leakage) versus the need for low spectral leakage associated with the phase slope estimation, is to use the average of the 2 lowest-order tapers to obtain the spectral values, while using the standard deviation of the 6 lowest-order tapers to estimate standard deviations on each phase point. Figure 4 shows an example spectral estimate using this method. Adaptive methods exist for more sophisticated multi-taper weighting schemes in the estimation of power spectra (Park et al., 1987), and such methods should be developed for cross-spectral phase estimation.

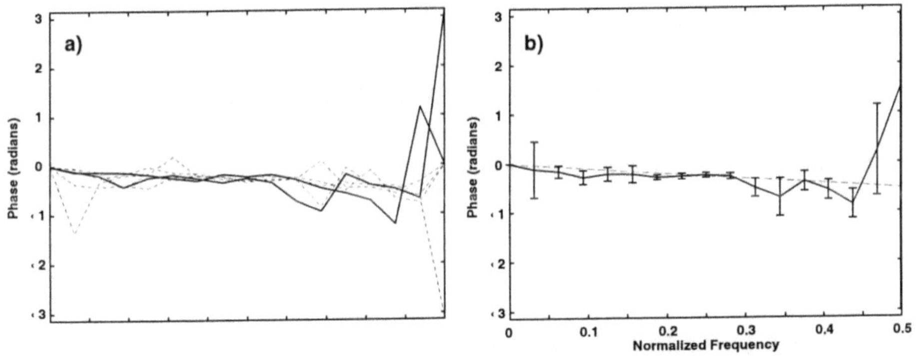

Figure 4. a) Cross-spectral phase estimates from 6 lowest-order eigenspectra for a synthetic pair of similar waveforms. The 2 lowest order phase estimates are shown as bold, solid lines; 4 higher order cross-spectral phase estimates are shown as dotted lines. In the phase slope estimation, the average of the 2 lowest order eigenspectra is used. b) Cross-spectral phase estimate for example event pair. The solid line represents the mean cross-spectral phase taken from the zero and first order eigenspectra. Standard deviations (vertical bars) are calculated from the 6 lowest-order eigenspectra, then mapped by the tangent function (9) to provide a conservative estimate of cross-spectral phase slope error. The zero-intercept phase slope (dashed line) is proportional to the subsample lag estimate for the waveform pair.

Prior to estimating the phase slope, we down-weight highly uncertain phase estimates (thus further reducing possible phase-slope flattening bias) by stretching the standard deviations of the multitaper estimates, σ_j, using the mapping

$$\sigma_j' = \tan(\sigma_j) \quad (-\pi/2 + \varepsilon < \sigma_j < \pi/2 - \varepsilon) \tag{10}$$

(we currently use $\varepsilon = 0.01$) and removing points from the phase slope estimation if σ_j is outside of the range specified in (10). We estimate the phase slope and its standard deviation using the L_2 (least-squares) zero-intercept linear regression for the P usable frequency points

$$\frac{d\phi}{df} = \frac{\sum_{j=1}^{P} \phi_j / \sigma_j'}{\sum_{j=1}^{P} f_j / \sigma_j'} \tag{11a}$$

$$\sigma\left(\frac{d\phi}{df}\right) = \frac{\left(\sum_{j=1}^{P} 1/\sigma_j'\right)^{1/2}}{\sum_{j=1}^{P} f_j / \sigma_j'} \tag{11b}$$

2.3 Solving for Pick Corrections from Interevent Lags

The desired N-vector of pick adjustments, $\mathbf{b}$, for N events is the solution to

$$\mathbf{G}\,\mathbf{b} = \mathbf{d} \tag{12}$$

where $\mathbf{d}$ is a Q (up to $N(N-1)/2$-length) vector of weighted inter-event lags with elements

$$d_i = \frac{l_{1,i} + l_{2,i}}{\sigma^{\dagger}_i} \tag{13}$$

where the first and second terms are the coarse and fine lag estimates, the standard deviation, $\sigma^{\dagger}_i$, is the quadrature sum of the fine and coarse lag standard deviation estimates, and the contribution from the fine lag estimate has been converted to time using Eq. 7. The system matrix, **G**, is a weighted first-difference operator on **b** of the form

$$\mathbf{G} = \begin{bmatrix} -1/\sigma^{\dagger}_{21} & 1/\sigma^{\dagger}_{21} & 0 & 0 & 0 & \cdots \\ -1/\sigma^{\dagger}_{31} & 0 & 1/\sigma^{\dagger}_{31} & 0 & 0 & \cdots \\ -1/\sigma^{\dagger}_{41} & 0 & 0 & 1/\sigma^{\dagger}_{41} & 0 & \cdots \\ \vdots & \vdots & \vdots & \vdots & \vdots & \ddots \\ 0 & -1/\sigma^{\dagger}_{32} & 1/\sigma^{\dagger}_{32} & 0 & 0 & \cdots \\ 0 & -1/\sigma^{\dagger}_{42} & 0 & 1/\sigma^{\dagger}_{42} & 0 & \cdots \\ 0 & -1/\sigma^{\dagger}_{52} & 0 & 0 & 1/\sigma^{\dagger}_{52} & \cdots \\ \vdots & \vdots & \vdots & \vdots & \vdots & \ddots \end{bmatrix} \tag{14}$$

where the double subscript notation refers to the indices of the two events being compared in each row constraint. Note that **G**, being solely a difference operator, does not constrain the mean value of the elements of **b**. This element of the solution space is typically set to zero, under the assumption that the mean pre-existing pick value is a good approximation to the first arrival. In cases where this is not true, it is desirable to use a master or stacked event to determine an appropriate offset.

It is advantageous to omit spurious constraints in Eq. 12 that may arise from low correlation event pairs. This can be done based on the value of the cross-correlation maximum or through a composite or other measure taking into account catalogue hypocenter distances or other information. A constraint system which is reduced in this manner can produce subpopulations of events that are not associated. The simplest example of this is an orphan event pair that does correlate well with any others. In such cases Eq. 12 can be conceptually split into subsystems and the unconstrained elements in the solution space relating these subsystems can be set to zero or to a master event-pair-derived value. The detailed issue of how best to generally cluster events in such situations is a broad one that is beyond the scope of this paper. One straightforward method is through single-correlation-link event association into equivalence classes which can be subjected to subsequent analysis (e.g., Aster and Scott, 1993), although there are many other schemes for forming equivalence classes of related objects (e.g., Carr et al., 1999; Davis, 1986; Ludwig and Reynolds, 1988); more research into the more general techniques for seismogram cluster characterization clearly remains to be done.

G is sparse, as only two elements in each row are nonzero; for a system composed of a single cluster its Q by N dimensions result in of order N^3

entries, of which only $2Q$ (of order N^2) are nonzero. We exploit this system sparseness to reduce vastly the necessary computer storage and solution time by parameterizing **G** by two Q-length integer vectors, A^- and A^+, which contain the row indices of the negative and positive entries for each constraint, and by a Q-length vector, $\sigma^\dagger$, containing the corresponding lag standard deviations $\sigma^\dagger{}_i$. This storage scheme is capable of easily representing very large (up to millions of elements) systems of the form of Eq. 14 within currently available workstation memory constraints.

Straightforward linear approaches to solving Eq. 14 for a least-squares (L_2) solution are via the normal equations, Cholesky factorization, or through singular value decomposition (e.g., Press et al., 1996). Such methods, however, require calculation of the (non-sparse) $\mathbf{G}^T\mathbf{G}$ (which contains of order N^4 entries) or other intermediate objects which ignore the considerable computational and memory advantages associated with the sparse storage of **G**. Additionally, the L_2 solution is vulnerable to large perturbations from interevent lag outliers, which can be expected to occur due to cycle-skipping or data glitches, excessive noise, or grossly inaccurate initial picks, all of which can result in miscorrelation of data segments.

To avoid storage limitations and to reduce outlier problems, we solve Eq. 12 by implementing iterative Polak-Ribiere conjugate gradient minimization (Polak, 1971; Press et al., 1996) programmed to operate efficiently with our (A^-, A^+) sparse storage scheme. This and other iterative minimization routines are easily implemented to obtain the more robust minimum 1-norm residual (L_1) solution, rather than the outlier-susceptible L_2 solution. For the L_1 solution, the functional to be minimized is

$$\mu^{(1)} = \sum_{i=1}^{Q} \frac{\left| d_i - d_{i,pred} \right|}{\sigma_i^\dagger} \tag{15}$$

and the gradient of $\mu^{(1)}$ at a general solution space point, $\boldsymbol{x}$, is

$$\nabla\mu^{(1)} = \mathrm{sgn}(\mathbf{G}\,\mathbf{x} - \mathbf{d}) \tag{16}$$

where the sgn function operates on each element of a vector, returning 1 if the argument is positive, -1 if the argument is negative, and 0 if the argument is zero. To avoid the slope discontinuity at the origin in Eq. 16, we set the gradient calculation to zero when an element becomes very small ($< 10^{-20}$).

From a probabilistic viewpoint, the L_1 misfit minimizing solution Eq. 15 is the maximum likelihood solution for exponentially distributed random variable data with mean $\bar{\mu}$ and standard deviation σ, which is characterized by the probability distribution function

$$P^{(1)}(x) = \frac{1}{2\sigma}\exp\left(-|x - \bar{\mu}|/\sigma\right) \tag{17}$$

rather than by the shorter-tailed Gaussian (L_2) data distribution function

$$P^{(2)}(x) = \frac{1}{\sigma\sqrt{2\pi}}\exp\left(-(x - \bar{v})^2/\left(2\sigma^2\right)\right) \tag{18}$$

Parker and McNutt (1980) describe the statistics of $\mu^{(1)}$ (Eq. 15) under assumed Gaussian data errors. We invoke this formalism as a quality-of-fit measure to assess statistical consistency between the L_1 misfit minimizing solution (Eq. 15) and the modeling assumption that the relative lag estimates and their standard deviations form a consistent set of first-difference constraints on **b**. The L_1 analogue to the L_2 (χ^2) q statistic for R=Q-N degrees of freedom is well-approximated by a third-moment expression, where the probability that a greater value of the random variable $\mu^{(1)}(R)$ than the *particular* misfit value,

$$y = \sum_{i=1}^{Q}\frac{\left|d_i - d_{i,pred}\right|}{\sigma_i^{\dagger}} \tag{19}$$

could have occurred is

$$q(y,R) = Prob(\mu^{(1)}(R) > x(y,R)) = S(x) - \frac{\gamma Z^{(2)}(x)}{6} \tag{20}$$

The first term on the right-hand side of Eq. 20 is the cumulative probability integral

$$S(x) = \frac{1}{\sigma_1(2\pi)^{1/2}}\int_{-\infty}^{x}\exp\left(-t^2/\left(2\sigma_1^{\,2}\right)\right)\mathrm{d}t \tag{21}$$

for a zero-mean Gaussian distribution with standard deviation

$$\sigma_1 = (1 - 2/\pi)\,R \tag{22}$$

and the second term is proportional to both the skewness, γ, of $\mu^{(1)}$,

$$\gamma = \frac{2-\pi/2}{(\pi/2-1)^{3/2}\, R^{1/2}} \tag{23}$$

and to

$$Z^{(2)}(x) = \frac{x^2-1}{(2\pi)^{1/2}} e^{-x^2/2} \tag{24}$$

where

$$x = \frac{y-\bar{\mu}}{\sigma_1} \tag{25}$$

and the mean value of the probability distribution function $\mu^{(1)}$ for R degrees of freedom is

$$\bar{\mu} = (2/\pi)^{1/2} R \tag{26}$$

Although the L_1 solution is considerably more robust than the L_2 solution in the presence of data outliers, we find that we can further improve the L_1 solution by conservatively rejecting outliers based on interim solution residual information and by successively re-solving the system. To do this, we first calculate the L_1 solution to Eq. 12 and its particular misfit measure, y (Eq. 19), using the full constraint system of Q relative lags. If there are data outliers, a large value of y will produce a highly unlikely value of $q(y,R)$ (Eq. 20 will be nearly zero). If q is very small, we successively cull constraints and data from $\boldsymbol{G}$, starting with the largest terms in y, and solving for a new value, y_i, after each ith constraint removal until we reach a conservative threshold for model-misfit consistency of $q(y_i, R_i) > 0.02$. For large systems, we incorporate a bisection method to do this more efficiently, where a larger number of constraints are discarded in each step; if the value of q is then higher than the selected limit, half of the discarded constraints are restored, we solve again, retest for q, and so forth. Generally a satisfactory value of q can be found in fewer than ten bisection steps. We then obtain a final pick adjustment solution for the reduced system and calculate 1-σ error bars on this solution via Monte Carlo propagation of data errors, where 50 solutions are found by perturbing the data using Gaussian noise consistent with the standard deviations for each lag constraint.

Figure 5 shows a simple synthetic illustration of the solution process for a small (N = 6) synthetic cluster, initially constrained by a full set of $Q = (6)(5)/2 = 15$ inter-event lag estimates. The true pattern of pick adjustments was chosen arbitrarily to be a zero-mean half-period sine function with an amplitude of 1 time unit. The 15 first-difference data points were randomized by adding a Gaussian error term with a standard deviation of 0.2 time units. Outliers were introduced to the system by adding large random terms to data points 3 and 7.

Figures 5a and 5b show the L_1 solution and data fit for the entire data and constraint set, where the recovery of true pick adjustments has been skewed by the data outliers, and the probability of a worse misfit is zero to single-precision. After automatically rejecting the two system constraints with the largest residual contributions, as described above, and re-solving the problem, we obtain a revised solution (Figure 5c) with an acceptable data misfit (Figure 5d) of $q \approx 0.06$ and an L_1 misfit improvement between solution and true model of 59% accompanied by generally reduced 1-σ error bars.

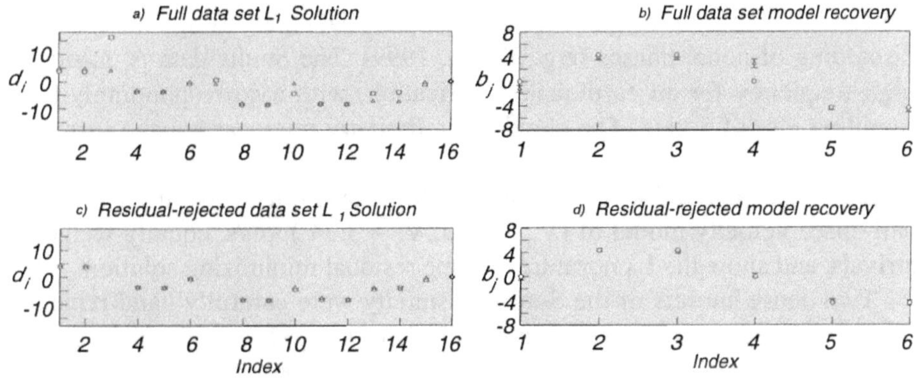

Figure 5. Example showing improved solution robustness for the pick correction system (Eq. 12) using the L_1 norm and iterative residual-based constraint rejection, and solving for a synthetic suite of pick adjustments, constrained by lag values with Gaussian noise. a) Full (Q=15; N=6) constraint solution (triangles) with Monte Carlo-estimated standard errors, and the true model (circles). Because of two large outlier data points inserted at indices 3 and 7, this solution is significantly in error. The sub-optimal model recovery is especially apparent in the pick adjustment solution for event 4, where the retrieved value departs from the true model by approximately four standard deviations. b) Data fit for the solution of (a). Predicted data are shown as triangles and the true data are shown as circles accompanied by standard errors. After two iterations of residual-based rejection and evaluation of the L_1 quality-of-fit statistic (Eq. 20), both outliers were identified and removed. c) Refined solution, showing improved model recovery. d) Data fit and standard deviation estimates for the residual-rejected data set.

2.4 Example Application

A rapidly growing worldwide digital waveform archive suggests a wide variety of applications which could be driven by automatic repicking algorithms. An instructive example is provided by microseismicity recorded in association with injection experiments at Soultz-sous-Forêts hot dry rock geothermal field in the Rhinegraben region of Alsace, France (Baria et al., 1999).

Between May and October, 1993, injection experiments were carried out at the reservoir, resulting in hydrofracturing and induced faulting that produced over 16,000 microearthquakes. The largest event had a magnitude of $M_L \approx 1.9$, while most were much smaller with inferred source dimensions of 1 m or less (Phillips, 2000). These earthquakes are more difficult to locate than those of many tectonic earthquake data sets, as the events were recorded on a sparse network consisting of just three deep borehole multicomponent seismometers and one downhole hydrophone (Baria et al., 1999; Phillips, 2000). In such a sparse network, one must use nearly all available phases to constrain hypocenters adequately, whereas a more dense station distribution might allow for more data selection, such as the discarding of nodal phases (e.g., Rubin, 1999). The Soultz data is unusually high-frequency for an earthquake application, with a correspondingly high sampling rate of 5 kHz. The cloud of preliminary microearthquake locations from initial analyst pick data and the seismometer locations are shown in Figure 6. In this and in all subsequent location Figures, we used Phillips' half-space velocity model of ($V_p = 5.85$, $V_s = 3.34$) km/s, equally-weighted arrivals, and show the L_2 norm travel-time residual minimizing solution.

Two dense subsets of the Soultz seismicity were carefully hand-repicked (Phillips, 2000), via visual cross-correlation to re-estimate P and S arrival times. Subsequent relocations show well-resolved, narrow, intersecting joint features with greatly enhanced interpretability over the initial diffuse event cloud. Such seismogenic images are of great value in discerning the geometry and estimating the efficiency of heat exchanger systems in geothermal reservoirs (e.g., Niitsuma et al., 1999). Hypocenters calculated using preliminary picks are shown in map view and in cross-section in Figures 7a and 7b for the 311-event "deep cluster", which occupies an approximately 65-m-radius spherical portion of the seismogenic cloud. Relocation of this seismicity using the Phillips repicks are shown in Figures 7c and 7d. Note the distinct, nearly orthogonal intersecting features apparent in the relocated hypocenters, thought to represent seismogenic joints in this small spatial sampling of the heat exchanger.

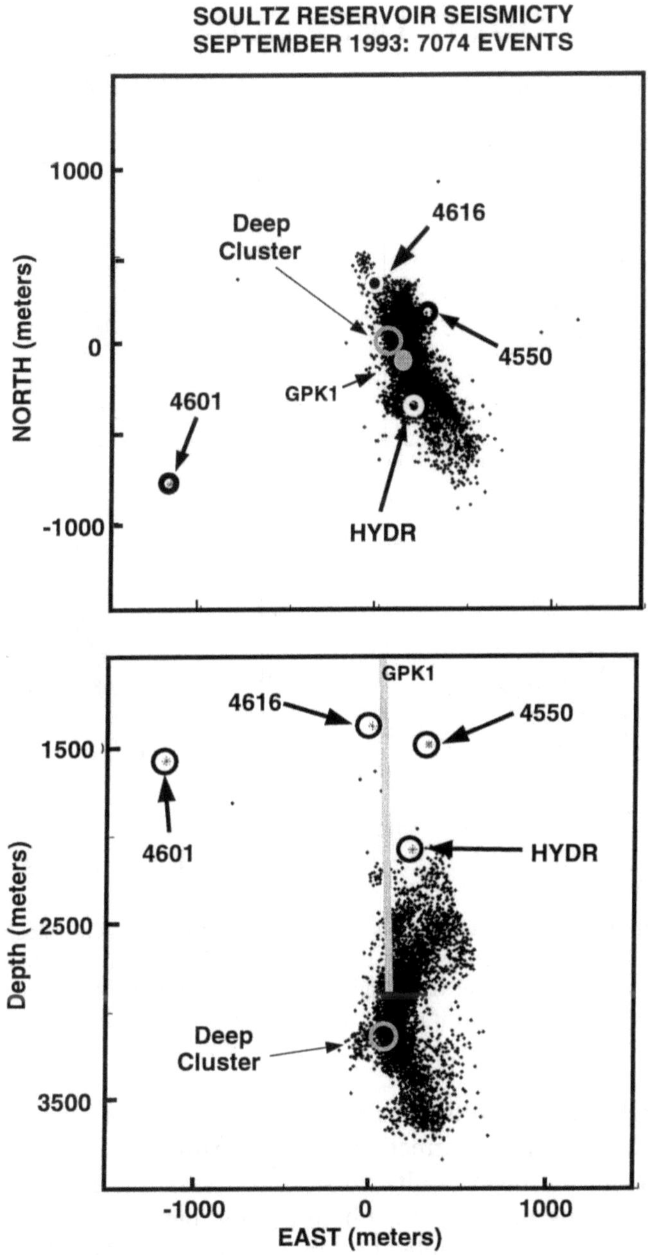

Figure 6. Soultz hot dry rock geothermal reservoir seismicity resulting from September 1993 injection experiments. a) Map view of the site, showing the cloud of induced microearthquakes, located using analyst picks of P and S arrivals at four borehole sensors (4550, 4601 and 4616 are four-component borehole seismometers; HYDR is a hydrophone), the GPK1 injection well, and the location of the Phillips (2000) deep seismic cluster. b) Cross-sectional view. After Phillips (2000).

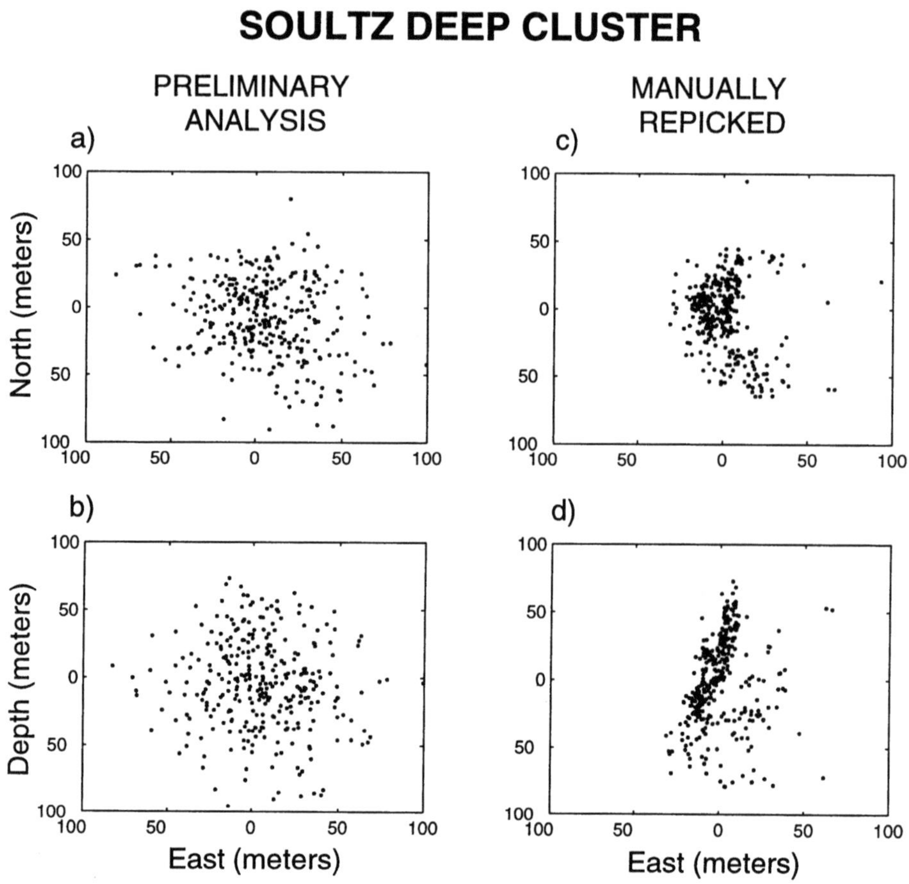

Figure 7. Deep cluster of 311 Soultz events. a) Map view of hypocenters calculated using analyst picks. b) East-West cross-section of hypocenters shown in (a). c) Map view of hypocenters relocated after manual repicking and location (Phillips, 2000). d) East-west cross-section of relocated events in (c). Note the collapsing of the diffuse cloud into two narrow, linear, intersecting features.

We show relocation results after applying the automatic lag estimation and conjugate gradient pick correction solver, starting with the preliminary analyst picks from the deep Soultz cluster, to assess the degree to which automatic reprocessing may approach high-quality manual repicking results. In this example, we use 64-sample (12.8 ms) segments of data for P phases, except at the relatively high-frequency station 4550 (32 samples or 6.4 ms), and 128-sample (25.6 ms) data segments for all S phases, except at the more distant, lower-frequency station 4601 (256 samples or 51.2 ms). Data segment lengths were selected, after viewing representative segments of the data, to accommodate approximately 1.5 full zero-crossing cycles for typical data. For this data set, we found this to be sufficiently long to span the range of analyst pick inconsistency, yet also sufficiently short to window highly similar waveform segments in the vicinity of the first breaks. Further development is desirable to make the choice of these window lengths more automatic. Figure 8 shows waveforms for a typical event from this cluster.

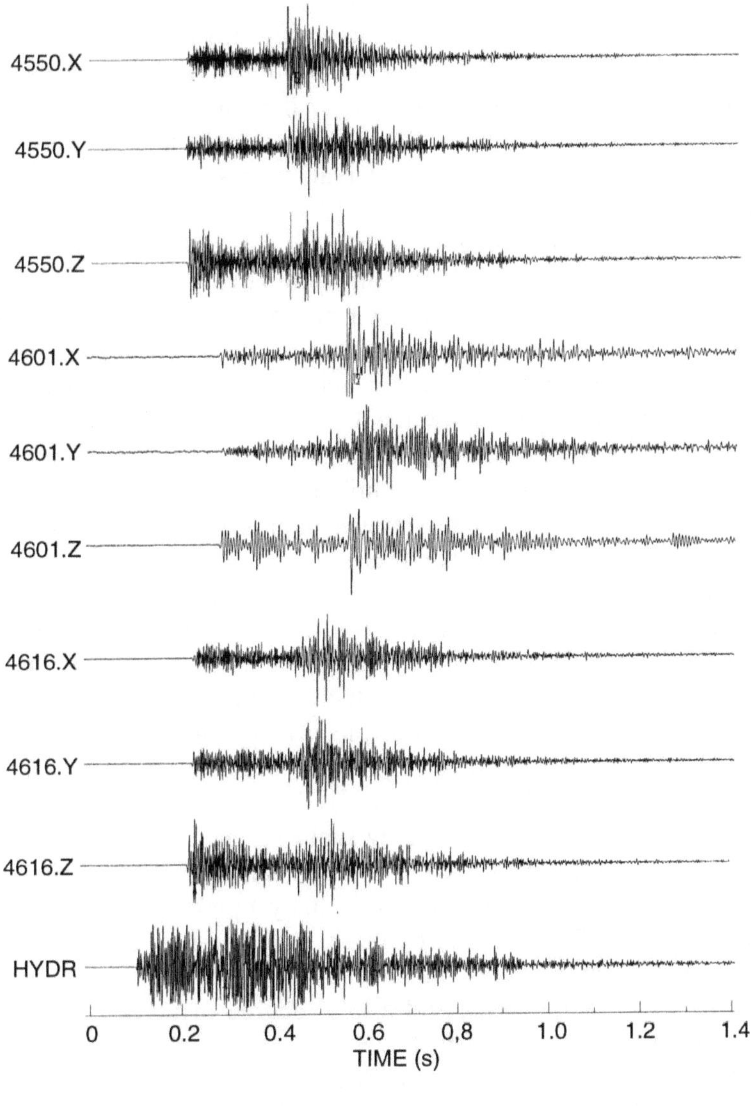

Figure 8. Example waveforms from a deep cluster event, in the $(x,y,z) = (E,N,V)$ coordinate system. Because the seismometers are all in deep boreholes, background noise is generally very low (e.g., Withers et al., 1996) and P-wave onsets are clear. S-waves arrival time relative lag determinations are noisier due to interference from the P-wave coda. Note the clipping on the hydrophone channel, which renders cross-correlation of S-waves problematic for that sensor.

The component rotation, cross-correlation, and pick-correction solving steps were carried out seven times; once for each P arrival at a single- or multi-component receiver and once for each S arrival at each of the multi-component seismometers. S arrivals at the hydrophone were not included in the preliminary phase catalog, so we are unable to apply the re-picker to raw hydrophone S-wave picks. Additionally, the hydrophone is severely underdamped and a large fraction of the waveforms are clipped at the time of the S arrival, which would disincline us from processing S arrivals at this channel due to the nonlinear corruption of the waveform. Phillips (2000), however, estimated S arrivals on the hydrophone channel for these events, and we therefore adopted these S-arrival estimates for the hydrophone channel.

Figure 9 shows an example event pair recorded at station 4550 (Figure 6), for which the P-wave arrival is being compared. In the upper panel we show the two waveforms aligned on their preliminary picks. Trace alignment following coarse cross-correlation is shown in the second panel. The third panel shows the cross-spectral phase as determined from multitaper spectral analysis. Vertical error bars show the 1-sigma standard deviations at each frequency bin mapped by the tangent mapping (9). The slope of the best-fit zero-intercept line provides the subsample lag adjustment for the event pair (7), which, added to the coarse (integer sample) lag, provides the final estimate of relative lag for these two traces.

Once all phases have been correlated for each station and a consistent solution has been obtained through the L_1-norm conjugate gradient solver, the new picks are entered into the trace headers and are used for event relocation or other applications. To assess the degree of improvement in pick consistency in going from the initial picks to our estimates, we show in Figures 10 - 13 the waveforms aligned on preliminary picks (*left*) and automatically determined repicks (*right*) for P and S arrivals, where individual traces are plotted as a gray-scale surface with dark peaks and white troughs. The scatter in waveform alignment in the first panels arises from the inconsistency among preliminary analyst picks, and manifests itself in the diffuse cloud of preliminary hypocenters (Figure 7). Note that, despite a few outliers, the automatic alignments show vastly improved consistency compared to the original picks, and a high degree of waveform similarity becomes evident among most of the traces.

Automatically relocated hypocenters for the deep cluster are shown in Figures 14c and 14d. For comparison, initial locations are shown in Figures 14a and 14b, and the relocations of Phillips from manually repicked phases are shown in Figure 14e and 14f. Linear trends associated with probable high permeability paths as noted by Phillips are readily discernible in the automatically repicked hypocenter locations. Figure 15 shows the hypocenters of Figure 14 in a perspective image chosen to emphasize the resolving of the two intersecting joint features. Substantial convergence of the automatic hypocenter estimates to those of Phillips is also demonstrated in Figure 16, where a median hypocenter discrepancy of 31 m for the preliminary analyst catalogue is reduced to 7 m for the automatically relocated catalogue.

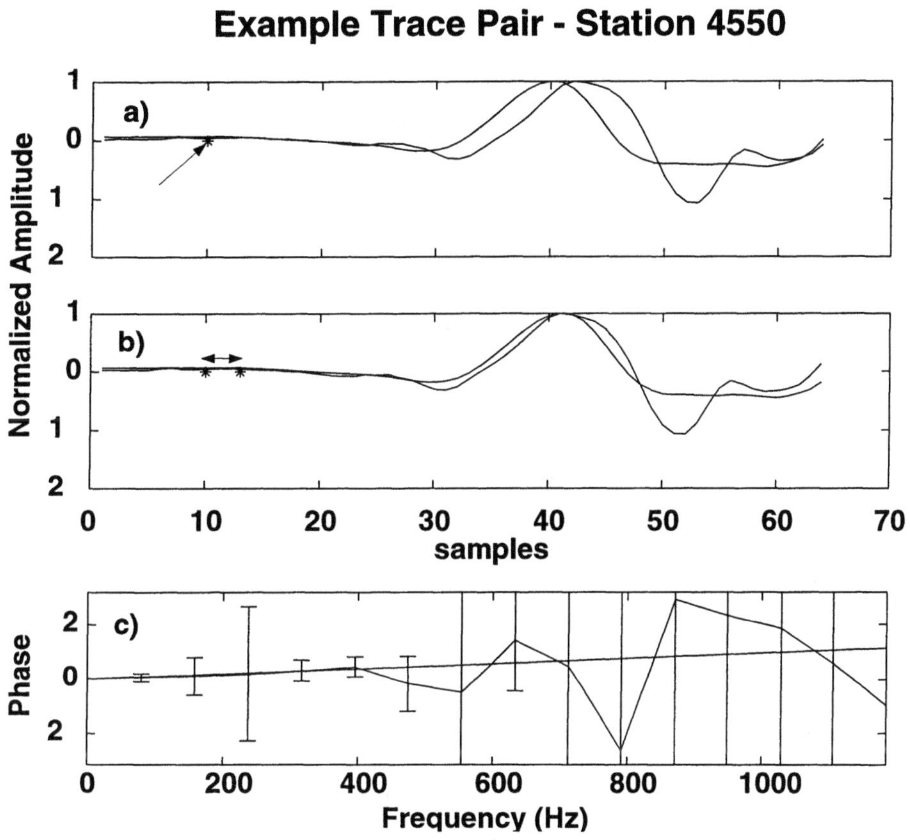

Figure 9. Example event pair from Soultz deep cluster, recorded at station 4550. a) Two 64-sample (12.8 ms) waveforms aligned on analyst picks (arrow), showing waveform misalignment due to analyst pick inconsistency. The waveforms shown are the projections the normalized multicomponent signals in the direction of maximum joint signal linearity (Eq. 2). b) Waveform alignment following coarse cross-correlation. The double-headed arrow indicates the 5-sample (1 ms) shift required to align these waveforms to the nearest integer sample. c) Mean cross-spectral phase as determined from the application of the two lowest-order eigentapers. Vertical error bars show 1-sigma standard deviations for each frequency bin (Eq. 9). The slope of the best-fit zero-intercept line scaled to time (Eq. 7), when added to the coarse (integer sample) lag, produces a total lag estimate for these two traces of 1.1 ms with a standard deviation of about 0.3 ms.

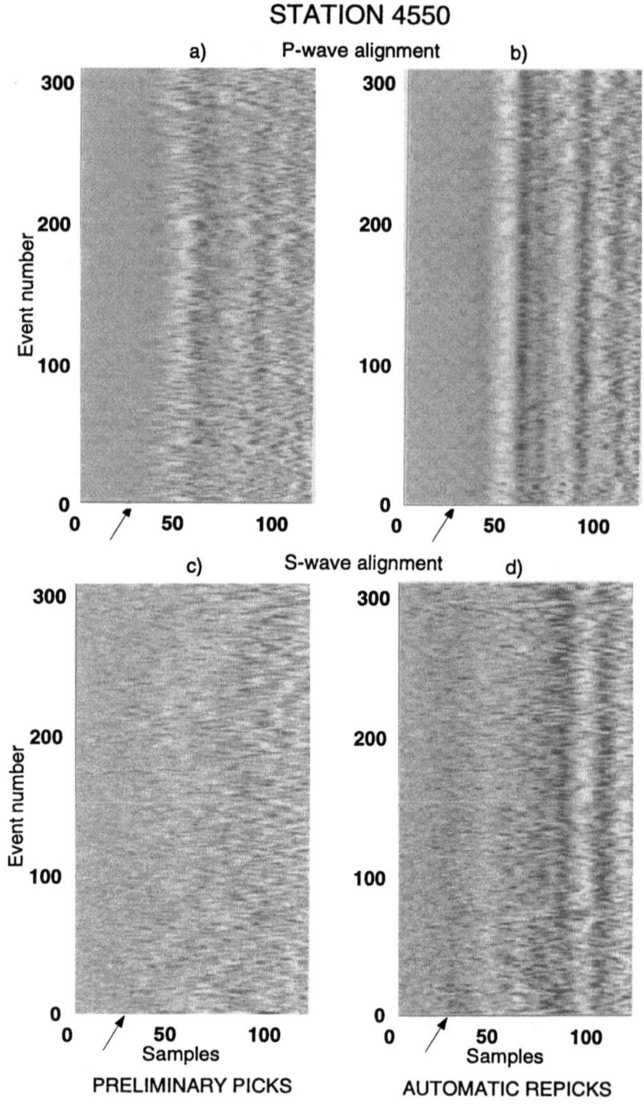

Figure 10. Waveform alignment plots for station 4550 (128 samples or 25.6 ms). a) Vertical (Z) component traces showing waveforms for Soultz deep cluster, aligned on analyst P-wave picks. Traces are plotted as a 3-dimensional shaded surface where peaks are white and troughs are black. Arrows indicate approximate pick times on which traces are aligned. b) Vertical (Z) component traces shown aligned on the automatically adjusted P-wave arrivals. Note the significant improvement in alignment of primary peaks and troughs. c) and d) East (X) component traces showing waveforms aligned on c) analyst S-wave picks and d) automatically adjusted S-wave picks. Although joint covariance matrix decomposition-based rotation (Eqs. 1-3) and adaptive prefiltering (Eq. 6) were employed in the correlation itself, we show unfiltered waveforms in the cardinal orientations similar to that which might be viewed in a typical analyst review situation. The RMS values of the zero-mean pick corrections are 6.3 samples (1.26 ms) for P waves and 20.8 samples (4.16 ms) for S waves.

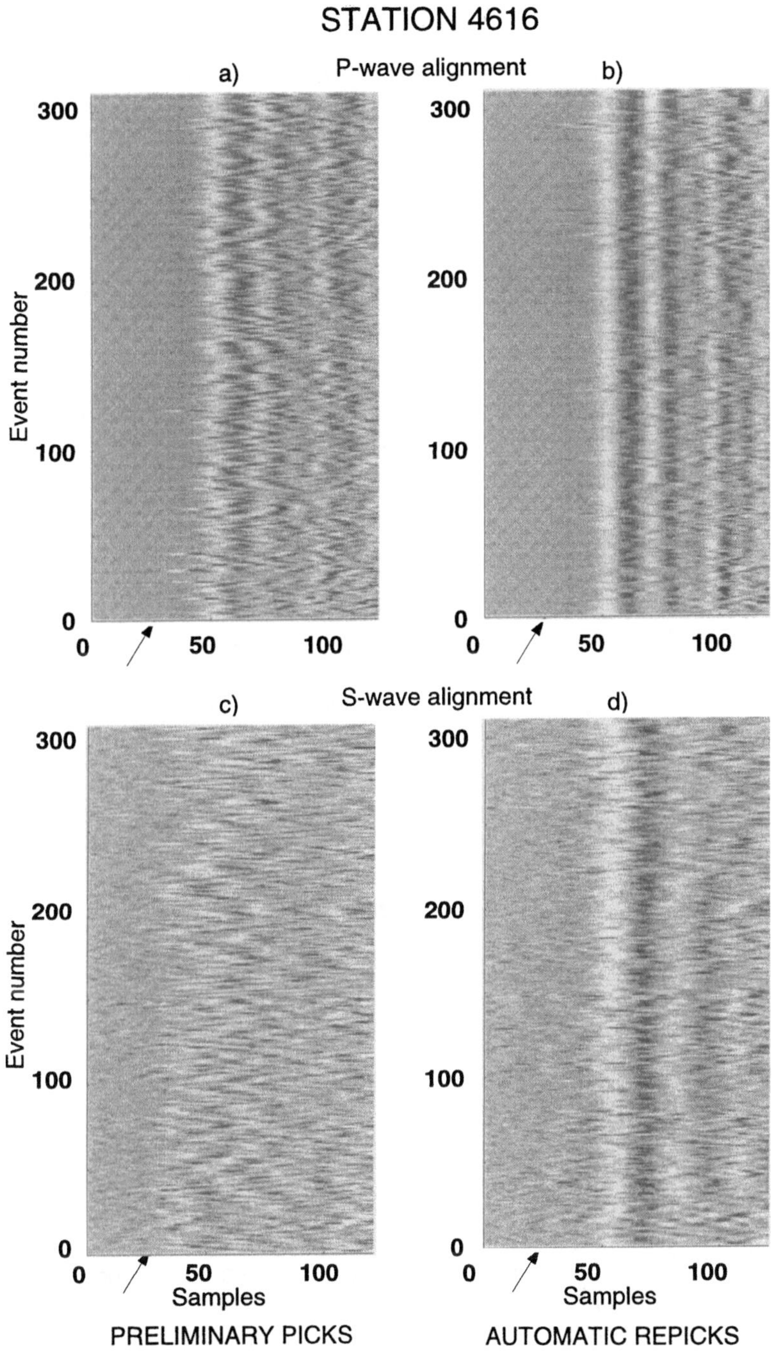

Figure 11. Waveform alignment plots for station 4616 in the format of Figure 10. The RMS values of the zero-mean pick corrections are 4.9 samples (1.0 ms) for P waves and 14.2 samples (2.84 ms) for S waves.

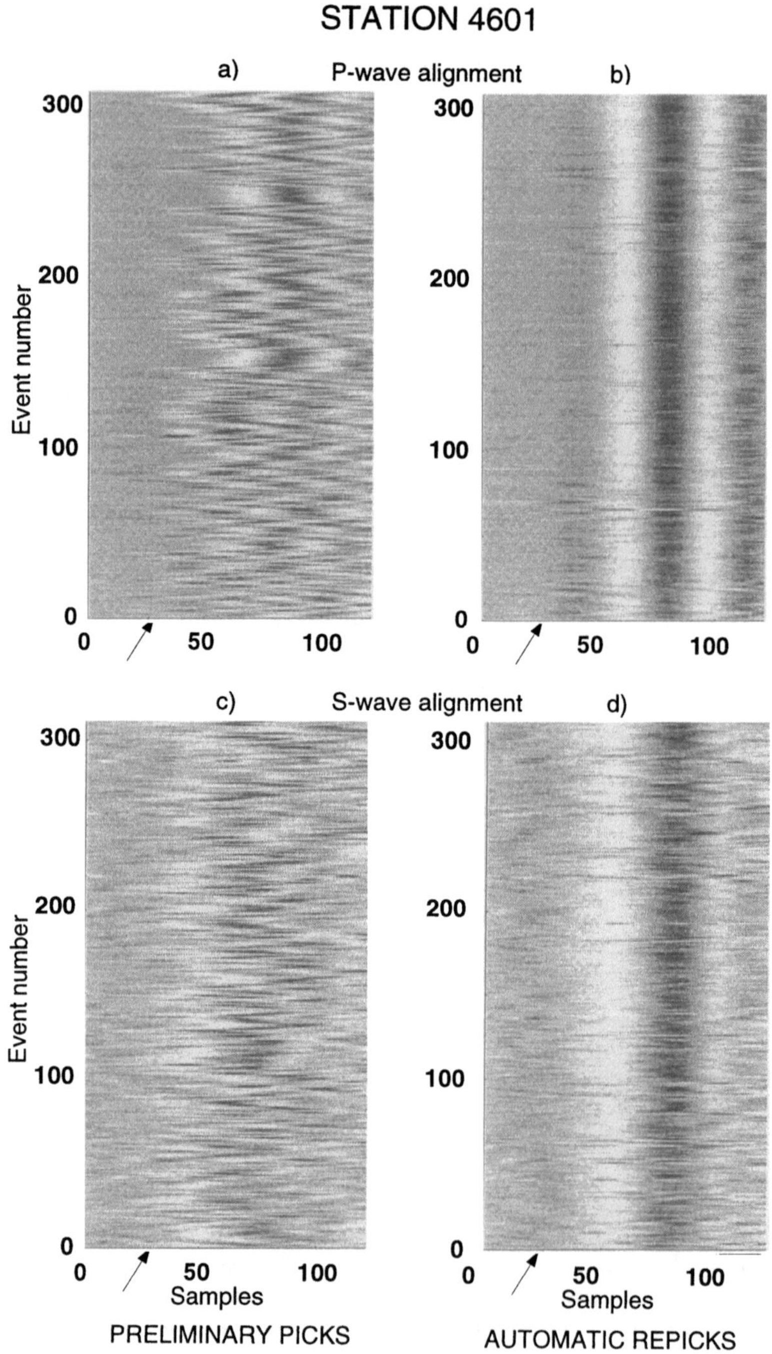

Figure 12. Waveform alignment plots for station 4601 in the format of Figure 10. The RMS values of the zero-mean pick corrections are 15.1 samples (3.02 ms) for P waves and 21.3 samples (4.27 ms) for S waves.

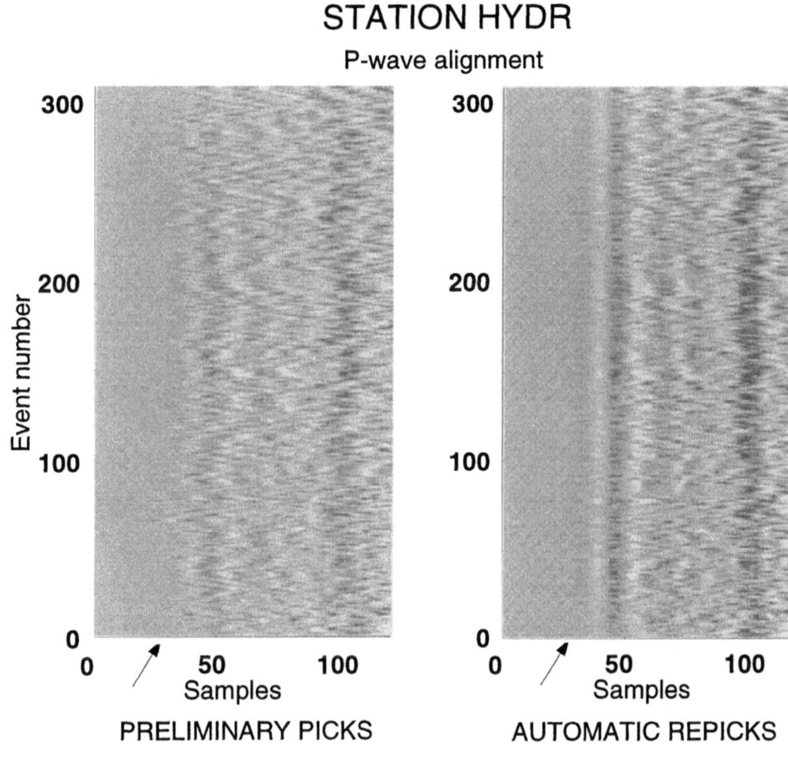

Figure 13. Waveform alignment plots for hydrophone station HYDR in the format of Figure 10. S-wave arrivals were not estimated due to clipping and the lack of analyst S-wave picks in the preliminary catalog. For the purpose of this comparative analysis the hydrophone S-wave manual picks of Phillips (2000) were adopted. The RMS value of the zero-mean pick correction is 4.1 samples (0.82 ms) for the P wave.

3. DISCUSSION

These results are not expected to perfectly reproduce those of Phillips for several reasons. First, the automatic algorithm rotates every event pair to optimize joint signal linearity, so the correlated waveforms will not be identical to the native components chosen for comparison in the manual repicking. Second, rather than identifying a particular peak or trough within the early part of the wave train for alignment (Phillips, 2000), the automatic algorithm correlates a fixed-length window of data which has had cross-coherency-weighted filtering applied. For some traces, Phillips applied an *a priori* bandpass filter prior to his single-component visual cross-correlation. Third, in the manual repicking, nodal or reverse-polarity arrivals were handled as special cases and first arrivals were visually estimated. In the

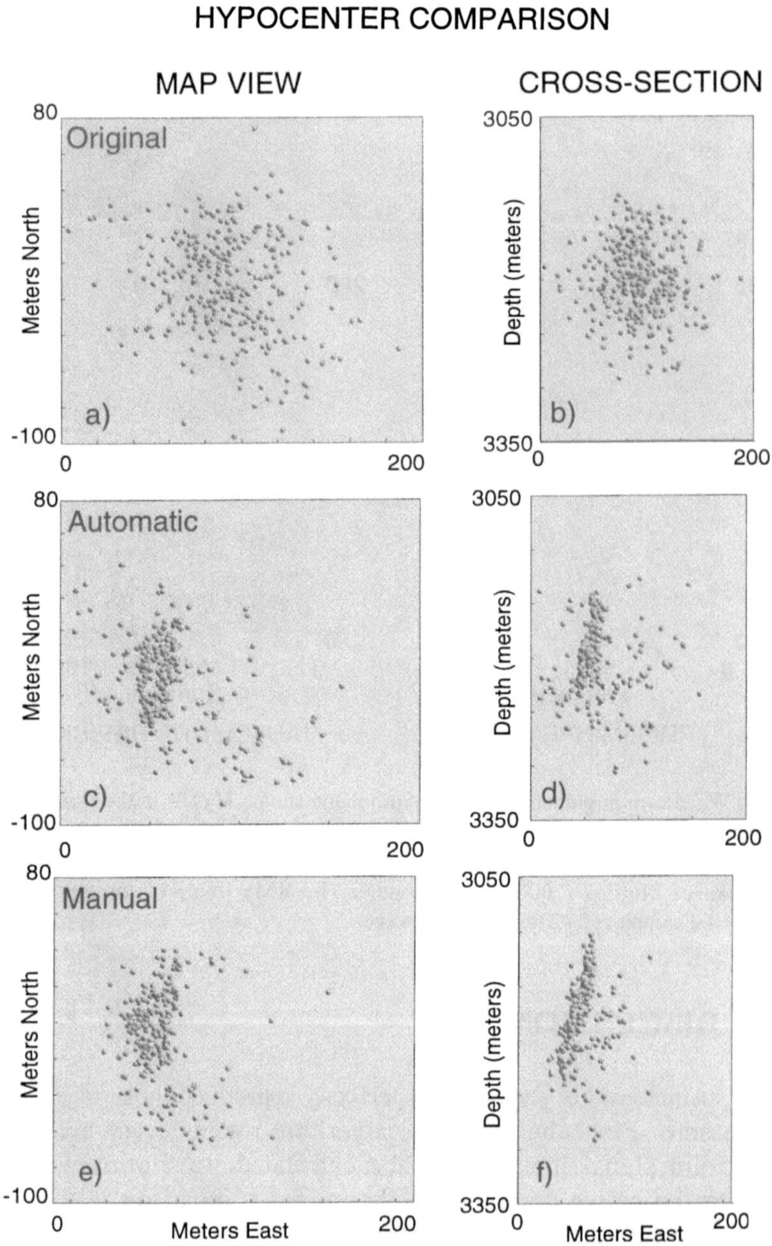

Figure 14. Soultz deep cluster relocation comparison. a) Hypocenters located using original analyst picks - map view. b) Cross section of hypocenters in (a). c) Hypocenters re-located using automatically adjusted picks. d) Cross-section of (c). e) Relocations obtained using Phillips (2000) manual repicks. f) Cross-section of hypocenters shown in (e). Note that the automatic repicks result in locations whose structure is similar to that obtained with the manual repicking.

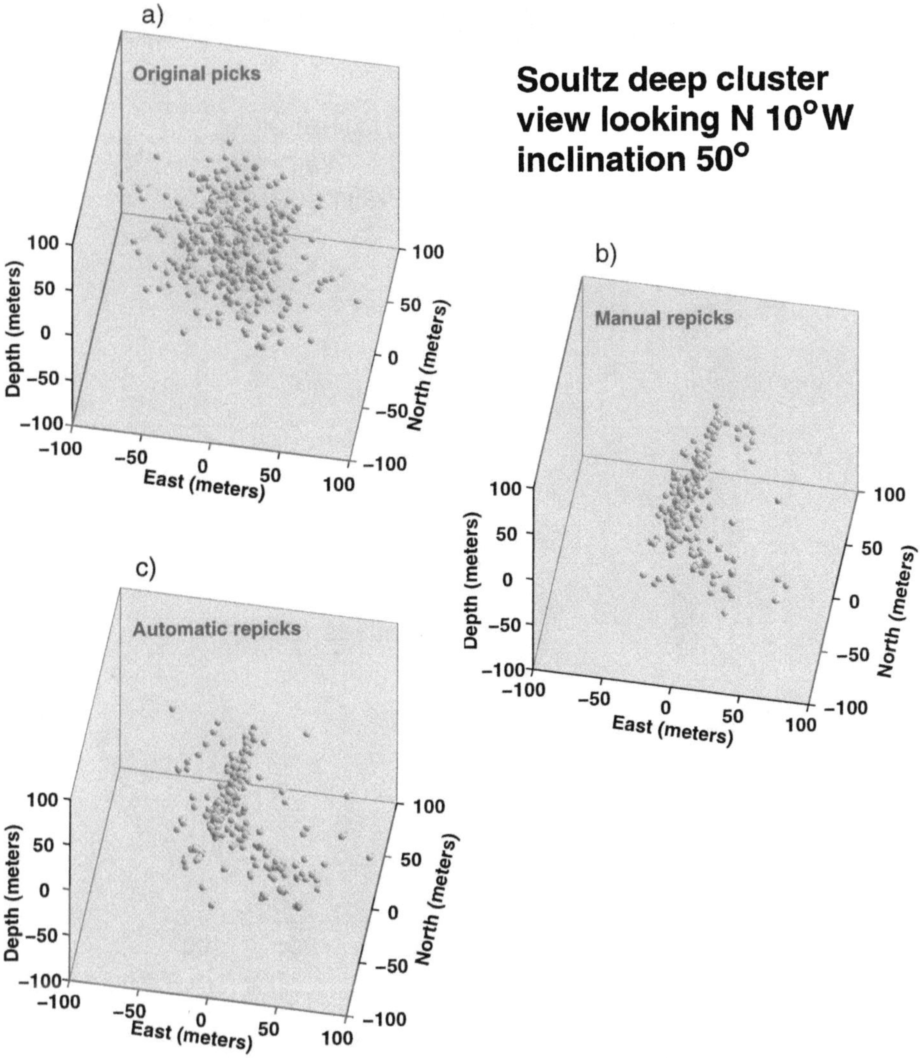

Figure 15. Perspective plot showing Soultz deep cluster events illustrated in map view and cross-section in Figure 14. View angle is looking 10 degrees West of North, from an inclination of 50 degrees above the horizontal. a) Original locations. b) Phillips' manually repicked event locations. c) Locations from automatically repicked phases. Note the two narrow, intersecting, nearly orthogonal features apparent in both (b) and (c) which are not discernible in the diffuse scatter of hypocenters shown in (a).

automatic approach, the data windows are cross-correlated with all other events and the relative quality of the lag adjustments is accommodated via application of the standard deviations within the system solution and subsequent residual based constraint rejection.

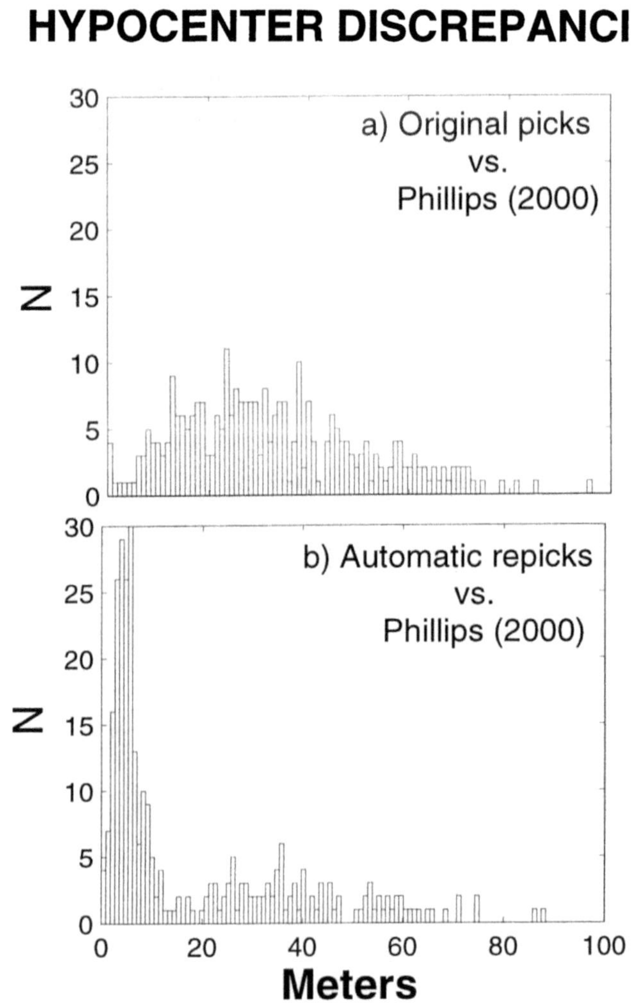

Figure 16. Discrepancy histograms (m) between hypocenter locations for Soultz deep cluster. a) Hypocenters calculated from original, analyst picks compared to the Phillips manual repicks. b) Hypocenters calculated from automatic repicks versus Phillips manual repicks. Median discrepancy distance relative to Phillips' locations is reduced from 31.3 m in (a) to 7.2 m in (b).

In the case where very small subsets of different events are involved, or in the handling of pathological cases involving nonlinear data distortion (e.g., noise spikes, data dropouts, clipping), large-scale automatic reprocessing will probably not provide the adaptability and flexibility of human analysis for some time, although the incorporation of a clustering algorithm within the flow of the algorithm, as discussed earlier, will assist in identifying and efficiently segregating such events and will enable the processing of more spatially extensive regions. We thus do not claim event-for-event superiority in general in this particular example. Rather, we demonstrate that the vastly superior efficiency of automatic processing can provide substantial improvement in locations which approach those obtained through painstaking visual methods for most events in a large catalog. This can be accomplished in an unbiased manner with a small fraction of the time and other resources required to manually or semi-automatically process the same events. We see no reason why this methodology should not ultimately be applicable to the largest data sets with dramatic results, although we note that this particular algorithm will benefit significantly from further developments, especially in adaptive filtering and event clustering.

In some situations, such as data sets with high levels of incoherent noise, automatic, spectral-based methods may be superior to the best practical human analysis in identifying similarities and providing consistent pick estimates. Limited analyst resources and the significant time investment required for consistent manual reprocessing of even a modest data set, such as our Soultz example, comprising a few hundred events and four sensors, suggest that adaptive, automatic techniques will be widely applied in future high-resolution imaging of seismogenic zones.

REFERENCES

Antolik, M., R. Nadeau, R. Aster, and T. McEvilly (1996) Differential analysis of coda Q using similar microearthquakes in seismic gaps. Part 2: Application to seismograms recorded by the Parkfield High Resolution Seismic Network, *Bull. Seism. Soc. Am.* **86**, 890-910.

Aster, R.C. and J. Scott (1993) Comprehensive characterization of waveform similarity in microearthquake data sets, *Bull. Seismol. Soc. Am.* **83**, 1307-1314.

Aster, R., P. Shearer, and J. Berger (1990) Quantitative measurements of shear-wave polarizations at the Anza seismic network, southern California – implications for shear-wave splitting and earthquake prediction, *J. Geophys. Res.* **95**, 12,449-12,474.

Aster, R., G. Slad, J. Henton, and M. Antolik (1996) Differential analysis of coda Q using similar microearthquakes in seismic gaps.Part 1: Techniques and application to seismograms recorded in the Anza seismic gap, southern California, *Bull. Seism. Soc. Am.* **86** 868-889.

Baria, R., J. Baumgartner, A. Gerard, R. Jung, and J. Garnish (1999) European HDR research programme at Soultz-sous-Forêts (France) 1987-1996, *Geothermics* **28**, 655-669.

Beroza, G., A. Cole, and W. Ellsworth (1995) Stability of coda wave attenuation during the Loma Prieta, California earthquake sequence, *J. Geophys. Res.* **100**, 3977-3987.

Carr, D., C. Young, J. Harris, R. Aster, and X. Zhang (1999) Cluster analysis for CTBT seismic event monitoring, *Seism Res. Lett.* **70**, 227-228.

Davis, J. (1986) *Statistics and data analysis in geology*, J. Wiley and Sons, New York.

Dodge, D.A., G.C. Beroza and W.L. Ellsworth (1995) Foreshock sequence of the 1992 Landers, California earthquake and its implications for earthquake nucleation, *J. Geophys. Res.* **100**, 9865-9880.

Fehler, M.C., L.S. House, and H. Kaieda (1987) Determining planes along which earthquakes occur: Method and application to earthquakes accompanying hydraulic fracturing, *J. Geophys. Res.* **92**, 9407-9414.

Fehler, M., W.S. Phillips, R. Jones, L. House, R. Aster, and C. Rowe (2000) A method for improving relative earthquake locations, *Bull. Seismol. Soc. Amer.*, submitted.

Fremont, M.-J. and S.D. Malone (1987) High precision relative locations of earthquakes at Mount St. Helens, Washington, *J. Geophys. Res.* **92**, 10,223-10,236.

Haase, J., P. Shearer, and R. Aster (1995) Constraints on temporal variations in velocity near Anza, California, from analysis of similar event pairs, *Bull Seismol. Soc. Am.* **85**, 194-206.

Goldstein, P., D. Dodge, M. Firpo, and S. Ruppert (1998) What's new in SAC2000? Enhanced processing and database access, *Seis. Res. Lett.* **69**, 202-205.

Got, J.-L., J. Frechet, and F.W. Klein (1994) Deep fault plane geometry inferred from multiplet relative relocation beneath the south flank of Kilauea, *J. Geophys. Res.* **99**, 15,375-15,386.

Gillard, D., A.M. Rubin, and P. Okubo (1996) Highly concentrated seismicity caused by deformation of Kilauea's deep magma system, *Nature* **384**, 343-346.

Lees, J.M. (1998) Multiplet analysis at Coso geothermal, *Bull. Seism. Soc. Am.* **88**, 1127-1143.

Ito, A. (1985) High resolution relative hypocenters of similar earthquakes by cross-spectral analysis method, *J. Phys. Earth* **33**, 279-294.

Jones, R.H. and R.C. Stewart (1997) A method for determining significant structures in a cloud of earthquakes, *J. Geophys. Res.* **102**, 8245-8254.

Lees, J.M. (1998) Multiplet analysis at Coso geothermal, *Bull. Seism. Soc. Am.* **88**, 1127-1143.

Ludwig, J. and J. Reynolds (1988) Statistical Ecology, J. Wiley and Sons, New York.

Nadeau, R., W. Foxall, and T. McEvilly (1995) Clustering and periodic recurrence of microearthquakes on the San Andreas Fault at Parkfield, California, *Science* **267**, 503-507.

Niitsuma et al. (1999) Current status of seismic and borehole measurements for HDR/HWR development, *Geothermics* **I28**, 475-490.

Park, J., C.R. Lindberg, and F.L. Vernon III (1987) Multitaper spectral analysis of high-frequency seismograms, *J. Geophys. Res.* **92**, 12,675-12,684.

Parker, R., and M. McNutt (1980) Statistics for the one-norm misfit measure, *J. Geophys. Res.* **85**, 4429-4430.

Phillips, W.S., L.S. House, and M.C. Fehler (1997) Detailed joint structure in a geothermal reservoir from studies of induced microearthquake clusters, *J. Geophys. Res.* **102**, 11,745-11,763.

Phillips, W.S. (2000) Precise microearthquake locations and fluid flow in the geothermal reservoir at Soultz-sous-Forêts, France, *Bull. Seismol. Soc. Am.*, in press.

Poupinet, G., W.L. Ellsworth, and J. Frechet (1984) Monitoring velocity variations in the crust using earthquake doublets: An application to the Calaveras Fault, California, *J. Geophys. Res.* **89**, 5719-5731.

Polak, E., (1971) *Computational Methods in Optimization*, Academic Press, New York.

Press, W.H., B.P. Flannery, S.A. Teukolsky, and W.T. Vetterling (1989) *Numerical Recipes in C*, Cambridge University Press.

Rubin, A.M., D. Gillard, and J.-L. Got (1998) A reinterpretation of seismicity associated with the January 1983 dike intrusion at Kilauea Volcano, Hawaii, *J. Geophys. Res.* **103**, 10,003-10,015.

Rubin, A.M., D. Gillard, and J.-L. Got (1999) Microseismic lineations along creeping faults, *Nature* **400**, 635-641.

Shearer, P.M. (1997) Improving local earthquake locations using the L1 norm and waveform cross correlation: Application to the Whittier Narrows, California, aftershock sequence, *J. Geophys, Res.* **102**, 8269-8283.

Shearer, P.M. (1998) Evidence from a cluster of small earthquakes for a fault at 18 km depth beneath Oak Ridge, Southern California, *Bull. Seism. Soc. Am.* **88**, 1327-1336.

Sheriff, R, and L. Geldart (1995) *Exploration Seismology*, Second Edition, Cambridge University Press, 575 pp.

Spudich, P. and T. Bostwick (1987) Studies of the seismic coda using an earthquake cluster as a deeply buried seismograph array, *J. Geophs. Res.* **92**, 10,526-10,546.

Thomson, J.D. (1982) Spectrum estimation and harmonic analysis, *Proceedings of the IEEE* **70**, 1055-1096.

Waldhauser, F., W. Ellsworth, and A. Cole. (1999) Slip-parallel seismic lineations on the northern Hayward Fault, California, *Geophys. Res. Lett.* **26**, 3525-3528.

Withers, M., R. Aster, C. Young, E. Chael (1996) High-frequency analysis of seismic background noise and signal-to-noise ratio near Datil, New Mexico, *Bull. Seism. Soc. Am.* **86**, 1507-1515.

Index

MODERN APPROACHES IN GEOPHYSICS

1. E.I. Galperin: *Vertical Seismic Profiling and Its Exploration Potential.* 1985 ISBN 90-277-1450-9
2. E.I. Galperin, I.L. Nersesov and R.M. Galperina: *Borehole Seismology.* 1986 ISBN 90-277-1967-5
3. Jean-Pierre Cordier: *Velocities in Reflection Seismology.* 1985 ISBN 90-277-2024-X
4. Gregg Parkes and Les Hatton: *The Marine Seismic Source.* 1986 ISBN 90-277-2228-5
5. Guust Nolet (ed.): *Seismic Tomography.* 1987 Hb: ISBN 90-277-2521-7; Pb: ISBN 90-277-2583-7
6. N.J. Vlaar, G. Nolet, M.J.R. Wortel and S.A.P.L. Cloetingh (eds.): *Mathematical Geophysics.* 1988 ISBN 90-277-2620-5
7. J. Bonnin, M. Cara, A. Cisternas and R. Fantechi (eds.): *Seismic Hazard in Mediterranean Regions.* 1988 ISBN 90-277-2779-1
8. Paul L. Stoffa (ed.): *Tau-p: A Plane Wave Approach to the Analysis of Seismic Data.* 1989 ISBN 0-7923-0038-6
9. V.I. Keilis-Borok (ed.): *Seismic Surface Waves in a Laterally Inhomogeneous Earth.* 1989 ISBN 0-7923-0044-0
10. V. Babuska and M. Cara: *Seismic Anisotropy in the Earth.* 1991 ISBN 0-7923-1321-6
11. A.I. Shemenda: *Subduction.* Insights from Physical Modeling. 1994 ISBN 0-7923-3042-0
12. O. Diachok, A. Caiti, P. Gerstoft and H. Schmidt (eds.): *Full Field Inversion Methods in Ocean and Seismo-Acoustics.* 1995 ISBN 0-7923-3459-0
13. M. Kelbert and I. Sazonov: *Pulses and Other Wave Processes in Fluids.* An Asymptotical Approach to Initial Problems. 1996 ISBN 0-7923-3928-2
14. B.E. Khesin, V.V. Alexeyev and L.V. Eppelbaum: *Interpretation of Geophysical Fields in Complicated Environments.* 1996 ISBN 0-7923-3964-9
15. F. Scherbaum: *Of Poles and Zeros.* Fundamentals of Digital Seismology. 1996 Hb: ISBN 0-7923-4012-4; Pb: ISBN 0-7923-4013-2
16. J. Koyama: *The Complex Faulting Process of Earthquakes.* 1997 ISBN 0-7923-4499-5
17. L. Tauxe: *Paleomagnetic Principles and Practice.* 1998 ISBN 0-7923-5258-0
18. C.H. Thurber and N. Rabinowitz (eds.): *Advances in Seismic Event Location.* 2000 ISBN 0-7923-6392-2

KLUWER ACADEMIC PUBLISHERS – DORDRECHT / BOSTON / LONDON

MIX
Papier aus verantwortungsvollen Quellen
Paper from responsible sources
FSC® C105338

If you have any concerns about our products,
you can contact us on
ProductSafety@springernature.com

In case Publisher is established outside the EU,
the EU authorized representative is:
Springer Nature Customer Service Center GmbH
Europaplatz 3, 69115 Heidelberg, Germany

Printed by Libri Plureos GmbH
in Hamburg, Germany